Undergraduate Texts in Mathematics

Undergraduate Texts in Mathematics

Undergraduate Texts in Mathematics are generally aimed at third- and fourth-year undergraduate mathematics students at North American universities. These texts strive to provide students and teachers with new perspectives and novel approaches. The books include motivation that guides the reader to an appreciation of interrelations among different aspects of the subject. They feature examples that illustrate key concepts as well as exercises that strengthen understanding.

More information about this series at http://www.springer.com/series/666

Stephen Abbott

Understanding Analysis

Second Edition

 Springer

Stephen Abbott
Department of Mathematics
Middlebury College
Middlebury, VT, USA

ISSN 0172-6056   ISSN 2197-5604   (electronic)
Undergraduate Texts in Mathematics
ISBN 978-1-4939-2711-1   ISBN 978-1-4939-2712-8   (eBook)
DOI 10.1007/978-1-4939-2712-8

Library of Congress Control Number: 2015937969

Mathematics Subject Classification (2010): 26-01

Springer New York Heidelberg Dordrecht London

Printed on acid-free paper

Springer Science+Business Media LLC New York is part of Springer Science+Business Media (www.springer.com)

Preface

My primary goal in writing *Understanding Analysis* was to create an elementary one-semester book that exposes students to the rich rewards inherent in taking a mathematically rigorous approach to the study of functions of a real variable. The aim of a course in real analysis should be to challenge and improve mathematical intuition rather than to verify it. There is a tendency, however, to center an introductory course too closely around the familiar theorems of the standard calculus sequence. Producing a rigorous argument that polynomials are continuous is good evidence for a well-chosen definition of continuity, but it is not the reason the subject was created and certainly not the reason it should be required study. By shifting the focus to topics where an untrained intuition is severely disadvantaged (e.g., rearrangements of infinite series, nowhere-differentiable continuous functions, Cantor sets), my intent is to bring an intellectual liveliness to this course by offering the beginning student access to some truly significant achievements of the subject.

The Main Objectives

Real analysis stands as a beacon of stability in the otherwise unpredictable evolution of the mathematics curriculum. Amid the various pedagogical revolutions in calculus, computing, statistics, and data analysis, nearly every undergraduate program continues to require at least one semester of real analysis. My own department once challenged this norm by creating a mathematical sciences track that allowed students to replace our two core proof-writing classes with electives in departments like physics and computer science. Within a few years, however, we concluded that the pieces did not hold together without a course in analysis. Analysis is, at once, a course in philosophy and applied mathematics. It is abstract and axiomatic in nature, but is engaged with the mathematics used by economists and engineers.

How then do we teach a successful course to students with such diverse interests and expectations? Our desire to make analysis required study for wider audiences must be reconciled with the fact that many students find the subject quite challenging and even a bit intimidating. One unfortunate resolution of this dilemma is to make the course easier by making it less interesting. The omitted material is inevitably what gives analysis its true flavor. A better solution is to find a way to make the more advanced topics accessible and worth the effort.

I see three essential goals that a semester of real analysis should try to meet:

1. Students need to be confronted with questions that expose the insufficiency of an informal understanding of the objects of calculus. The need for a more rigorous study should be carefully motivated.

2. Having seen mainly intuitive or heuristic arguments, students need to learn what constitutes a rigorous mathematical proof and how to write one.

3. Most importantly, there needs to be significant reward for the difficult work of firming up the logical structure of limits. Specifically, real analysis should not be just an elaborate reworking of standard introductory calculus. Students should be exposed to the tantalizing complexities of the real line, to the subtleties of different flavors of convergence, and to the intellectual delights hidden in the paradoxes of the infinite.

The philosophy of *Understanding Analysis* is to focus attention on questions that give analysis its inherent fascination. Does the Cantor set contain any irrational numbers? Can the set of points where a function is discontinuous be arbitrary? Are derivatives continuous? Are derivatives integrable? Is an infinitely differentiable function necessarily the limit of its Taylor series? In giving these topics center stage, the hard work of a rigorous study is justified by the fact that *they are inaccessible without it.*

The Audience

This book is an introductory text. The only prerequisite is a robust understanding of the results from single-variable calculus. The theorems of linear algebra are not needed, but the exposure to abstract arguments and proof writing that usually comes with this course would be a valuable asset. Complex numbers are never used.

The proofs in *Understanding Analysis* are written with the beginning student firmly in mind. Brevity and other stylistic concerns are postponed in favor of including a significant level of detail. Most proofs come with a generous amount of discussion about the context of the argument. What should the proof entail? Which definitions are relevant? What is the overall strategy? Whenever there is a choice, efficiency is traded for an opportunity to reinforce some previously learned technique. Especially familiar or predictable arguments are often deferred to the exercises.

The search for recurring ideas exists at the proof-writing level and also on the larger expository level. I have tried to give the course a narrative tone by picking up on the unifying themes of approximation and the transition from the finite to the infinite. Often when we ask a question in analysis the answer is "sometimes." Can the order of a double summation be exchanged? Is term-by-term differentiation of an infinite series allowed? By focusing on this recurring pattern, each successive topic builds on the intuition of the previous one. The questions seem more natural, and a coherent story emerges from what might otherwise appear as a long list of theorems and proofs.

This book always emphasizes core ideas over generality, and it makes no effort to be a complete, deductive catalog of results. It is designed to capture the intellectual imagination. Those who become interested are then exceptionally well prepared for a second course starting from complex-valued functions on more general spaces, while those content with a single semester come away with a strong sense of the essence and purpose of real analysis.

The Structure of the Book

Although the book finds its way to some sophisticated results, the main body of each chapter consists of a lean and focused treatment of the core topics that make up the center of most courses in analysis. Fundamental results about completeness, compactness, sequential and functional limits, continuity, uniform convergence, differentiation, and integration are all incorporated.

What is specific here is where the emphasis is placed. In the chapter on integration, for instance, the exposition revolves around deciphering the relationship between continuity and the Riemann integral. Enough properties of the integral are obtained to justify a proof of the Fundamental Theorem of Calculus, but the theme of the chapter is the pursuit of a characterization of integrable functions in terms of continuity. Whether or not Lebesgue's measure-zero criterion is treated, framing the material in this way is still valuable because it is the questions that are important. Mathematics is not a static discipline. Students should be aware of the historical reasons for the creation of the mathematics they are learning and by extension realize that there is no last word on the subject. In the case of integration, this point is made explicitly by including some relatively modern developments on the generalized Riemann integral in the additional topics of the last chapter.

The structure of the chapters has the following distinctive features.

Discussion Sections: Each chapter begins with the discussion of some motivating examples and open questions. The tone in these discussions is intentionally informal, and full use is made of familiar functions and results from calculus. The idea is to freely explore the terrain, providing context for the upcoming definitions and theorems. After these exploratory introductions, the tone of the writing changes, and the treatment becomes rigorously tight but still not overly formal. With the questions in place, the need for the ensuing development of the material is well motivated and the payoff is in sight.

Project Sections: The penultimate section of each chapter (the final section is a short epilogue) is written with the exercises incorporated into the exposition. Proofs are outlined but not completed, and additional exercises are included to elucidate the material being discussed. The sections are written as self-guided tutorials, but they can also be the subject of lectures. I typically use them in place of a final examination, and they work especially well as collaborative assignments that can culminate in a class presentation. The body of each chapter contains the necessary tools, so there is some satisfaction in letting the students use their newly acquired skills to ferret out for themselves answers to questions that have been driving the exposition.

Building a Course

Although this book was originally designed for a 12–14-week semester, it has been used successfully in any number of formats including independent study. The dependence of the sections follows the natural ordering, but there is some flexibility as to what can be treated and omitted.

- The introductory discussions to each chapter can be the subject of lecture, assigned as reading, omitted, or substituted with something preferable. There are no theorems proved here that show up later in the text. I do develop some important examples in these introductions (the Cantor set, Dirichlet's nowhere-continuous function) that probably need to find their way into discussions at some point.

- Chapter 3, Basic Topology of **R**, is much longer than it needs to be. All that is required by the ensuing chapters are fundamental results about open and closed sets and a thorough understanding of sequential compactness. The characterization of compactness using open covers as well as the section on perfect and connected sets are included for their own intrinsic interest. They are not, however, crucial to any future proofs. The one exception to this is a presentation of the Intermediate Value Theorem (IVT) as a special case of the preservation of connected sets by continuous functions. To keep connectedness truly optional, I have included two direct proofs of IVT based on completeness results from Chapter 1.

- All the project sections (1.6, 2.8, 3.5, 4.6, 5.4, 6.7, 7.6, 8.1–8.6) are optional in the sense that no results in later chapters depend on material in these sections. The six topics covered in Chapter 8 are also written in this tutorial-style format, where the exercises make up a significant part of the development. The only one of these sections that might benefit from a lecture is the unit on Fourier series, which is a bit longer than the others.

Changes in the Second Edition

In light of the encouraging feedback—especially from students—I decided not to attempt any major alterations to the central narrative of the text as it was set out in the original edition. Some longer sections have been edited down, or in one case split in two, and the unit on Taylor series is now part of the core material of Chapter 6 instead of being relegated to the closing project section. In contrast to the main body of the book, significant effort has gone into revising the exercises and projects. There are roughly 150 new exercises in this edition alongside 200 or so of what I feel are the most effective problems from the first edition. Some of these introduce new ideas not covered in the chapters (e.g., Euler's constant, infinite products, inverse functions), but the majority are designed to kindle debates about the major ideas under discussion in what I hope are engaging ways. There are ample propositions to prove but also a good supply of Moore-method type exercises that require assessing the validity of various conjectures, deciphering invented definitions, or searching for examples that may not exist.

The introductory discussion to Chapter 6 is new and tells the story of how Euler's deft and audacious manipulations of power series led to a computation of $\sum 1/n^2$. Providing a proper proof for Euler's sum is the topic of one of three new project sections. The other two are a treatment of the Weierstrass Approximation Theorem and an exploration of how to best extend the domain of the factorial function to all of $\mathbf{R}$. Each of these three topics represents a seminal achievement in the history of analysis, but my decision to include them has as much to do with the associated ideas that accompany the main proofs. For the Weierstrass Approximation Theorem, the particular argument that I chose relies on Taylor series and a deep understanding of uniform convergence, making it an ideal project to conclude Chapter 6. The journey to a proper definition of $x!$ allowed me to include a short unit on improper integrals and a proof of Leibniz's rule for differentiating under the integral sign. The accompanying topics for the project on Euler's sum are an analysis of the integral remainder formula for Taylor series and a proof of Wallis's famous product formula for π. Yes these are challenging arguments but they are also beautiful ideas. Returning to the thesis of this text, it is my conviction that encounters with results like these make the task of learning analysis less daunting and more meaningful. They make the epsilons matter.

Acknowledgements

I never met Robert Bartle, although it seems like I did. As a student and a young professor, I spent many hours learning and teaching analysis from his books. I did eventually correspond with him back in 2000 while working on the first edition of this text because I wanted to include a project based on his article, "Return to the Riemann Integral." Professor Bartle was gracious and helpful, even though he was editing his own competing text to include the same material. In September 2003, Robert Bartle died following a long battle with cancer at the age of 76. The section his article inspired on the Generalized Riemann integral continues to be one of my favorite projects to assign, but it is fair to say that Professor Bartle's lucid mathematical writing has been a source of inspiration for the entire text.

My long and winding journey to find an elegant proof of Euler's sum constructed only from theorems in the first seven chapters in this text came to a happy conclusion in Peter Duren's recently published *Invitation to Classical Analysis*. A treasure trove of fascinating topics that have been largely excised from the undergraduate curriculum, Duren's book is remarkable in part for how much he accomplishes without the use of Lebesgue's theory. T.W. Körner's wonderfully opinionated *A Companion to Analysis* is another engaging read that inspired a few of the new exercises in this edition. *Analysis by Its History*, by E. Hairer and G. Wanner, and *A Radical Approach to Real Analysis*, by David Bressoud, were both cited in the acknowledgements of the first edition as sources for many of the historical anecdotes that permeate the text. Since then, Professor Bressoud has published a sequel, *A Radical Approach to Lebesgue's Theory of Integration*, which I heartily recommend.

The significant contributions of Benjamin Lotto, Loren Pitt, and Paul Humke to the content of the first edition warrant a second nod in these acknowledgements. As for the new edition, Dan Velleman taught from a draft of the text and provided much helpful feedback. Whatever problems still remain are likely places where I stubbornly did not follow Dan's advice. Back in 2001, Steve Kennedy penned a review of *Understanding Analysis* which I am sure enhanced the audience for this book. His kind assessment nevertheless included a number of worthy suggestions for improvement, most of which I have incorporated. I should also acknowledge Fernando Gouvea as the one who suggested that the series of articles by David Fowler on the factorial function would fit well with the themes of this book. The result is Section 8.4.

I would like to express my continued appreciation to the staff at Springer, and in particular to Marc Strauss and Eugene Ha for their support and unwavering faith in the merits of this project. The large email file of thoughtful suggestions from users of the book is too long to enumerate, but perhaps this is the place to say that I continue to welcome comments from readers, even moderately disgruntled ones. The most gratifying aspect of authoring the first edition is the sense of being connected to the larger mathematical community and of being an active participant in it.

The margins of my original copy of *Understanding Analysis* are filled with vestiges of my internal debates about what to revise, what to preserve, and what to discard. The final decisions I made are the result of 15 years of classroom experiments with the text, and it is comforting to report that the main body of the book has weathered the test of time with only a modest tune-up. On a similarly positive note, the original dedication of this book to my wife Katy is another feature of the first edition that has required no additional editing.

Middlebury, VT, USA   Stephen Abbott
March 2015

Contents

Chapter 1

The Real Numbers

1.1 Discussion: The Irrationality of $\sqrt{2}$

Toward the end of his distinguished career, the renowned British mathematician G.H. Hardy eloquently laid out a justification for a life of studying mathematics in *A Mathematician's Apology*, an essay first published in 1940. At the center of Hardy's defense is the thesis that mathematics is an aesthetic discipline. For Hardy, the applied mathematics of engineers and economists held little charm. "Real mathematics," as he referred to it, "must be justified as art if it can be justified at all."

To help make his point, Hardy includes two theorems from classical Greek mathematics, which, in his opinion, possess an elusive kind of beauty that, although difficult to define, is easy to recognize. The first of these results is Euclid's proof that there are an infinite number of prime numbers. The second result is the discovery, attributed to the school of Pythagoras from around 500 B.C., that $\sqrt{2}$ is irrational. It is this second theorem that demands our attention. (A course in number theory would focus on the first.) The argument uses only arithmetic, but its depth and importance cannot be overstated. As Hardy says, "[It] is a 'simple' theorem, simple both in idea and execution, but there is no doubt at all about [it being] of the highest class. [It] is as fresh and significant as when it was discovered—two thousand years have not written a wrinkle on [it]."

Theorem 1.1.1. *There is no rational number whose square is 2.*

Proof. A rational number is any number that can be expressed in the form p/q, where p and q are integers. Thus, what the theorem asserts is that no matter how p and q are chosen, it is never the case that $(p/q)^2 = 2$. The line of attack is indirect, using a type of argument referred to as a proof by contradiction. The idea is to assume that there *is* a rational number whose square is 2 and then proceed along logical lines until we reach a conclusion that is unacceptable.

© Springer Science+Business Media New York 2015
S. Abbott, *Understanding Analysis*, Undergraduate Texts
in Mathematics, DOI 10.1007/978-1-4939-2712-8_1

At this point, we will be forced to retrace our steps and reject the erroneous assumption that some rational number squared is equal to 2. In short, we will prove that the theorem is true by demonstrating that it cannot be false.

And so assume, for contradiction, that there exist integers p and q satisfying

(1)
$$\left(\frac{p}{q}\right)^2 = 2.$$

We may also assume that p and q have no common factor, because, if they had one, we could simply cancel it out and rewrite the fraction in lowest terms. Now, equation (1) implies

(2)
$$p^2 = 2q^2.$$

From this, we can see that the integer p^2 is an even number (it is divisible by 2), and hence p must be even as well because the square of an odd number is odd. This allows us to write $p = 2r$, where r is also an integer. If we substitute $2r$ for p in equation (2), then a little algebra yields the relationship

$$2r^2 = q^2.$$

But now the absurdity is at hand. This last equation implies that q^2 is even, and hence q must also be even. Thus, we have shown that p and q are both even (i.e., divisible by 2) when they were originally assumed to have no common factor. From this logical impasse, we can only conclude that equation (1) *cannot* hold for any integers p and q, and thus the theorem is proved. □

A component of Hardy's definition of beauty in a mathematical theorem is that the result have lasting and serious implications for a network of other mathematical ideas. In this case, the ideas under assault were the Greeks' understanding of the relationship between geometric *length* and arithmetic *number*. Prior to the preceding discovery, it was an assumed and commonly used fact that, given two line segments $\overline{AB}$ and $\overline{CD}$, it would always be possible to find a third line segment whose length divides evenly into the first two. In modern terminology, this is equivalent to asserting that the length of $\overline{CD}$ is a rational multiple of the length of $\overline{AB}$. Looking at the diagonal of a unit square (Fig. 1.1), it now followed (using the Pythagorean Theorem) that this was not always the case. Because the Pythagoreans implicitly interpreted number to mean rational number, they were forced to accept that number was a strictly weaker notion than length.

Rather than abandoning arithmetic in favor of geometry (as the Greeks seem to have done), our resolution to this limitation is to strengthen the concept of number by moving from the rational numbers to a larger number system. From a modern point of view, this should seem like a familiar and somewhat natural phenomenon. We begin with the *natural numbers*

$$\mathbf{N} = \{1, 2, 3, 4, 5, \ldots\}.$$

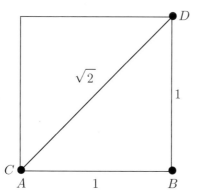

Figure 1.1: $\sqrt{2}$ EXISTS AS A GEOMETRIC LENGTH.

The influential German mathematician Leopold Kronecker (1823–1891) once asserted that "The natural numbers are the work of God. All of the rest is the work of mankind." Debating the validity of this claim is an interesting conversation for another time. For the moment, it at least provides us with a place to start. If we restrict our attention to the natural numbers **N**, then we can perform addition perfectly well, but we must extend our system to the *integers*

$$\mathbf{Z} = \{\ldots, -3, -2, -1, 0, 1, 2, 3, \ldots\}$$

if we want to have an additive identity (zero) and the additive inverses necessary to define subtraction. The next issue is multiplication and division. The number 1 acts as the multiplicative identity, but in order to define division we need to have multiplicative inverses. Thus, we extend our system again to the *rational numbers*

$$\mathbf{Q} = \left\{ \text{all fractions } \frac{p}{q} \text{ where } p \text{ and } q \text{ are integers with } q \neq 0 \right\}.$$

Taken together, the properties of **Q** discussed in the previous paragraph essentially make up the definition of what is called a *field*. More formally stated, a field is any set where addition and multiplication are well-defined operations that are commutative, associative, and obey the familiar distributive property $a(b+c) = ab + ac$. There must be an additive identity, and every element must have an additive inverse. Finally, there must be a multiplicative identity, and multiplicative inverses must exist for all nonzero elements of the field. Neither **Z** nor **N** is a field. The finite set $\{0, 1, 2, 3, 4\}$ is a field when addition and multiplication are computed modulo 5. This is not immediately obvious but makes an interesting exercise.

The set **Q** also has a natural *order* defined on it. Given any two rational numbers r and s, exactly one of the following is true:

$$r < s, \quad r = s, \quad \text{or} \quad r > s.$$

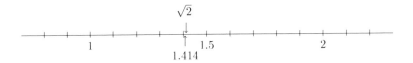

Figure 1.2: Approximating $\sqrt{2}$ with rational numbers.

This ordering is transitive in the sense that if $r < s$ and $s < t$, then $r < t$, so we are conveniently led to a mental picture of the rational numbers as being laid out from left to right along a number line. Unlike $\mathbf{Z}$, there are no intervals of empty space. Given any two rational numbers $r < s$, the rational number $(r+s)/2$ sits halfway in between, implying that the rational numbers are densely nestled together.

With the field properties of $\mathbf{Q}$ allowing us to safely carry out the algebraic operations of addition, subtraction, multiplication, and division, let's remind ourselves just what it is that $\mathbf{Q}$ is lacking. By Theorem 1.1.1, it is apparent that we cannot always take square roots. The problem, however, is actually more fundamental than this. Using only rational numbers, it is possible to *approximate* $\sqrt{2}$ quite well (Fig. 1.2). For instance, $1.414^2 = 1.999396$. By adding more decimal places to our approximation, we can get even closer to a value for $\sqrt{2}$, but, even so, we are now well aware that there is a "hole" in the rational number line where $\sqrt{2}$ ought to be. Of course, there are quite a few other holes—at $\sqrt{3}$ and $\sqrt{5}$, for example. Returning to the dilemma of the ancient Greek mathematicians, if we want every length along the number line to correspond to an actual number, then another extension to our number system is in order. Thus, to the chain $\mathbf{N} \subseteq \mathbf{Z} \subseteq \mathbf{Q}$ we append the *real numbers* $\mathbf{R}$.

The question of how to actually construct $\mathbf{R}$ from $\mathbf{Q}$ is rather complicated business. It is discussed in Section 1.3, and then again in more detail in Section 8.6. For the moment, it is not too inaccurate to say that $\mathbf{R}$ is obtained by filling in the gaps in $\mathbf{Q}$. Wherever there is a hole, a new *irrational* number is defined and placed into the ordering that already exists on $\mathbf{Q}$. The real numbers are then the union of these irrational numbers together with the more familiar rational ones. What properties does the set of irrational numbers have? How do the sets of rational and irrational numbers fit together? Is there a kind of symmetry between the rationals and the irrationals, or is there some sense in which we can argue that one type of real number is more common than the other? The one method we have seen so far for generating examples of irrational numbers is through square roots. Not too surprisingly, other roots such as $\sqrt[3]{2}$ or $\sqrt[5]{3}$ are most often irrational. Can all irrational numbers be expressed as algebraic combinations of nth roots and rational numbers, or are there still other irrational numbers beyond those of this form?

1.2 Some Preliminaries

The vocabulary necessary for the ensuing development comes from set theory and the theory of functions. This should be familiar territory, but a brief review of the terminology is probably a good idea, if only to establish some agreed-upon notation.

Sets

Intuitively speaking, a *set* is any collection of objects. These objects are referred to as the *elements* of the set. For our purposes, the sets in question will most often be sets of real numbers, although we will also encounter sets of functions and, on a few occasions, sets whose elements are other sets.

Given a set A, we write $x \in A$ if x (whatever it may be) is an element of A. If x is not an element of A, then we write $x \notin A$. Given two sets A and B, the *union* is written $A \cup B$ and is defined by asserting that

$$x \in A \cup B \text{ provided that } x \in A \text{ or } x \in B \text{ (or potentially both)}.$$

The *intersection* $A \cap B$ is the set defined by the rule

$$x \in A \cap B \text{ provided } x \in A \text{ and } x \in B.$$

Example 1.2.1. (i) There are many acceptable ways to assert the contents of a set. In the previous section, the set of natural numbers was defined by listing the elements: $\mathbf{N} = \{1, 2, 3, \ldots\}$.

(ii) Sets can also be described in words. For instance, we can define the set E to be the collection of even natural numbers.

(iii) Sometimes it is more efficient to provide a kind of rule or algorithm for determining the elements of a set. As an example, let

$$S = \{r \in \mathbf{Q} : r^2 < 2\}.$$

Read aloud, the definition of S says, "Let S be the set of all rational numbers whose squares are less than 2." It follows that $1 \in S$, $4/3 \in S$, but $3/2 \notin S$ because $9/4 \geq 2$.

Using the previously defined sets to illustrate the operations of intersection and union, we observe that

$$\mathbf{N} \cup E = \mathbf{N}, \quad \mathbf{N} \cap E = E, \quad \mathbf{N} \cap S = \{1\}, \text{ and } E \cap S = \emptyset.$$

The set $\emptyset$ is called the *empty set* and is understood to be the set that contains no elements. An equivalent statement would be to say that E and S are *disjoint*.

A word about the equality of two sets is in order (since we have just used the notion). The *inclusion* relationship $A \subseteq B$ or $B \supseteq A$ is used to indicate that every element of A is also an element of B. In this case, we say A is a *subset* of B, or B *contains* A. To assert that $A = B$ means that $A \subseteq B$ and $B \subseteq A$. Put another way, A and B have exactly the same elements.

Quite frequently in the upcoming chapters, we will want to apply the union and intersection operations to infinite collections of sets.

Example 1.2.2. Let

$$
\begin{aligned}
A_1 &= \mathbf{N} = \{1, 2, 3, \ldots\}, \\
A_2 &= \{2, 3, 4, \ldots\}, \\
A_3 &= \{3, 4, 5, \ldots\},
\end{aligned}
$$

and, in general, for each $n \in \mathbf{N}$, define the set

$$
A_n = \{n, n+1, n+2, \ldots\}.
$$

The result is a nested chain of sets

$$
A_1 \supseteq A_2 \supseteq A_3 \supseteq A_4 \supseteq \cdots,
$$

where each successive set is a subset of all the previous ones. Notationally,

$$
\bigcup_{n=1}^{\infty} A_n, \quad \bigcup_{n \in \mathbf{N}} A_n, \quad \text{or} \quad A_1 \cup A_2 \cup A_3 \cup \cdots
$$

are all equivalent ways to indicate the set whose elements consist of any element that appears in at least one particular A_n. Because of the nested property of this particular collection of sets, it is not too hard to see that

$$
\bigcup_{n=1}^{\infty} A_n = A_1.
$$

The notion of intersection has the same kind of natural extension to infinite collections of sets. For this example, we have

$$
\bigcap_{n=1}^{\infty} A_n = \emptyset.
$$

Let's be sure we understand why this is the case. Suppose we had some natural number m that we thought might actually satisfy $m \in \bigcap_{n=1}^{\infty} A_n$. What this would mean is that $m \in A_n$ for *every* A_n in our collection of sets. Because m is not an element of A_{m+1}, no such m exists and the intersection is empty.

As mentioned, most of the sets we encounter will be sets of real numbers. Given $A \subseteq \mathbf{R}$, the *complement* of A, written A^c, refers to the set of all elements of $\mathbf{R}$ not in A. Thus, for $A \subseteq \mathbf{R}$,

$$
A^c = \{x \in \mathbf{R} : x \notin A\}.
$$

A few times in our work to come, we will refer to De Morgan's Laws, which state that
$$(A \cap B)^c = A^c \cup B^c \quad \text{and} \quad (A \cup B)^c = A^c \cap B^c.$$
Proofs of these statements are discussed in Exercise 1.2.5.

Admittedly, there is something imprecise about the definition of set presented at the beginning of this discussion. The defining sentence begins with the phrase "Intuitively speaking," which might seem an odd way to embark on a course of study that purportedly intends to supply a rigorous foundation for the theory of functions of a real variable. In some sense, however, this is unavoidable. Each repair of one level of the foundation reveals something below it in need of attention. The theory of sets has been subjected to intense scrutiny over the past century precisely because so much of modern mathematics rests on this foundation. But such a study is really only advisable once it is understood why our naive impression about the behavior of sets is insufficient. For the direction in which we are heading, this will not happen, although an indication of some potential pitfalls is given in Section 1.7.

Functions

Definition 1.2.3. Given two sets A and B, a *function* from A to B is a rule or mapping that takes each element $x \in A$ and associates with it a single element of B. In this case, we write $f : A \to B$. Given an element $x \in A$, the expression $f(x)$ is used to represent the element of B associated with x by f. The set A is called the *domain* of f. The *range* of f is not necessarily equal to B but refers to the subset of B given by $\{y \in B : y = f(x) \text{ for some } x \in A\}$.

This definition of function is more or less the one proposed by Peter Lejeune Dirichlet (1805–1859) in the 1830s. Dirichlet was a German mathematician who was one of the leaders in the development of the rigorous approach to functions that we are about to undertake. His main motivation was to unravel the issues surrounding the convergence of Fourier series. Dirichlet's contributions figure prominently in Section 8.5, where an introduction to Fourier series is presented, but we will also encounter his name in several earlier chapters along the way. What is important at the moment is that we see how Dirichlet's definition of function liberates the term from its interpretation as a type of "formula." In the years leading up to Dirichlet's time, the term "function" was generally understood to refer to algebraic entities such as $f(x) = x^2 + 1$ or $g(x) = \sqrt{x^4 + 4}$. Definition 1.2.3 allows for a much broader range of possibilities.

Example 1.2.4. In 1829, Dirichlet proposed the unruly function

$$g(x) = \begin{cases} 1 & \text{if } x \in \mathbf{Q} \\ 0 & \text{if } x \notin \mathbf{Q}. \end{cases}$$

The domain of g is all of $\mathbf{R}$, and the range is the set $\{0, 1\}$. There is no single formula for g in the usual sense, and it is quite difficult to graph this function (see Section 4.1 for a rough attempt), but it certainly qualifies as a function according to the criterion in Definition 1.2.3. As we study the theoretical nature of continuous, differentiable, or integrable functions, examples such as this one will provide us with an invaluable testing ground for the many conjectures we encounter.

Example 1.2.5 (Triangle Inequality). The *absolute value function* is so important that it merits the special notation $|x|$ in place of the usual $f(x)$ or $g(x)$. It is defined for every real number via the piecewise definition

$$|x| = \begin{cases} x & \text{if } x \geq 0 \\ -x & \text{if } x < 0. \end{cases}$$

With respect to multiplication and division, the absolute value function satisfies

(i) $|ab| = |a||b|$ and

(ii) $|a + b| \leq |a| + |b|$

for all choices of a and b. Verifying these properties (Exercise 1.2.6) is just a matter of examining the different cases that arise when a, b, and $a+b$ are positive and negative. Property (ii) is called the *triangle inequality*. This innocuous looking inequality turns out to be fantastically important and will be frequently employed in the following way. Given three real numbers $a, b,$ and c, we certainly have

$$|a - b| = |(a - c) + (c - b)|.$$

By the triangle inequality,

$$|(a - c) + (c - b)| \leq |a - c| + |c - b|,$$

so we get

(1) $$|a - b| \leq |a - c| + |c - b|.$$

Now, the expression $|a - b|$ is equal to $|b - a|$ and is best understood as the *distance* between the points a and b on the number line. With this interpretation, equation (1) makes the plausible statement that the distance from a to b is less than or equal to the distance from a to c plus the distance from c to b. Pretending for a moment that these are points in the plane (instead of on the real line), it should be evident why this is referred to as the "triangle inequality."

Logic and Proofs

Writing rigorous mathematical proofs is a skill best learned by doing, and there is plenty of on-the-job training just ahead. As Hardy indicates, there is an artistic quality to mathematics of this type, which may or may not come easily, but that is not to say that anything especially mysterious is happening. A proof is an essay of sorts. It is a set of carefully crafted directions, which, when followed, should leave the reader absolutely convinced of the truth of the proposition in

question. To achieve this, the steps in a proof must follow logically from previous steps or be justified by some other agreed-upon set of facts. In addition
to being valid, these steps must also fit coherently together to form a cogent
argument. Mathematics has a specialized vocabulary, to be sure, but that does
not exempt a good proof from being written in grammatically correct English.

The one proof we have seen at this point (to Theorem 1.1.1) uses an indirect
strategy called *proof by contradiction*. This powerful technique will be employed
a number of times in our upcoming work. Nevertheless, most proofs are direct.
(It also bears mentioning that using an indirect proof when a direct proof is
available is generally considered bad form.) A direct proof begins from some
valid statement, most often taken from the theorem's hypothesis, and then proceeds through rigorously logical deductions to a demonstration of the theorem's
conclusion. As we saw in Theorem 1.1.1, an indirect proof always begins by
negating what it is we would like to prove. This is not always as easy to do as it
may sound. The argument then proceeds until (hopefully) a logical contradiction with some other accepted fact is uncovered. Many times, this accepted fact
is part of the hypothesis of the theorem. When the contradiction is with the
theorem's hypothesis, we technically have what is called a *contrapositive* proof.

The next proposition illustrates a number of the issues just discussed and
introduces a few more.

Theorem 1.2.6. *Two real numbers a and b are equal if and only if for every
real number $\epsilon > 0$ it follows that $|a - b| < \epsilon$.*

Proof. There are two key phrases in the statement of this proposition that
warrant special attention. One is "for every," which will be addressed in a
moment. The other is "if and only if." To say "if and only if" in mathematics
is an economical way of stating that the proposition is true in two directions.
In the forward direction, we must prove the statement:

($\Rightarrow$) *If $a = b$, then for every real number $\epsilon > 0$ it follows that $|a - b| < \epsilon$.*

We must also prove the converse statement:

($\Leftarrow$) *If for every real number $\epsilon > 0$ it follows that $|a - b| < \epsilon$, then we must
have $a = b$.*

For the proof of the first statement, there is really not much to say. If $a = b$,
then $|a - b| = 0$, and so certainly $|a - b| < \epsilon$ no matter what $\epsilon > 0$ is chosen.

For the second statement, we give a proof by contradiction. The conclusion
of the proposition in this direction states that $a = b$, so we assume that $a \neq b$.
Heading off in search of a contradiction brings us to a consideration of the phrase
"for every $\epsilon > 0$." Some equivalent ways to state the hypothesis would be to
say that "for all possible choices of $\epsilon > 0$" or "no matter how $\epsilon > 0$ is selected,
it is always the case that $|a - b| < \epsilon$." But assuming $a \neq b$ (as we are doing at
the moment), the choice of

$$\epsilon_0 = |a - b| > 0$$

poses a serious problem. We are assuming that $|a - b| < \epsilon$ is true for *every* $\epsilon > 0$, so this must certainly be true of the particular ϵ_0 just defined. However, the statements

$$|a - b| < \epsilon_0 \quad \text{and} \quad |a - b| = \epsilon_0$$

cannot both be true. This contradiction means that our initial assumption that $a \neq b$ is unacceptable. Therefore, $a = b$, and the indirect proof is complete. □

One of the most fundamental skills required for reading and writing analysis proofs is the ability to confidently manipulate the quantifying phrases "for all" and "there exists." Significantly more attention will be given to this issue in many upcoming discussions.

Induction

One final trick of the trade, which will arise with some frequency, is the use of *induction* arguments. Induction is used in conjunction with the natural numbers $\mathbf{N}$ (or sometimes with the set $\mathbf{N} \cup \{0\}$). The fundamental principle behind induction is that if S is some subset of $\mathbf{N}$ with the property that

(i) S contains 1 and

(ii) whenever S contains a natural number n, it also contains $n + 1$,

then it must be that $S = \mathbf{N}$. As the next example illustrates, this principle can be used to define sequences of objects as well as to prove facts about them.

Example 1.2.7. Let $x_1 = 1$, and for each $n \in \mathbf{N}$ define

$$x_{n+1} = (1/2)x_n + 1.$$

Using this rule, we can compute $x_2 = (1/2)(1) + 1 = 3/2$, $x_3 = 7/4$, and it is immediately apparent how this leads to a definition of x_n for all $n \in \mathbf{N}$.

The sequence just defined appears at the outset to be increasing. For the terms computed, we have $x_1 \leq x_2 \leq x_3$. Let's use induction to prove that this trend continues; that is, let's show

(2) $$x_n \leq x_{n+1}$$

for all values of $n \in \mathbf{N}$.

For $n = 1$, $x_1 = 1$ and $x_2 = 3/2$, so that $x_1 \leq x_2$ is clear. Now, we want to show that

if we have $x_n \leq x_{n+1}$, then it follows that $x_{n+1} \leq x_{n+2}$.

Think of S as the set of natural numbers for which the claim in equation (2) is true. We have shown that $1 \in S$. We are now interested in showing that if $n \in S$, then $n+1 \in S$ as well. Starting from the induction hypothesis $x_n \leq x_{n+1}$, we can multiply across the inequality by $1/2$ and add 1 to get

$$\frac{1}{2}x_n + 1 \leq \frac{1}{2}x_{n+1} + 1,$$

which is precisely the desired conclusion $x_{n+1} \leq x_{n+2}$. By induction, the claim is proved for all $n \in \mathbf{N}$.

Any discussion about why induction is a valid argumentative technique immediately opens up a box of questions about how we understand the natural numbers. Earlier, in Section 1.1, we avoided this issue by referencing Kronecker's famous comment that the natural numbers are somehow divinely given. Although we will not improve on this explanation here, it should be pointed out that a more atheistic and mathematically satisfying approach to $\mathbf{N}$ is possible from the point of view of axiomatic set theory. This brings us back to a recurring theme of this chapter. Pedagogically speaking, the foundations of mathematics are best learned and appreciated in a kind of reverse order. A rigorous study of the natural numbers and the theory of sets is certainly recommended, but only after we have an understanding of the subtleties of the real number system. It is this latter topic that is the business of real analysis.

Exercises

Exercise 1.2.1. (a) Prove that $\sqrt{3}$ is irrational. Does a similar argument work to show $\sqrt{6}$ is irrational?

(b) Where does the proof of Theorem 1.1.1 break down if we try to use it to prove $\sqrt{4}$ is irrational?

Exercise 1.2.2. Show that there is no rational number r satisfying $2^r = 3$.

Exercise 1.2.3. Decide which of the following represent true statements about the nature of sets. For any that are false, provide a specific example where the statement in question does not hold.

(a) If $A_1 \supseteq A_2 \supseteq A_3 \supseteq A_4 \cdots$ are all sets containing an infinite number of elements, then the intersection $\bigcap_{n=1}^{\infty} A_n$ is infinite as well.

(b) If $A_1 \supseteq A_2 \supseteq A_3 \supseteq A_4 \cdots$ are all finite, nonempty sets of real numbers, then the intersection $\bigcap_{n=1}^{\infty} A_n$ is finite and nonempty.

(c) $A \cap (B \cup C) = (A \cap B) \cup C$.

(d) $A \cap (B \cap C) = (A \cap B) \cap C$.

(e) $A \cap (B \cup C) = (A \cap B) \cup (A \cap C)$.

Exercise 1.2.4. Produce an infinite collection of sets $A_1, A_2, A_3, \ldots$ with the property that every A_i has an infinite number of elements, $A_i \cap A_j = \emptyset$ for all $i \neq j$, and $\bigcup_{i=1}^{\infty} A_i = \mathbf{N}$.

Exercise 1.2.5 (De Morgan's Laws). Let A and B be subsets of $\mathbf{R}$.

(a) If $x \in (A \cap B)^c$, explain why $x \in A^c \cup B^c$. This shows that $(A \cap B)^c \subseteq A^c \cup B^c$.

(b) Prove the reverse inclusion $(A \cap B)^c \supseteq A^c \cup B^c$, and conclude that $(A \cap B)^c = A^c \cup B^c$.

(c) Show $(A \cup B)^c = A^c \cap B^c$ by demonstrating inclusion both ways.

Exercise 1.2.6.   (a) Verify the triangle inequality in the special case where a and b have the same sign.

(b) Find an efficient proof for all the cases at once by first demonstrating $(a + b)^2 \leq (|a| + |b|)^2$.

(c) Prove $|a - b| \leq |a - c| + |c - d| + |d - b|$ for all a, b, c, and d.

(d) Prove $||a| - |b|| \leq |a - b|$. (The unremarkable identity $a = a - b + b$ may be useful.)

Exercise 1.2.7. Given a function f and a subset A of its domain, let $f(A)$ represent the range of f over the set A; that is, $f(A) = \{f(x) : x \in A\}$.

(a) Let $f(x) = x^2$. If $A = [0, 2]$ (the closed interval $\{x \in \mathbf{R} : 0 \leq x \leq 2\}$) and $B = [1, 4]$, find $f(A)$ and $f(B)$. Does $f(A \cap B) = f(A) \cap f(B)$ in this case? Does $f(A \cup B) = f(A) \cup f(B)$?

(b) Find two sets A and B for which $f(A \cap B) \neq f(A) \cap f(B)$.

(c) Show that, for an arbitrary function $g : \mathbf{R} \to \mathbf{R}$, it is always true that $g(A \cap B) \subseteq g(A) \cap g(B)$ for all sets $A, B \subseteq \mathbf{R}$.

(d) Form and prove a conjecture about the relationship between $g(A \cup B)$ and $g(A) \cup g(B)$ for an arbitrary function g.

Exercise 1.2.8. Here are two important definitions related to a function $f :$ $A \to B$. The function f is *one-to-one* (1–1) if $a_1 \neq a_2$ in A implies that $f(a_1) \neq f(a_2)$ in B. The function f is *onto* if, given any $b \in B$, it is possible to find an element $a \in A$ for which $f(a) = b$.

Give an example of each or state that the request is impossible:

(a) $f : \mathbf{N} \to \mathbf{N}$ that is 1–1 but not onto.

(b) $f : \mathbf{N} \to \mathbf{N}$ that is onto but not 1–1.

(c) $f : \mathbf{N} \to \mathbf{Z}$ that is 1–1 and onto.

Exercise 1.2.9. Given a function $f : D \to \mathbf{R}$ and a subset $B \subseteq \mathbf{R}$, let $f^{-1}(B)$ be the set of all points from the domain D that get mapped into B; that is, $f^{-1}(B) = \{x \in D : f(x) \in B\}$. This set is called the *preimage* of B.

(a) Let $f(x) = x^2$. If A is the closed interval $[0, 4]$ and B is the closed interval $[-1, 1]$, find $f^{-1}(A)$ and $f^{-1}(B)$. Does $f^{-1}(A \cap B) = f^{-1}(A) \cap f^{-1}(B)$ in this case? Does $f^{-1}(A \cup B) = f^{-1}(A) \cup f^{-1}(B)$?

(b) The good behavior of preimages demonstrated in (a) is completely general. Show that for an arbitrary function $g : \mathbf{R} \to \mathbf{R}$, it is always true that $g^{-1}(A \cap B) = g^{-1}(A) \cap g^{-1}(B)$ and $g^{-1}(A \cup B) = g^{-1}(A) \cup g^{-1}(B)$ for all sets $A, B \subseteq \mathbf{R}$.

Exercise 1.2.10. Decide which of the following are true statements. Provide a short justification for those that are valid and a counterexample for those that are not:

(a) Two real numbers satisfy $a < b$ if and only if $a < b + \epsilon$ for every $\epsilon > 0$.

(b) Two real numbers satisfy $a < b$ if $a < b + \epsilon$ for every $\epsilon > 0$.

(c) Two real numbers satisfy $a \leq b$ if and only if $a < b + \epsilon$ for every $\epsilon > 0$.

Exercise 1.2.11. Form the logical negation of each claim. One trivial way to do this is to simply add "It is not the case that. . . " in front of each assertion. To make this interesting, fashion the negation into a positive statement that avoids using the word "not" altogether. In each case, make an intuitive guess as to whether the claim or its negation is the true statement.

(a) For all real numbers satisfying $a < b$, there exists an $n \in \mathbf{N}$ such that $a + 1/n < b$.

(b) There exists a real number $x > 0$ such that $x < 1/n$ for all $n \in \mathbf{N}$.

(c) Between every two distinct real numbers there is a rational number.

Exercise 1.2.12. Let $y_1 = 6$, and for each $n \in \mathbf{N}$ define $y_{n+1} = (2y_n - 6)/3$.

(a) Use induction to prove that the sequence satisfies $y_n > -6$ for all $n \in \mathbf{N}$.

(b) Use another induction argument to show the sequence $(y_1, y_2, y_3, \ldots)$ is decreasing.

Exercise 1.2.13. For this exercise, assume Exercise 1.2.5 has been successfully completed.

(a) Show how induction can be used to conclude that
$$(A_1 \cup A_2 \cup \cdots \cup A_n)^c = A_1^c \cap A_2^c \cap \cdots \cap A_n^c$$
for any finite $n \in \mathbf{N}$.

(b) It is tempting to appeal to induction to conclude
$$\left(\bigcup_{i=1}^{\infty} A_i \right)^c = \bigcap_{i=1}^{\infty} A_i^c,$$
but induction does not apply here. Induction is used to prove that a particular statement holds for every value of $n \in \mathbf{N}$, but this does not imply the validity of the infinite case. To illustrate this point, find an example of a collection of sets $B_1, B_2, B_3, \ldots$ where $\bigcap_{i=1}^{n} B_i \neq \emptyset$ is true for every $n \in \mathbf{N}$, but $\bigcap_{i=1}^{\infty} B_i \neq \emptyset$ fails.

(c) Nevertheless, the infinite version of De Morgan's Law stated in (b) is a valid statement. Provide a proof that does not use induction.

1.3   The Axiom of Completeness

What exactly is a real number? In Section 1.1, we got as far as saying that the set $\mathbf{R}$ of real numbers is an extension of the rational numbers $\mathbf{Q}$ in which there are no holes or gaps. We want every length along the number line—such as $\sqrt{2}$—to correspond to a real number and vice versa.

We are going to improve on this definition, but as we do so, it is important to keep in mind our earlier acknowledgment that whatever precise statements we formulate will necessarily rest on other unproven assumptions or undefined terms. At some point, we must draw a line and confess that this is what we have decided to accept as a reasonable place to start. Naturally, there is some debate about where this line should be drawn. One way to view the mathematics of the 19th and 20th centuries is as a stalwart attempt to move this line further and further back toward some unshakable foundation. The majority of the material covered in this book is attributable to the mathematicians working in the early and middle parts of the 1800s. Augustin Louis Cauchy (1789–1857), Bernhard Bolzano (1781–1848), Niels Henrik Abel (1802–1829), Peter Lejeune Dirichlet, Karl Weierstrass (1815–1897), and Bernhard Riemann (1826–1866) all figure prominently in the discovery of the theorems that follow. But here is the interesting point. Nearly all of this work was done using intuitive assumptions about the nature of $\mathbf{R}$ quite similar to our own informal understanding at this point. Eventually, enough scrutiny was directed at the detailed structure of $\mathbf{R}$ so that, in the 1870s, a handful of ways to rigorously *construct* $\mathbf{R}$ from $\mathbf{Q}$ were proposed.

Following this historical model, our own rigorous construction of $\mathbf{R}$ from $\mathbf{Q}$ is postponed until Section 8.6. By this point, the need for such a construction will be more justified and easier to appreciate. In the meantime, we have many proofs to write, so it is important to lay down, as explicitly as possible, the assumptions that we intend to make about the real numbers.

An Initial Definition for R

First, $\mathbf{R}$ is a set containing $\mathbf{Q}$. The operations of addition and multiplication on $\mathbf{Q}$ extend to all of $\mathbf{R}$ in such a way that every element of $\mathbf{R}$ has an additive inverse and every nonzero element of $\mathbf{R}$ has a multiplicative inverse. Echoing the discussion in Section 1.1, we assume $\mathbf{R}$ is a *field*, meaning that addition and multiplication of real numbers are commutative, associative, and the distributive property holds. This allows us to perform all of the standard algebraic manipulations that are second nature to us. We also assume that the familiar properties of the ordering on $\mathbf{Q}$ extend to all of $\mathbf{R}$. Thus, for example, such deductions as "If $a < b$ and $c > 0$, then $ac < bc$" will be carried out freely without much comment. To summarize the situation in the official terminology

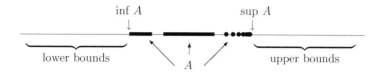

Figure 1.3: Definition of sup A **and** inf A.

of the subject, we assume that $\mathbf{R}$ is an *ordered field*, which contains $\mathbf{Q}$ as a subfield. (A rigorous definition of "ordered field" is presented in Section 8.6.)

This brings us to the final, and most distinctive, assumption about the real number system. We must find some way to clearly articulate what we mean by insisting that $\mathbf{R}$ does not contain the gaps that permeate $\mathbf{Q}$. Because this is the defining difference between the rational numbers and the real numbers, we will be excessively precise about how we phrase this assumption, hereafter referred to as the *Axiom of Completeness*.

Axiom of Completeness. *Every nonempty set of real numbers that is bounded above has a least upper bound.*

Now, what exactly does this mean?

Least Upper Bounds and Greatest Lower Bounds

Let's first state the relevant definitions, and then look at some examples.

Definition 1.3.1. A set $A \subseteq \mathbf{R}$ is *bounded above* if there exists a number $b \in \mathbf{R}$ such that $a \leq b$ for all $a \in A$. The number b is called an *upper bound* for A.

Similarly, the set A is *bounded below* if there exists a *lower bound* $l \in \mathbf{R}$ satisfying $l \leq a$ for every $a \in A$.

Definition 1.3.2. A real number s is the *least upper bound* for a set $A \subseteq \mathbf{R}$ if it meets the following two criteria:

(i) s is an upper bound for A;

(ii) if b is any upper bound for A, then $s \leq b$.

The least upper bound is also frequently called the *supremum* of the set A. Although the notation $s = \text{lub } A$ is sometimes used, we will always write $s = \sup A$ for the least upper bound.

The *greatest lower bound* or *infimum* for A is defined in a similar way (Exercise 1.3.1) and is denoted by inf A (Fig. 1.3).

Although a set can have a host of upper bounds, it can have only one *least* upper bound. If s_1 and s_2 are both least upper bounds for a set A, then by property (ii) in Definition 1.3.2 we can assert $s_1 \leq s_2$ and $s_2 \leq s_1$. The conclusion is that $s_1 = s_2$ and least upper bounds are unique.

Example 1.3.3. Let

$$A = \left\{ \frac{1}{n} : n \in N \right\} = \left\{ 1, \frac{1}{2}, \frac{1}{3}, \dots \right\}.$$

The set A is bounded above and below. Successful candidates for an upper bound include 3, 2, and 3/2. For the least upper bound, we claim $\sup A = 1$. To argue this rigorously using Definition 1.3.2, we need to verify that properties (i) and (ii) hold. For (i), we just observe that $1 \geq 1/n$ for all choices of $n \in \mathbf{N}$. To verify (ii), we begin by assuming we are in possession of some other upper bound b. Because $1 \in A$ and b is an upper bound for A, we must have $1 \leq b$. This is precisely what property (ii) asks us to show.

Although we do not quite have the tools we need for a rigorous proof (see Theorem 1.4.2), it should be somewhat apparent that $\inf A = 0$.

An important lesson to take from Example 1.3.3 is that $\sup A$ and $\inf A$ may or may not be elements of the set A. This issue is tied to understanding the crucial difference between the maximum and the supremum (or the minimum and the infimum) of a given set.

Definition 1.3.4. A real number a_0 is a *maximum* of the set A if a_0 is an element of A and $a_0 \geq a$ for all $a \in A$. Similarly, a number a_1 is a *minimum* of A if $a_1 \in A$ and $a_1 \leq a$ for every $a \in A$.

Example 1.3.5. To belabor the point, consider the open interval

$$(0, 2) = \{ x \in \mathbf{R} : 0 < x < 2 \},$$

and the closed interval

$$[0, 2] = \{ x \in \mathbf{R} : 0 \leq x \leq 2 \}.$$

Both sets are bounded above (and below), and both have the same least upper bound, namely 2. It is *not* the case, however, that both sets have a maximum. A maximum is a specific type of upper bound that is required to be an element of the set in question, and the open interval $(0, 2)$ does not possess such an element. Thus, the supremum can exist and not be a maximum, but when a maximum exists, then it is also the supremum.

Let's turn our attention back to the Axiom of Completeness. Although we can see now that not every nonempty bounded set contains a maximum, the Axiom of Completeness asserts that every such set does have a least upper bound. We are not going to prove this. An *axiom* in mathematics is an accepted assumption, to be used without proof. Preferably, an axiom should be an elementary statement about the system in question that is so fundamental that it seems to need no justification. Perhaps the Axiom of Completeness fits this description, and perhaps it does not. Before deciding, let's remind ourselves why *it is not a valid statement about* $\mathbf{Q}$.

Example 1.3.6. Consider again the set

$$S = \{r \in \mathbf{Q} : r^2 < 2\},$$

and pretend for the moment that our world consists only of rational numbers. The set S is certainly bounded above. Taking $b = 2$ works, as does $b = 3/2$. But notice what happens as we go in search of the *least* upper bound. (It may be useful here to know that the decimal expansion for $\sqrt{2}$ begins $1.4142\ldots$.) We might try $b = 142/100$, which is indeed an upper bound, but then we discover that $b = 1415/1000$ is an upper bound that is smaller still. Is there a smallest one?

In the rational numbers, there is not. In the real numbers, there is. Back in $\mathbf{R}$, the Axiom of Completeness states that we may set $\alpha = \sup S$ and be confident that such a number exists. In the next section, we will prove that $\alpha^2 = 2$. But according to Theorem 1.1.1, this implies α is not a rational number. If we are restricting our attention to only rational numbers, then α is not an allowable option for $\sup S$, and the search for a least upper bound goes on indefinitely. Whatever rational upper bound is discovered, it is always possible to find one smaller.

The tools needed to carry out the computations described in Example 1.3.6 depend on results about how $\mathbf{Q}$ and $\mathbf{N}$ fit inside of $\mathbf{R}$. These are discussed in the next section. In the meantime, it is possible to prove some intuitive algebraic properties of least upper bounds just using the definition.

Example 1.3.7. Let $A \subseteq \mathbf{R}$ be nonempty and bounded above, and let $c \in \mathbf{R}$. Define the set $c + A$ by

$$c + A = \{c + a : a \in A\}.$$

Then $\sup(c + A) = c + \sup A$.

To properly verify this we focus separately on each part of Definition 1.3.2. Setting $s = \sup A$, we see that $a \leq s$ for all $a \in A$, which implies $c + a \leq c + s$ for all $a \in A$. Thus, $c + s$ is an upper bound for $c + A$ and condition (i) is verified.

For (ii), let b be an arbitrary upper bound for $c + A$; i.e., $c + a \leq b$ for all $a \in A$. This is equivalent to $a \leq b - c$ for all $a \in A$, from which we conclude that $b - c$ is an upper bound for A. Because s is the *least* upper bound of A, $s \leq b - c$, which can be rewritten as $c + s \leq b$. This verifies part (ii) of Definition 1.3.2, and we conclude $\sup(c + A) = c + \sup A$.

There is an equivalent and useful way of characterizing least upper bounds. As the previous example illustrates, Definition 1.3.2 of the supremum has two parts. Part (i) says that $\sup A$ must be an upper bound, and part (ii) states that it must be the smallest one. The following lemma offers an alternative way to restate part (ii).

Lemma 1.3.8. *Assume $s \in \mathbf{R}$ is an upper bound for a set $A \subseteq \mathbf{R}$. Then, $s = \sup A$ if and only if, for every choice of $\epsilon > 0$, there exists an element $a \in A$ satisfying $s - \epsilon < a$.*

Proof. Here is a short rephrasing of the lemma: Given that s is an upper bound, s is the least upper bound if and only if any number smaller than s is not an upper bound. Putting it this way almost qualifies as a proof, but we will expand on what exactly is being said in each direction.

($\Rightarrow$) For the forward direction, we assume $s = \sup A$ and consider $s - \epsilon$, where $\epsilon > 0$ has been arbitrarily chosen. Because $s - \epsilon < s$, part (ii) of Definition 1.3.2 implies that $s - \epsilon$ is *not* an upper bound for A. If this is the case, then there must be some element $a \in A$ for which $s - \epsilon < a$ (because otherwise $s - \epsilon$ would be an upper bound). This proves the lemma in one direction.

($\Leftarrow$) Conversely, assume s is an upper bound with the property that no matter how $\epsilon > 0$ is chosen, $s - \epsilon$ is no longer an upper bound for A. Notice that what this implies is that if b is any number less than s, then b is not an upper bound. (Just let $\epsilon = s - b$.) To prove that $s = \sup A$, we must verify part (ii) of Definition 1.3.2. (Read it again.) Because we have just argued that any number smaller than s cannot be an upper bound, it follows that if b is some other upper bound for A, then $s \leq b$. $\square$

It is certainly the case that all of our conclusions to this point about least upper bounds have analogous versions for greatest lower bounds. The Axiom of Completeness does not explicitly assert that a nonempty set bounded below has an infimum, but this is because we do not need to assume this fact as part of the axiom. Using the Axiom of Completeness, there are several ways to prove that greatest lower bounds exist for nonempty bounded sets. One such proof is explored in Exercise 1.3.3.

Exercises

Exercise 1.3.1.   (a) Write a formal definition in the style of Definition 1.3.2 for the *infimum* or *greatest lower bound* of a set.

(b) Now, state and prove a version of Lemma 1.3.8 for greatest lower bounds.

Exercise 1.3.2. Give an example of each of the following, or state that the request is impossible.

(a) A set B with $\inf B \geq \sup B$.

(b) A finite set that contains its infimum but not its supremum.

(c) A bounded subset of $\mathbf{Q}$ that contains its supremum but not its infimum.

Exercise 1.3.3.   (a) Let A be nonempty and bounded below, and define $B = \{b \in \mathbf{R} : b$ is a lower bound for $A\}$. Show that $\sup B = \inf A$.

(b) Use (a) to explain why there is no need to assert that greatest lower bounds exist as part of the Axiom of Completeness.

Exercise 1.3.4. Let $A_1, A_2, A_3, \ldots$ be a collection of nonempty sets, each of which is bounded above.

(a) Find a formula for $\sup(A_1 \cup A_2)$. Extend this to $\sup\left(\bigcup_{k=1}^{n} A_k\right)$.

(b) Consider $\sup\left(\bigcup_{k=1}^{\infty} A_k\right)$. Does the formula in (a) extend to the infinite case?

Exercise 1.3.5. As in Example 1.3.7, let $A \subseteq \mathbf{R}$ be nonempty and bounded above, and let $c \in \mathbf{R}$. This time define the set $cA = \{ca : a \in A\}$.

(a) If $c \geq 0$, show that $\sup(cA) = c \sup A$.

(b) Postulate a similar type of statement for $\sup(cA)$ for the case $c < 0$.

Exercise 1.3.6. Given sets A and B, define $A+B = \{a+b : a \in A \text{ and } b \in B\}$. Follow these steps to prove that if A and B are nonempty and bounded above then $\sup(A+B) = \sup A + \sup B$.

(a) Let $s = \sup A$ and $t = \sup B$. Show $s + t$ is an upper bound for $A + B$.

(b) Now let u be an arbitrary upper bound for $A + B$, and temporarily fix $a \in A$. Show $t \leq u - a$.

(c) Finally, show $\sup(A + B) = s + t$.

(d) Construct another proof of this same fact using Lemma 1.3.8.

Exercise 1.3.7. Prove that if a is an upper bound for A, and if a is also an element of A, then it must be that $a = \sup A$.

Exercise 1.3.8. Compute, without proofs, the suprema and infima (if they exist) of the following sets:

(a) $\{m/n : m, n \in \mathbf{N} \text{ with } m < n\}$.

(b) $\{(-1)^m/n : m, n \in \mathbf{N}\}$.

(c) $\{n/(3n + 1) : n \in \mathbf{N}\}$.

(d) $\{m/(m + n) : m, n \in \mathbf{N}\}$.

Exercise 1.3.9. (a) If $\sup A < \sup B$, show that there exists an element $b \in B$ that is an upper bound for A.

(b) Give an example to show that this is not always the case if we only assume $\sup A \leq \sup B$.

Exercise 1.3.10 (Cut Property). The *Cut Property* of the real numbers is the following:

If A and B are nonempty, disjoint sets with $A \cup B = \mathbf{R}$ and $a < b$ for all $a \in A$ and $b \in B$, then there exists $c \in \mathbf{R}$ such that $x \leq c$ whenever $x \in A$ and $x \geq c$ whenever $x \in B$.

(a) Use the Axiom of Completeness to prove the Cut Property.

(b) Show that the implication goes the other way; that is, assume **R** possesses the Cut Property and let E be a nonempty set that is bounded above. Prove $\sup E$ exists.

(c) The punchline of parts (a) and (b) is that the Cut Property could be used in place of the Axiom of Completeness as the fundamental axiom that distinguishes the real numbers from the rational numbers. To drive this point home, give a concrete example showing that the Cut Property is not a valid statement when **R** is replaced by **Q**.

Exercise 1.3.11. Decide if the following statements about suprema and infima are true or false. Give a short proof for those that are true. For any that are false, supply an example where the claim in question does not appear to hold.

(a) If A and B are nonempty, bounded, and satisfy $A \subseteq B$, then $\sup A \leq \sup B$.

(b) If $\sup A < \inf B$ for sets A and B, then there exists a $c \in \mathbf{R}$ satisfying $a < c < b$ for all $a \in A$ and $b \in B$.

(c) If there exists a $c \in \mathbf{R}$ satisfying $a < c < b$ for all $a \in A$ and $b \in B$, then $\sup A < \inf B$.

1.4   Consequences of Completeness

The first application of the Axiom of Completeness is a result that may look like a more natural way to mathematically express the sentiment that the real line contains no gaps.

Theorem 1.4.1 (Nested Interval Property). *For each $n \in \mathbf{N}$, assume we are given a closed interval $I_n = [a_n, b_n] = \{x \in \mathbf{R} : a_n \leq x \leq b_n\}$. Assume also that each I_n contains I_{n+1}. Then, the resulting nested sequence of closed intervals*

$$I_1 \supseteq I_2 \supseteq I_3 \supseteq I_4 \supseteq \cdots$$

has a nonempty intersection; that is, $\bigcap_{n=1}^{\infty} I_n \neq \emptyset$.

Proof. In order to show that $\bigcap_{n=1}^{\infty} I_n$ is not empty, we are going to use the Axiom of Completeness (AoC) to produce a single real number x satisfying $x \in I_n$ for every $n \in \mathbf{N}$. Now, AoC is a statement about bounded sets, and the one we want to consider is the set

$$A = \{a_n : n \in \mathbf{N}\}$$

of left-hand endpoints of the intervals.

Because the intervals are nested, we see that every b_n serves as an upper bound for A. Thus, we are justified in setting

$$x = \sup A.$$

Now, consider a particular $I_n = [a_n, b_n]$. Because x is an upper bound for A, we have $a_n \le x$. The fact that each b_n is an upper bound for A and that x is the least upper bound implies $x \le b_n$.

Altogether then, we have $a_n \le x \le b_n$, which means $x \in I_n$ for every choice of $n \in \mathbf{N}$. Hence, $x \in \bigcap_{n=1}^{\infty} I_n$, and the intersection is not empty. □

The Density of Q in R

The set $\mathbf{Q}$ is an extension of $\mathbf{N}$, and $\mathbf{R}$ in turn is an extension of $\mathbf{Q}$. The next few results indicate how $\mathbf{N}$ and $\mathbf{Q}$ sit inside of $\mathbf{R}$.

Theorem 1.4.2 (Archimedean Property). (i) *Given any number $x \in \mathbf{R}$, there exists an $n \in \mathbf{N}$ satisfying $n > x$.*

(ii) *Given any real number $y > 0$, there exists an $n \in \mathbf{N}$ satisfying $1/n < y$.*

Proof. Part (i) of the proposition states that $\mathbf{N}$ is not bounded above. There has never been any doubt about the truth of this, and it could be reasonably argued that we should not have to prove it at all, especially in light of the fact that we have decided to take other familiar properties of $\mathbf{N}$, $\mathbf{Z}$, and $\mathbf{Q}$ as given.

The counterargument is that there is still a great deal of mystery about what the real numbers actually are. What we have said so far is that $\mathbf{R}$ is an extension of $\mathbf{Q}$ that maintains the algebraic and order properties of the rationals but also possesses the least upper bound property articulated in the Axiom of Completeness. In the absence of any other information about $\mathbf{R}$, we have to consider the possibility that in extending $\mathbf{Q}$ we unwittingly acquired some new numbers that are upper bounds for $\mathbf{N}$. In fact, as disorienting as it may sound, there *are* ordered field extensions of $\mathbf{Q}$ that include "numbers" bigger than every natural number. Theorem 1.4.2 asserts that the real numbers do not contain such exotic creatures. The Axiom of Completeness, which we adopted to patch up the holes in $\mathbf{Q}$, carries with it the implication that $\mathbf{N}$ is an unbounded subset of $\mathbf{R}$.

And so to the proof. Assume, for contradiction, that $\mathbf{N}$ *is* bounded above. By the Axiom of Completeness (AoC), $\mathbf{N}$ should then have a least upper bound, and we can set $\alpha = \sup \mathbf{N}$. If we consider $\alpha - 1$, then we no longer have an upper bound (see Lemma 1.3.8), and therefore there exists an $n \in \mathbf{N}$ satisfying $\alpha - 1 < n$. But this is equivalent to $\alpha < n + 1$. Because $n + 1 \in \mathbf{N}$, we have a contradiction to the fact that α is supposed to be an upper bound for $\mathbf{N}$. (Notice that the contradiction here depends only on AoC and the fact that $\mathbf{N}$ is closed under addition.)

Part (ii) follows from (i) by letting $x = 1/y$. □

This familiar property of $\mathbf{N}$ is the key to an extremely important fact about how $\mathbf{Q}$ fits inside of $\mathbf{R}$.

Theorem 1.4.3 (Density of Q in R). *For every two real numbers a and b with $a < b$, there exists a rational number r satisfying $a < r < b$.*

Proof. A rational number is a quotient of integers, so we must produce $m \in \mathbf{Z}$ and $n \in \mathbf{N}$ so that

(1) $$a < \frac{m}{n} < b.$$

The first step is to choose the denominator n large enough so that consecutive increments of size $1/n$ are too close together to "step over" the interval (a, b).

Using the Archimedean Property (Theorem 1.4.2), we may pick $n \in \mathbf{N}$ large enough so that

(2) $$\frac{1}{n} < b - a.$$

Inequality (1) (which we are trying to prove) is equivalent to $na < m < nb$. With n already chosen, the idea now is to choose m to be the smallest integer greater than na. In other words, pick $m \in \mathbf{Z}$ so that

$$m - 1 \overset{(3)}{\leq} na \overset{(4)}{<} m.$$

Now, inequality (4) immediately yields $a < m/n$, which is half of the battle. Keeping in mind that inequality (2) is equivalent to $a < b - 1/n$, we can use (3) to write

$$
\begin{aligned}
m &\leq\ na + 1 \\
&<\ n\left(b - \frac{1}{n}\right) + 1 \\
&=\ nb.
\end{aligned}
$$

Because $m < nb$ implies $m/n < b$, we have $a < m/n < b$, as desired. $\square$

Theorem 1.4.3 is paraphrased by saying that $\mathbf{Q}$ is *dense* in $\mathbf{R}$. Without working too hard, we can use this result to show that the irrational numbers are dense in $\mathbf{R}$ as well.

Corollary 1.4.4. *Given any two real numbers $a < b$, there exists an irrational number t satisfying $a < t < b$.*

Proof. Exercise 1.4.5. $\square$

The Existence of Square Roots

It is time to tend to some unfinished business left over from Example 1.3.6 and this chapter's opening discussion.

Theorem 1.4.5. *There exists a real number $\alpha \in \mathbf{R}$ satisfying $\alpha^2 = 2$.*

Proof. After reviewing Example 1.3.6, consider the set

$$T = \{t \in \mathbf{R} : t^2 < 2\}$$

and set $\alpha = \sup T$. We are going to prove $\alpha^2 = 2$ by ruling out the possibilities $\alpha^2 < 2$ and $\alpha^2 > 2$. Keep in mind that there are two parts to the definition of $\sup T$, and they will both be important. (This always happens when a supremum is used in an argument.) The strategy is to demonstrate that $\alpha^2 < 2$ violates the fact that α is an upper bound for T, and $\alpha^2 > 2$ violates the fact that it is the least upper bound.

Let's first see what happens if we assume $\alpha^2 < 2$. In search of an element of T that is larger than α, write

$$
\begin{aligned}
\left(\alpha + \frac{1}{n}\right)^2 &= \alpha^2 + \frac{2\alpha}{n} + \frac{1}{n^2} \\
&< \alpha^2 + \frac{2\alpha}{n} + \frac{1}{n} \\
&= \alpha^2 + \frac{2\alpha + 1}{n}.
\end{aligned}
$$

But now assuming $\alpha^2 < 2$ gives us a little space in which to fit the $(2\alpha + 1)/n$ term and keep the total less than 2. Specifically, choose $n_0 \in \mathbf{N}$ large enough so that

$$\frac{1}{n_0} < \frac{2 - \alpha^2}{2\alpha + 1}.$$

This implies $(2\alpha + 1)/n_0 < 2 - \alpha^2$, and consequently that

$$\left(\alpha + \frac{1}{n_0}\right)^2 < \alpha^2 + (2 - \alpha^2) = 2.$$

Thus, $\alpha + 1/n_0 \in T$, contradicting the fact that α is an upper bound for T. We conclude that $\alpha^2 < 2$ cannot happen.

Now, what about the case $\alpha^2 > 2$? This time, write

$$
\begin{aligned}
\left(\alpha - \frac{1}{n}\right)^2 &= \alpha^2 - \frac{2\alpha}{n} + \frac{1}{n^2} \\
&> \alpha^2 - \frac{2\alpha}{n}.
\end{aligned}
$$

The remainder of the argument is requested in Exercise 1.4.7. □

A small modification of this proof can be made to show that $\sqrt{x}$ exists for any $x \geq 0$. A formula for expanding $(\alpha + 1/n)^m$ called the binomial formula can be used to show that $\sqrt[m]{x}$ exists for arbitrary values of $m \in \mathbf{N}$.

Exercises

Exercise 1.4.1. Recall that $\mathbf{I}$ stands for the set of irrational numbers.

(a) Show that if $a, b \in \mathbf{Q}$, then ab and $a + b$ are elements of $\mathbf{Q}$ as well.

(b) Show that if $a \in \mathbf{Q}$ and $t \in \mathbf{I}$, then $a + t \in \mathbf{I}$ and $at \in \mathbf{I}$ as long as $a \neq 0$.

(c) Part (a) can be summarized by saying that $\mathbf{Q}$ is closed under addition and multiplication. Is $\mathbf{I}$ closed under addition and multiplication? Given two irrational numbers s and t, what can we say about $s + t$ and st?

Exercise 1.4.2. Let $A \subseteq \mathbf{R}$ be nonempty and bounded above, and let $s \in \mathbf{R}$ have the property that for all $n \in \mathbf{N}$, $s + \frac{1}{n}$ is an upper bound for A and $s - \frac{1}{n}$ is not an upper bound for A. Show $s = \sup A$.

Exercise 1.4.3. Prove that $\bigcap_{n=1}^{\infty}(0, 1/n) = \emptyset$. Notice that this demonstrates that the intervals in the Nested Interval Property must be closed for the conclusion of the theorem to hold.

Exercise 1.4.4. Let $a < b$ be real numbers and consider the set $T = \mathbf{Q} \cap [a, b]$. Show $\sup T = b$.

Exercise 1.4.5. Using Exercise 1.4.1, supply a proof for Corollary 1.4.4 by considering the real numbers $a - \sqrt{2}$ and $b - \sqrt{2}$.

Exercise 1.4.6. Recall that a set B is *dense* in $\mathbf{R}$ if an element of B can be found between any two real numbers $a < b$. Which of the following sets are dense in $\mathbf{R}$? Take $p \in \mathbf{Z}$ and $q \in \mathbf{N}$ in every case.

(a) The set of all rational numbers p/q with $q \leq 10$.

(b) The set of all rational numbers p/q with q a power of 2.

(c) The set of all rational numbers p/q with $10|p| \geq q$.

Exercise 1.4.7. Finish the proof of Theorem 1.4.5 by showing that the assumption $\alpha^2 > 2$ leads to a contradiction of the fact that $\alpha = \sup T$.

Exercise 1.4.8. Give an example of each or state that the request is impossible. When a request is impossible, provide a compelling argument for why this is the case.

(a) Two sets A and B with $A \cap B = \emptyset$, $\sup A = \sup B$, $\sup A \notin A$ and $\sup B \notin B$.

(b) A sequence of nested open intervals $J_1 \supseteq J_2 \supseteq J_3 \supseteq \cdots$ with $\bigcap_{n=1}^{\infty} J_n$ nonempty but containing only a finite number of elements.

(c) A sequence of nested unbounded closed intervals $L_1 \supseteq L_2 \supseteq L_3 \supseteq \cdots$ with $\bigcap_{n=1}^{\infty} L_n = \emptyset$. (An unbounded closed interval has the form $[a, \infty) = \{x \in R : x \geq a\}$.)

(d) A sequence of closed bounded (not necessarily nested) intervals $I_1, I_2, I_3, \ldots$ with the property that $\bigcap_{n=1}^{N} I_n \neq \emptyset$ for all $N \in \mathbf{N}$, but $\bigcap_{n=1}^{\infty} I_n = \emptyset$.

1.5 Cardinality

The applications of the Axiom of Completeness to this point have basically served to restore our confidence in properties we already felt we knew about the real number system. One final consequence of completeness that we are about to explore is of a very different nature and, on its own, represents an astounding intellectual discovery. The traditional way that mathematics gets done is by one mathematician modifying and expanding on the work of those who came before. This model does not seem to apply to Georg Cantor (1845–1918), at least with regard to his work on the theory of infinite sets.

At the moment, we have an image of $\mathbf{R}$ as consisting of rational and irrational numbers, continuously packed together along the real line. We have seen that both $\mathbf{Q}$ and $\mathbf{I}$ (the set of irrationals) are dense in $\mathbf{R}$, meaning that in every interval (a, b) there exist rational and irrational numbers alike. Mentally, there is a temptation to think of $\mathbf{Q}$ and $\mathbf{I}$ as being intricately mixed together in equal proportions, but this turns out not to be the case. In a way that Cantor made precise, the irrational numbers far outnumber the rational numbers in making up the real line.

1–1 Correspondence

The term *cardinality* is used in mathematics to refer to the size of a set. The cardinalities of finite sets can be compared simply by attaching a natural number to each set. The set of Snow White's dwarfs is smaller than the set of United States Supreme Court Justices because 7 is less than 9. But how might we draw this same conclusion without referring to any numbers? Cantor's idea was to attempt to put the sets into a 1–1 correspondence with each other. There are fewer dwarfs than Justices because, if the dwarfs were all simultaneously appointed to the bench, there would still be two empty chairs to fill. On the other hand, the cardinality of the Supreme Court is the same as the cardinality of the set of fielders on a baseball team. This is because, when the judges take the field, it is possible to arrange them so that there is exactly one judge at every position.

The advantage of this method of comparing the sizes of sets is that it works equally well on sets that are infinite.

Definition 1.5.1. A function $f : A \to B$ is *one-to-one* (1–1) if $a_1 \neq a_2$ in A implies that $f(a_1) \neq f(a_2)$ in B. The function f is *onto* if, given any $b \in B$, it is possible to find an element $a \in A$ for which $f(a) = b$.

A function $f : A \to B$ that is both 1–1 and onto provides us with exactly what we mean by a 1–1 correspondence between two sets. The property of being 1–1 means that no two elements of A correspond to the same element of B (no two judges are playing the same position), and the property of being onto ensures that every element of B corresponds to something in A (there is a judge at every position).

Definition 1.5.2. The set A *has the same cardinality as* B if there exists $f : A \to B$ that is 1–1 and onto. In this case, we write $A \sim B$.

Example 1.5.3.   (i) If we let $E = \{2, 4, 6, \ldots\}$ be the set of even natural numbers, then we can show $\mathbf{N} \sim E$. To see why, let $f : \mathbf{N} \to E$ be given by $f(n) = 2n$.

$$
\begin{array}{ccccccc}
\mathbf{N}: & 1 & 2 & 3 & 4 & \cdots & n & \cdots \\
 & \updownarrow & \updownarrow & \updownarrow & \updownarrow & \cdots & \updownarrow & \\
E: & 2 & 4 & 6 & 8 & \cdots & 2n & \cdots
\end{array}
$$

It is certainly true that E is a proper subset of $\mathbf{N}$, and for this reason it may seem logical to say that E is a "smaller" set than $\mathbf{N}$. This is one way to look at it, but it represents a point of view that is heavily biased from an overexposure to finite sets. The definition of cardinality is quite specific, and from this point of view E and $\mathbf{N}$ are equivalent.

(ii) To make this point again, note that although $\mathbf{N}$ is contained in $\mathbf{Z}$ as a proper subset, we can show $\mathbf{N} \sim \mathbf{Z}$. This time let

$$
f(n) = \begin{cases} (n-1)/2 & \text{if } n \text{ is odd} \\ -n/2 & \text{if } n \text{ is even.} \end{cases}
$$

The important details to verify are that f does not map any two natural numbers to the same element of $\mathbf{Z}$ (f is 1–1) and that every element of $\mathbf{Z}$ gets "hit" by something in $\mathbf{N}$ (f is onto).

$$
\begin{array}{ccccccccc}
\mathbf{N}: & 1 & 2 & 3 & 4 & 5 & 6 & 7 & \cdots \\
 & \updownarrow & \updownarrow & \updownarrow & \updownarrow & \updownarrow & \updownarrow & \updownarrow & \\
\mathbf{Z}: & 0 & -1 & 1 & -2 & 2 & -3 & 3 & \cdots
\end{array}
$$

Example 1.5.4. A little calculus (which we will not supply) shows that the function $f(x) = x/(x^2 - 1)$ takes the interval $(-1, 1)$ onto $\mathbf{R}$ in a 1–1 fashion (Fig. 1.4). Thus $(-1, 1) \sim \mathbf{R}$. In fact, $(a, b) \sim \mathbf{R}$ for any interval (a, b).

Countable Sets

Definition 1.5.5. A set A is *countable* if $\mathbf{N} \sim A$. An infinite set that is not countable is called an *uncountable* set.

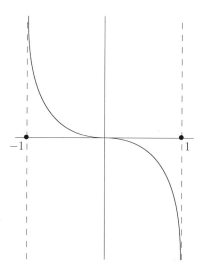

Figure 1.4: $(-1, 1) \sim \mathbf{R}$ **using** $f(x) = x/(x^2 - 1)$.

From Example 1.5.3, we see that both E and $\mathbf{Z}$ are countable sets. Putting a set into a 1–1 correspondence with $\mathbf{N}$, in effect, means putting all of the elements into an infinitely long list or sequence. Looking at Example 1.5.3, we can see that this was quite easy to do for E and required only a modest bit of shuffling for the set $\mathbf{Z}$. A natural question arises as to whether *all* infinite sets are countable. Given some infinite set such as $\mathbf{Q}$ or $\mathbf{R}$, it might seem as though, with enough cleverness, we should be able to fit all the elements of our set into a single list (i.e., into a correspondence with $\mathbf{N}$). After all, this list is infinitely long so there should be plenty of room. But alas, as Hardy remarks, "[The mathematician's] subject is the most curious of all—there is none in which truth plays such odd pranks."

Theorem 1.5.6. (i) *The set* $\mathbf{Q}$ *is countable.* (ii) *The set* $\mathbf{R}$ *is uncountable.*

Proof. (i) Set $A_1 = \{0\}$ and for each $n \geq 2$, let A_n be the set given by

$$A_n = \left\{ \pm \frac{p}{q} : \text{ where } p, q \in \mathbf{N} \text{ are in lowest terms with } p + q = n \right\}.$$

The first few of these sets look like

$$A_1 = \{0\}, \quad A_2 = \left\{ \frac{1}{1}, \frac{-1}{1} \right\}, \quad A_3 = \left\{ \frac{1}{2}, \frac{-1}{2}, \frac{2}{1}, \frac{-2}{1} \right\},$$

$$A_4 = \left\{ \frac{1}{3}, \frac{-1}{3}, \frac{3}{1}, \frac{-3}{1} \right\}, \quad \text{and} \quad A_5 = \left\{ \frac{1}{4}, \frac{-1}{4}, \frac{2}{3}, \frac{-2}{3}, \frac{3}{2}, \frac{-3}{2}, \frac{4}{1}, \frac{-4}{1} \right\}.$$

The crucial observation is that each A_n is *finite* and every rational number appears in exactly one of these sets. Our 1–1 correspondence with $\mathbf{N}$ is then achieved by consecutively listing the elements in each A_n.

$$\mathbf{N}: \quad 1 \quad 2 \quad 3 \quad 4 \quad 5 \quad 6 \quad 7 \quad 8 \quad 9 \quad 10 \quad 11 \quad 12 \quad \cdots$$

$$\updownarrow \ \updownarrow \ \updownarrow \ \updownarrow \ \updownarrow \ \updownarrow \ \updownarrow \ \updownarrow \ \updownarrow \ \updownarrow \ \updownarrow \ \updownarrow$$

$$\mathbf{Q}: \quad 0 \quad \tfrac{1}{1} \quad -\tfrac{1}{1} \quad \tfrac{1}{2} \quad -\tfrac{1}{2} \quad \tfrac{2}{1} \quad -\tfrac{2}{1} \quad \tfrac{1}{3} \quad -\tfrac{1}{3} \quad \tfrac{3}{1} \quad -\tfrac{3}{1} \quad \tfrac{1}{4} \quad \cdots$$

$$\underbrace{\qquad}_{A_1} \underbrace{\qquad}_{A_2} \underbrace{\qquad\qquad}_{A_3} \underbrace{\qquad\qquad\qquad}_{A_4}$$

Admittedly, writing an explicit formula for this correspondence would be an awkward task, and attempting to do so is not the best use of time. What matters is that we see why every rational number appears in the correspondence exactly once. Given, say, $22/7$, we have that $22/7 \in A_{29}$. Because the set of elements in $A_1, \ldots, A_{28}$ is finite, we can be confident that $22/7$ eventually gets included in the sequence. The fact that this line of reasoning applies to any rational number p/q is our proof that the correspondence is onto. To verify that it is 1–1, we observe that the sets A_n were constructed to be disjoint so that no rational number appears twice. This completes the proof of (i).

(ii) The second statement of Theorem 1.5.6 is the truly unexpected part, and its proof is done by contradiction. Assume that there *does* exist a 1–1, onto function $f : \mathbf{N} \to \mathbf{R}$. Again, what this suggests is that it is possible to enumerate the elements of $\mathbf{R}$. If we let $x_1 = f(1)$, $x_2 = f(2)$, and so on, then our assumption that f is onto means that we can write

$$(1) \qquad\qquad \mathbf{R} = \{x_1, x_2, x_3, x_4, \ldots\}$$

and be confident that every real number appears somewhere on the list. We will now use the Nested Interval Property (Theorem 1.4.1) to produce a real number that is not there.

Let I_1 be a closed interval that *does not* contain x_1. Next, let I_2 be a closed interval, contained in I_1, which does not contain x_2. The existence of such an I_2 is easy to verify. Certainly I_1 contains two smaller *disjoint* closed intervals, and x_2 can only be in one of these. In general, given an interval I_n, construct I_{n+1} to satisfy

(i) $I_{n+1} \subseteq I_n$ and

(ii) $x_{n+1} \notin I_{n+1}$.

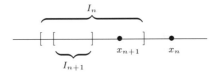

We now consider the intersection $\bigcap_{n=1}^{\infty} I_n$. If x_{n_0} is some real number from the list in (1), then we have $x_{n_0} \notin I_{n_0}$, and it follows that

$$x_{n_0} \notin \bigcap_{n=1}^{\infty} I_n.$$

Now, we are assuming that the list in (1) contains every real number, and this leads to the conclusion that

$$\bigcap_{n=1}^{\infty} I_n = \emptyset.$$

However, the Nested Interval Property (NIP) asserts that $\bigcap_{n=1}^{\infty} I_n \neq \emptyset$. By NIP, there is at least one $x \in \bigcap_{n=1}^{\infty} I_n$ that, consequently, cannot be on the list in (1). This contradiction means that such an enumeration of $\mathbf{R}$ is impossible, and we conclude that $\mathbf{R}$ is an *uncountable* set. $\qquad \square$

What exactly should we make of this discovery? It is an important exercise to show that any subset of a countable set must be either countable or finite. This should not be too surprising. If a set can be arranged into a single list, then deleting some elements from this list results in another (shorter, and potentially terminating) list. This means that countable sets are the smallest type of infinite set. Anything smaller is either still countable or finite.

The force of Theorem 1.5.6 is that the cardinality of $\mathbf{R}$ is, informally speaking, a larger type of infinity. The real numbers so outnumber the natural numbers that there is no way to map $\mathbf{N}$ onto $\mathbf{R}$. No matter how we attempt this, there are always real numbers to spare. The set $\mathbf{Q}$, on the other hand, is countable. As far as infinite sets are concerned, this is as small as it gets. What does this imply about the set $\mathbf{I}$ of irrational numbers? By imitating the demonstration that $\mathbf{N} \sim \mathbf{Z}$, we can prove that the union of two countable sets must be countable. Because $\mathbf{R} = \mathbf{Q} \cup \mathbf{I}$, it follows that $\mathbf{I}$ cannot be countable because otherwise $\mathbf{R}$ would be. The inescapable conclusion is that, despite the fact that we have encountered so few of them, the irrational numbers form a far greater subset of $\mathbf{R}$ than $\mathbf{Q}$.

The properties of countable sets described in this discussion are useful for a few exercises in upcoming chapters. For easier reference, we state them as some final propositions and outline their proofs in the exercises that follow.

Theorem 1.5.7. *If $A \subseteq B$ and B is countable, then A is either countable or finite.*

Theorem 1.5.8. (i) *If $A_1, A_2, \ldots A_m$ are each countable sets, then the union $A_1 \cup A_2 \cup \cdots \cup A_m$ is countable.*

(ii) *If A_n is a countable set for each $n \in \mathbf{N}$, then $\bigcup_{n=1}^{\infty} A_n$ is countable.*

Exercises

Exercise 1.5.1. Finish the following proof for Theorem 1.5.7.

Assume B is a countable set. Thus, there exists $f : \mathbf{N} \to B$, which is 1–1 and onto. Let $A \subseteq B$ be an infinite subset of B. We must show that A is countable.

Let $n_1 = \min\{n \in \mathbf{N} : f(n) \in A\}$. As a start to a definition of $g : \mathbf{N} \to A$, set $g(1) = f(n_1)$. Show how to inductively continue this process to produce a 1–1 function g from $\mathbf{N}$ onto A.

Exercise 1.5.2. Review the proof of Theorem 1.5.6, part (ii) showing that $\mathbf{R}$ is uncountable, and then find the flaw in the following erroneous proof that $\mathbf{Q}$ is uncountable:

Assume, for contradiction, that $\mathbf{Q}$ is countable. Thus we can write $\mathbf{Q} = \{r_1, r_2, r_3, \ldots\}$ and, as before, construct a nested sequence of closed intervals with $r_n \notin I_n$. Our construction implies $\bigcap_{n=1}^{\infty} I_n = \emptyset$ while NIP implies $\bigcap_{n=1}^{\infty} I_n \neq \emptyset$. This contradiction implies $\mathbf{Q}$ must therefore be uncountable.

Exercise 1.5.3. Use the following outline to supply proofs for the statements in Theorem 1.5.8.

(a) First, prove statement (i) for two countable sets, A_1 and A_2. Example 1.5.3 (ii) may be a useful reference. Some technicalities can be avoided by first replacing A_2 with the set $B_2 = A_2 \backslash A_1 = \{x \in A_2 : x \notin A_1\}$. The point of this is that the union $A_1 \cup B_2$ is equal to $A_1 \cup A_2$ and the sets A_1 and B_2 are disjoint. (What happens if B_2 is finite?)

Now, explain how the more general statement in (i) follows.

(b) Explain why induction *cannot* be used to prove part (ii) of Theorem 1.5.8 from part (i).

(c) Show how arranging $\mathbf{N}$ into the two-dimensional array

1	3	6	10	15	$\cdots$
2	5	9	14	$\cdots$	
4	8	13	$\cdots$		
7	12	$\cdots$			
11	$\cdots$				
$\vdots$					

leads to a proof of Theorem 1.5.8 (ii).

Exercise 1.5.4.   (a) Show $(a, b) \sim \mathbf{R}$ for any interval (a, b).

(b) Show that an unbounded interval like $(a, \infty) = \{x : x > a\}$ has the same cardinality as $\mathbf{R}$ as well.

(c) Using open intervals makes it more convenient to produce the required 1–1, onto functions, but it is not really necessary. Show that $[0, 1) \sim (0, 1)$ by exhibiting a 1–1 onto function between the two sets.

Exercise 1.5.5.   (a) Why is $A \sim A$ for every set A?

(b) Given sets A and B, explain why $A \sim B$ is equivalent to asserting $B \sim A$.

(c) For three sets A, B, and C, show that $A \sim B$ and $B \sim C$ implies $A \sim C$. These three properties are what is meant by saying that $\sim$ is an *equivalence relation*.

Exercise 1.5.6.   (a) Give an example of a countable collection of disjoint open intervals.

(b) Give an example of an uncountable collection of disjoint open intervals, or argue that no such collection exists.

Exercise 1.5.7. Consider the open interval (0,1), and let S be the set of points in the open unit square; that is, $S = \{(x, y) : 0 < x, y < 1\}$.

(a) Find a 1–1 function that maps $(0, 1)$ into, but not necessarily onto, S. (This is easy.)

(b) Use the fact that every real number has a decimal expansion to produce a 1–1 function that maps S into $(0, 1)$. Discuss whether the formulated function is onto. (Keep in mind that any terminating decimal expansion such as .235 represents the same real number as .234999....)

The Schröder–Bernstein Theorem discussed in Exercise 1.5.11 can now be applied to conclude that $(0, 1) \sim S$.

Exercise 1.5.8. Let B be a set of positive real numbers with the property that adding together any finite subset of elements from B always gives a sum of 2 or less. Show B must be finite or countable.

Exercise 1.5.9. A real number $x \in \mathbf{R}$ is called *algebraic* if there exist integers $a_0, a_1, a_2, \ldots, a_n \in \mathbf{Z}$, not all zero, such that

$$a_n x^n + a_{n-1} x^{n-1} + \cdots + a_1 x + a_0 = 0.$$

Said another way, a real number is algebraic if it is the root of a polynomial with integer coefficients. Real numbers that are not algebraic are called *transcendental* numbers. Reread the last paragraph of Section 1.1. The final question posed here is closely related to the question of whether or not transcendental numbers exist.

(a) Show that $\sqrt{2}$, $\sqrt[3]{2}$, and $\sqrt{3} + \sqrt{2}$ are algebraic.

(b) Fix $n \in \mathbf{N}$, and let A_n be the algebraic numbers obtained as roots of polynomials with integer coefficients that have degree n. Using the fact that every polynomial has a finite number of roots, show that A_n is countable.

(c) Now, argue that the set of all algebraic numbers is countable. What may we conclude about the set of transcendental numbers?

Exercise 1.5.10.   (a) Let $C \subseteq [0, 1]$ be uncountable. Show that there exists $a \in (0, 1)$ such that $C \cap [a, 1]$ is uncountable.

(b) Now let A be the set of all $a \in (0, 1)$ such that $C \cap [a, 1]$ is uncountable, and set $\alpha = \sup A$. Is $C \cap [\alpha, 1]$ an uncountable set?

(c) Does the statement in (a) remain true if "uncountable" is replaced by "infinite"?

Exercise 1.5.11 (Schröder–Bernstein Theorem). Assume there exists a
1–1 function $f : X \to Y$ and another 1–1 function $g : Y \to X$. Follow the steps
to show that there exists a 1–1, onto function $h : X \to Y$ and hence $X \sim Y$.

The strategy is to partition X and Y into components

$$X = A \cup A' \qquad \text{and} \qquad Y = B \cup B'$$

with $A \cap A' = \emptyset$ and $B \cap B' = \emptyset$, in such a way that f maps A onto B, and g
maps B' onto A'.

(a) Explain how achieving this would lead to a proof that $X \sim Y$.

(b) Set $A_1 = X \backslash g(Y) = \{x \in X : x \notin g(Y)\}$ (what happens if $A_1 = \emptyset$?) and
inductively define a sequence of sets by letting $A_{n+1} = g(f(A_n))$. Show
that $\{A_n : n \in \mathbf{N}\}$ is a pairwise disjoint collection of subsets of X, while
$\{f(A_n) : n \in \mathbf{N}\}$ is a similar collection in Y.

(c) Let $A = \bigcup_{n=1}^{\infty} A_n$ and $B = \bigcup_{n=1}^{\infty} f(A_n)$. Show that f maps A onto B.

(d) Let $A' = X \backslash A$ and $B' = Y \backslash B$. Show g maps B' onto A'.

1.6   Cantor's Theorem

Cantor's work into the theory of infinite sets extends far beyond the conclusions
of Theorem 1.5.6. Although initially resisted, his creative and relentless assault
in this area eventually produced a revolution in set theory and a paradigm shift
in the way mathematicians came to understand the infinite.

Cantor's Diagonalization Method

Cantor published his discovery that $\mathbf{R}$ is uncountable in 1874. Although it
has some modern polish on it, the argument presented in Theorem 1.5.6 (ii)
is actually quite similar to the one Cantor originally found. In 1891, Cantor
offered another proof of this same fact that is startling in its simplicity. It
relies on decimal representations for real numbers, which we will accept and use
without any formal definitions.

Theorem 1.6.1. *The open interval* $(0, 1) = \{x \in \mathbf{R} : 0 < x < 1\}$ *is
uncountable.*

Exercise 1.6.1. Show that $(0, 1)$ is uncountable if and only if $\mathbf{R}$ is uncountable.
This shows that Theorem 1.6.1 is equivalent to Theorem 1.5.6.

Proof. As with Theorem 1.5.6, we proceed by contradiction and assume that
there does exist a function $f : \mathbf{N} \to (0, 1)$ that is 1–1 and onto. For each $m \in \mathbf{N}$,
$f(m)$ is a real number between 0 and 1, and we represent it using the decimal
notation

$$f(m) = .a_{m1}a_{m2}a_{m3}a_{m4}a_{m5}\cdots .$$

What is meant here is that for each $m, n \in \mathbf{N}$, a_{mn} is the digit from the set $\{0, 1, 2, \ldots, 9\}$ that represents the nth digit in the decimal expansion of $f(m)$. The 1–1 correspondence between $\mathbf{N}$ and $(0, 1)$ can be summarized in the doubly indexed array

$\mathbf{N}$		$(0,1)$								
1	$\longleftrightarrow$	$f(1)$	$=$	$.a_{11}$	a_{12}	a_{13}	a_{14}	a_{15}	a_{16}	$\cdots$
2	$\longleftrightarrow$	$f(2)$	$=$	$.a_{21}$	$\mathbf{a_{22}}$	a_{23}	a_{24}	a_{25}	a_{26}	$\cdots$
3	$\longleftrightarrow$	$f(3)$	$=$	$.a_{31}$	a_{32}	$\mathbf{a_{33}}$	a_{34}	a_{35}	a_{36}	$\cdots$
4	$\longleftrightarrow$	$f(4)$	$=$	$.a_{41}$	a_{42}	a_{43}	$\mathbf{a_{44}}$	a_{45}	a_{46}	$\cdots$
5	$\longleftrightarrow$	$f(5)$	$=$	$.a_{51}$	a_{52}	a_{53}	a_{54}	$\mathbf{a_{55}}$	a_{56}	$\cdots$
6	$\longleftrightarrow$	$f(6)$	$=$	$.a_{61}$	a_{62}	a_{63}	a_{64}	a_{65}	$\mathbf{a_{66}}$	$\cdots$
$\vdots$		$\vdots$		$\vdots$	$\vdots$	$\vdots$	$\vdots$	$\vdots$	$\vdots$	$\ddots$

The key assumption about this correspondence is that *every* real number in $(0, 1)$ is assumed to appear somewhere on the list.

Now for the pearl of the argument. Define a real number $x \in (0, 1)$ with the decimal expansion $x = .b_1 b_2 b_3 b_4 \ldots$ using the rule

$$b_n = \begin{cases} 2 & \text{if } a_{nn} \neq 2 \\ 3 & \text{if } a_{nn} = 2. \end{cases}$$

Let's be clear about this. To compute the digit b_1, we look at the digit a_{11} in the upper left-hand corner of the array. If $a_{11} = 2$, then we choose $b_1 = 3$; otherwise, we set $b_1 = 2$.

Exercise 1.6.2. (a) Explain why the real number $x = .b_1 b_2 b_3 b_4 \ldots$ cannot be $f(1)$.

(b) Now, explain why $x \neq f(2)$, and in general why $x \neq f(n)$ for any $n \in \mathbf{N}$.

(c) Point out the contradiction that arises from these observations and conclude that $(0, 1)$ is uncountable. $\square$

Exercise 1.6.3. Supply rebuttals to the following complaints about the proof of Theorem 1.6.1.

(a) Every rational number has a decimal expansion, so we could apply this same argument to show that the set of rational numbers between 0 and 1 is uncountable. However, because we know that any subset of $\mathbf{Q}$ must be countable, the proof of Theorem 1.6.1 must be flawed.

(b) Some numbers have *two* different decimal representations. Specifically, any decimal expansion that terminates can also be written with repeating 9's. For instance, $1/2$ can be written as $.5$ or as $.4999\ldots$. Doesn't this cause some problems?

Exercise 1.6.4. Let S be the set consisting of all sequences of 0's and 1's. Observe that S is not a particular sequence, but rather a large set whose elements are sequences: namely,

$$S = \{(a_1, a_2, a_3, \ldots) : a_n = 0 \text{ or } 1\}.$$

As an example, the sequence $(1, 0, 1, 0, 1, 0, 1, 0, \ldots)$ is an element of S, as is the sequence $(1, 1, 1, 1, 1, 1, \ldots)$.

   Give a rigorous argument showing that S is uncountable.

   Having distinguished between the countable infinity of $\mathbf{N}$ and the uncountable infinity of $\mathbf{R}$, a new question that occupied Cantor was whether or not there existed an infinity "above" that of $\mathbf{R}$. This is logically treacherous territory. The same care we gave to defining the relationship "has the same cardinality as" needs to be given to defining relationships such as "has cardinality greater than" or "has cardinality less than or equal to." Nevertheless, without getting too weighed down with formal definitions, one gets a very clear sense from the next result that there is a hierarchy of infinite sets that continues well beyond the continuum of $\mathbf{R}$.

Power Sets and Cantor's Theorem

Given a set A, the *power set* $P(A)$ refers to the collection of all subsets of A. It is important to understand that $P(A)$ is itself considered a set whose elements are the different possible subsets of A.

Exercise 1.6.5.   (a) Let $A = \{a, b, c\}$. List the eight elements of $P(A)$. (Do not forget that $\emptyset$ is considered to be a subset of every set.)

(b) If A is finite with n elements, show that $P(A)$ has 2^n elements.

Exercise 1.6.6.   (a) Using the particular set $A = \{a, b, c\}$, exhibit two different 1–1 mappings from A into $P(A)$.

(b) Letting $C = \{1, 2, 3, 4\}$, produce an example of a 1–1 map $g : C \to P(C)$.

(c) Explain why, in parts (a) and (b), it is impossible to construct mappings that are *onto*.

   Cantor's Theorem states that the phenomenon in Exercise 1.6.6 holds for infinite sets as well as finite sets. Whereas mapping A *into* $P(A)$ is quite effortless, finding an *onto* map is impossible.

Theorem 1.6.2 (Cantor's Theorem). *Given any set A, there does not exist a function $f : A \to P(A)$ that is onto.*

Proof. This proof, like the others of its kind, is indirect. Thus, assume, for contradiction, that $f : A \to P(A)$ is onto. Unlike the usual situation in which we have sets of numbers for the domain and range, f is a correspondence between a set and its power set. For each element $a \in A$, $f(a)$ is a particular *subset* of A.

The assumption that f is onto means that every subset of A appears as $f(a)$ for some $a \in A$. To arrive at a contradiction, we will produce a subset $B \subseteq A$ that is not equal to $f(a)$ for any $a \in A$.

Construct B using the following rule. For each element $a \in A$, consider the subset $f(a)$. This subset of A may contain the element a or it may not. This depends on the function f. If $f(a)$ does not contain a, then we include a in our set B. More precisely, let

$$B = \{a \in A : a \notin f(a)\}.$$

Exercise 1.6.7. Return to the particular functions constructed in Exercise 1.6.6 and construct the subset B that results using the preceding rule. In each case, note that B is not in the range of the function used.

We now focus on the general argument. Because we have assumed that our function $f : A \to P(A)$ is onto, it must be that $B = f(a')$ for some $a' \in A$. The contradiction arises when we consider whether or not a' is an element of B.

Exercise 1.6.8. (a) First, show that the case $a' \in B$ leads to a contradiction.

(b) Now, finish the argument by showing that the case $a' \notin B$ is equally unacceptable. $\qquad\square$

To get an initial sense of its broad significance, let's apply this result to the set of natural numbers. Cantor's Theorem states that there is no onto function from $\mathbf{N}$ to $P(\mathbf{N})$; in other words, the power set of the natural numbers is uncountable. How does the cardinality of this newly discovered uncountable set compare to the uncountable set of real numbers?

Exercise 1.6.9. Using the various tools and techniques developed in the last two sections (including the exercises from Section 1.5), give a compelling argument showing that $P(\mathbf{N}) \sim \mathbf{R}$.

Exercise 1.6.10. As a final exercise, answer each of the following by establishing a 1–1 correspondence with a set of known cardinality.

(a) Is the set of all functions from $\{0, 1\}$ to $\mathbf{N}$ countable or uncountable?

(b) Is the set of all functions from $\mathbf{N}$ to $\{0, 1\}$ countable or uncountable?

(c) Given a set B, a subset $\mathcal{A}$ of $P(B)$ is called an *antichain* if no element of $\mathcal{A}$ is a subset of any other element of $\mathcal{A}$. Does $P(\mathbf{N})$ contain an uncountable antichain?

1.7   Epilogue

The relationship of having the same cardinality is an *equivalence relation* (see Exercise 1.5.5), meaning, roughly, that all of the sets in the mathematical universe can be organized into disjoint groups according to their size. Two sets appear in the same group, or *equivalence class*, if and only if they have the same cardinality. Thus, $\mathbf{N}$, $\mathbf{Z}$, and $\mathbf{Q}$ are grouped together in one class with all of the other countable sets, whereas $\mathbf{R}$ is in another class that includes the intervals (a, b) as well as $P(\mathbf{N})$. One implication of Cantor's Theorem is that $P(\mathbf{R})$—the set of all subsets of $\mathbf{R}$—is in a different class from $\mathbf{R}$, and there is no reason to stop here. The set of subsets of $P(\mathbf{R})$—namely $P(P(\mathbf{R}))$—is in yet another class, and this process continues indefinitely.

Having divided the universe of sets into disjoint groups, it would be convenient to attach a "number" to each collection which could be used the way natural numbers are used to refer to the sizes of finite sets. Given a set X, there exists something called the *cardinal number* of X, denoted $\operatorname{card} X$, which behaves very much in this fashion. For instance, two sets X and Y satisfy $\operatorname{card} X = \operatorname{card} Y$ if and only if $X \sim Y$. (Rigorously defining $\operatorname{card} X$ requires some significant set theory. One way this is done is to define $\operatorname{card} X$ to be a very particular set that can always be uniquely found in the same equivalence class as X.)

Looking back at Cantor's Theorem, we get the strong sense that there is an *order* on the sizes of infinite sets that should be reflected in our new cardinal number system. Specifically, if it is possible to map a set X *into* Y in a 1–1 fashion, then we want $\operatorname{card} X \leq \operatorname{card} Y$. Writing the strict inequality $\operatorname{card} X < \operatorname{card} Y$ should indicate that it is possible to map X into Y but that it is not the case that $X \sim Y$. Restated in this notation, Cantor's Theorem states that for every set A, $\operatorname{card} A < \operatorname{card} P(A)$.

There are some significant details to work out. A kind of metaphysical problem arises when we realize that an implication of Cantor's Theorem is that there can be no "largest" set. A declaration such as, "Let U be the set of all possible things," is paradoxical because we immediately get that $\operatorname{card} U < \operatorname{card} P(U)$ and thus the set U does not contain everything it was advertised to hold. Issues such as this one are ultimately resolved by imposing some restrictions on what can qualify as a set. As set theory was formalized, the axioms had to be crafted so that objects such as U are simply not allowed. A more down-to-earth problem in need of attention is demonstrating that our definition of "$\leq$" between cardinal numbers really is an ordering. This involves showing that cardinal numbers possess a property analogous to real numbers, which states that if $\operatorname{card} X \leq \operatorname{card} Y$ and $\operatorname{card} Y \leq \operatorname{card} X$, then $\operatorname{card} X = \operatorname{card} Y$. In the end, this boils down to proving that if there exists $f : X \to Y$ that is 1–1, and if there exists $g : Y \to X$ that is 1–1, then it is possible to find a function $h : X \to Y$ that is both 1–1 and onto. A proof of this fact eluded Cantor but was eventually supplied independently by Ernst Schröder (in 1896) and Felix Bernstein (in 1898). An argument for the Schröder–Bernstein Theorem is outlined in Exercise 1.5.11.

There was another deep problem stemming from the budding theory of cardinal numbers that occupied Cantor and which was not resolved during his lifetime. Because of the importance of countable sets, the symbol $\aleph_0$ ("aleph naught") is frequently used for card $\mathbf{N}$. The subscript "0" is appropriate when we remember that countable sets are the smallest type of infinite set. In terms of cardinal numbers, if card $X < \aleph_0$, then X is finite. Thus, $\aleph_0$ is the smallest infinite cardinal number. The cardinality of $\mathbf{R}$ is also significant enough to deserve the special designation $c = \text{card}\,\mathbf{R} = \text{card}(0,1)$. The content of Theorems 1.5.6 and 1.6.1 is that $\aleph_0 < c$. The question that plagued Cantor was whether there were any cardinal numbers strictly in between these two. Put another way, does there exist a set $A \subseteq \mathbf{R}$ with card $\mathbf{N} < \text{card}\,A < \text{card}\,\mathbf{R}$? Cantor was of the opinion that no such set existed. In the ordering of cardinal numbers, he conjectured, c was the immediate successor of $\aleph_0$.

Cantor's "continuum hypothesis," as it came to be called, was one of the most famous mathematical challenges of the past century. Its unexpected resolution came in two parts. In 1940, the German logician and mathematician Kurt Gödel demonstrated that, using only the agreed-upon set of axioms of set theory, there was no way to disprove the continuum hypothesis. In 1963, Paul Cohen successfully showed that, under the same rules, it was also impossible to prove this conjecture. Taken together, what these two discoveries imply is that the continuum hypothesis is undecidable. It can be accepted or rejected as a statement about the nature of infinite sets, and in neither case will any logical contradictions arise.

The mention of Kurt Gödel brings to mind a final comment about the significance of Cantor's work. Gödel is best known for his "Incompleteness Theorems," which pertain to the strength of axiomatic systems in general. What Gödel showed was that any consistent axiomatic system created to study arithmetic was necessarily destined to be "incomplete" in the sense that there would always be true statements that the system of axioms would be too weak to prove. At the heart of Gödel's very complicated proof is a type of manipulation closely related to what is happening in the proofs of Theorems 1.6.1 and 1.6.2. Variations of Cantor's proof methods can also be found in the limitative results of computer science. The "halting problem" asks, loosely, whether some general algorithm exists that can look at every program and decide if that program eventually terminates. The proof that no such algorithm exists uses a diagonalization-type construction at the core of the argument. The main point to make is that not only are the implications of Cantor's theorems profound but the argumentative techniques are as well. As a more immediate example of this phenomenon, the diagonalization method is used again in Chapter 6—in a constructive way—as a crucial step in the proof of the Arzela–Ascoli Theorem.

Chapter 2

Sequences and Series

2.1 Discussion: Rearrangements of Infinite Series

Consider the infinite series

$$\sum_{n=1}^{\infty} \frac{(-1)^{n+1}}{n} = 1 - \frac{1}{2} + \frac{1}{3} - \frac{1}{4} + \frac{1}{5} - \frac{1}{6} + \frac{1}{7} - \frac{1}{8} + \cdots.$$

If we naively begin adding from the left-hand side, we get a sequence of what are called *partial sums*. In other words, let s_n equal the sum of the first n terms of the series, so that $s_1 = 1$, $s_2 = 1/2$, $s_3 = 5/6$, $s_4 = 7/12$, and so on. One immediate observation is that the successive sums oscillate in a progressively narrower space. The odd sums decrease ($s_1 > s_3 > s_5 > \ldots$) while the even sums increase ($s_2 < s_4 < s_6 < \ldots$).

$$s_2 < s_4 < s_6 < \cdots S \cdots < s_5 < s_3 < s_1$$

It seems reasonable—and we will soon prove—that the sequence (s_n) eventually hones in on a value, call it S, where the odd and even partial sums "meet." At this moment, we cannot compute S precisely, but we know it falls somewhere between $7/12$ and $5/6$. Summing a few hundred terms reveals that $S \approx .69$. Whatever its value, there is now an overwhelming temptation to write

(1) $$S = 1 - \frac{1}{2} + \frac{1}{3} - \frac{1}{4} + \frac{1}{5} - \frac{1}{6} + \frac{1}{7} - \frac{1}{8} + \cdots$$

© Springer Science+Business Media New York 2015
S. Abbott, *Understanding Analysis*, Undergraduate Texts
in Mathematics, DOI 10.1007/978-1-4939-2712-8_2

meaning, perhaps, that if we could indeed add up *all* infinitely many of these numbers, then the sum would *equal* S. A more familiar example of an equation of this type might be

$$2 = 1 + \frac{1}{2} + \frac{1}{4} + \frac{1}{8} + \frac{1}{16} + \frac{1}{32} + \frac{1}{64} + \cdots,$$

the only difference being that in the second equation we have a more recognizable value for the sum.

But now for the crux of the matter. The symbols $+$, $-$, and $=$ in the preceding equations are deceptively familiar notions being used in a very unfamiliar way. The crucial question is whether or not properties of addition and equality that are well understood for finite sums remain valid when applied to infinite objects such as equation (1). The answer, as we are about to witness, is somewhat ambiguous.

Treating equation (1) in a standard algebraic way, let's multiply through by $1/2$ and add it back to equation (1):

$$\frac{1}{2}S = \quad \frac{1}{2} \quad - \frac{1}{4} \quad + \frac{1}{6} \quad - \frac{1}{8} \quad + \frac{1}{10} \quad - \frac{1}{12} \quad + \cdots$$

$$+ \; S = 1 - \frac{1}{2} + \frac{1}{3} - \frac{1}{4} + \frac{1}{5} - \frac{1}{6} + \frac{1}{7} - \frac{1}{8} + \frac{1}{9} - \frac{1}{10} + \frac{1}{11} - \frac{1}{12} + \frac{1}{13} - \cdots$$

$$(2) \quad \frac{3}{2}S = 1 \quad + \frac{1}{3} - \frac{1}{2} + \frac{1}{5} \quad + \frac{1}{7} - \frac{1}{4} + \frac{1}{9} \quad + \frac{1}{11} - \frac{1}{6} + \frac{1}{13} \quad \cdots$$

Now, look carefully at the result. The sum in equation (2) consists *precisely* of the same terms as those in the original equation (1), only in a different order. Specifically, the series in (2) is a rearrangement of (1) where we list the first two positive terms $(1 + \frac{1}{3})$ followed by the first negative term $(-\frac{1}{2})$, followed by the next two positive terms $(\frac{1}{5} + \frac{1}{7})$ and then the next negative term $(-\frac{1}{4})$. Continuing this, it is apparent that every term in (2) appears in (1) and vice versa. The rub comes when we realize that equation (2) asserts that the sum of these rearranged, but otherwise unaltered, numbers is equal to $3/2$ its original value. Indeed, adding a few hundred terms of equation (2) produces partial sums in the neighborhood of 1.03. Addition, in this infinite setting, is not commutative!

Let's look at a similar rearrangement of the series

$$\sum_{n=0}^{\infty}(-1/2)^{n}.$$

This series is geometric with first term 1 and common ratio $r = -1/2$. Using the formula $1/(1-r)$ for the sum of a geometric series (Example 2.7.5), we get

$$1 - \frac{1}{2} + \frac{1}{4} - \frac{1}{8} + \frac{1}{16} - \frac{1}{32} + \frac{1}{64} - \frac{1}{128} + \frac{1}{256} \cdots = \frac{1}{1-(-\frac{1}{2})} = \frac{2}{3}.$$

This time, some computational experimentation with the "two positives, one negative" rearrangement

$$1 + \frac{1}{4} - \frac{1}{2} + \frac{1}{16} + \frac{1}{64} - \frac{1}{8} + \frac{1}{256} + \frac{1}{1024} - \frac{1}{32} \cdots$$

yields partial sums quite close to $2/3$. The sum of the first 30 terms, for instance, equals .666667. Infinite addition is commutative in some instances but not in others.

Far from being a charming theoretical oddity of infinite series, this phenomenon can be the source of great consternation in many applied situations. How, for instance, should a double summation over two index variables be defined? Let's say we are given a *grid* of real numbers $\{a_{ij} : i, j \in \mathbf{N}\}$, where $a_{ij} = 1/2^{j-i}$ if $j > i$, $a_{ij} = -1$ if $j = i$, and $a_{ij} = 0$ if $j < i$.

$$\begin{bmatrix} -1 & \frac{1}{2} & \frac{1}{4} & \frac{1}{8} & \frac{1}{16} & \cdots \\ 0 & -1 & \frac{1}{2} & \frac{1}{4} & \frac{1}{8} & \cdots \\ 0 & 0 & -1 & \frac{1}{2} & \frac{1}{4} & \cdots \\ 0 & 0 & 0 & -1 & \frac{1}{2} & \cdots \\ 0 & 0 & 0 & 0 & -1 & \cdots \\ \vdots & \vdots & \vdots & \vdots & \vdots & \ddots \end{bmatrix}$$

We would like to attach a mathematical meaning to the summation

$$\sum_{i,j=1}^{\infty} a_{ij}$$

whereby we intend to include every term in the preceding array in the total. One natural idea is to temporarily fix i and sum across each row. A moment's reflection (and a fact about geometric series) shows that each row sums to 0. Summing the sums of the rows, we get

$$\sum_{i,j=1}^{\infty} a_{ij} = \sum_{i=1}^{\infty} \left(\sum_{j=1}^{\infty} a_{ij} \right) = \sum_{i=1}^{\infty} (0) = 0.$$

We could just as easily have decided to fix j and sum down each column first. In this case, we have

$$\sum_{i,j=1}^{\infty} a_{ij} = \sum_{j=1}^{\infty} \left(\sum_{i=1}^{\infty} a_{ij} \right) = \sum_{j=1}^{\infty} \left(\frac{-1}{2^{j-1}} \right) = -2.$$

Changing the order of the summation changes the value of the sum! One common way that double sums arise (although not this particular one) is from the multiplication of two series. There is a natural desire to write

$$\left(\sum a_i \right) \left(\sum b_j \right) = \sum_{i,j} a_i b_j,$$

except that the expression on the right-hand side makes no sense at the moment.

It is the pathologies that give rise to the need for rigor. A satisfying resolution to the questions raised will require that we be absolutely precise about what we mean as we manipulate these infinite objects. It may seem that progress is slow at first, but that is because we do not want to fall into the trap of letting the biases of our intuition corrupt our arguments. Rigorous proofs are meant to be a check on intuition, and in the end we will see that they vastly improve our mental picture of the mathematical infinite.

As a final example, consider something as intuitively fundamental as the associative property of addition applied to the series $\sum_{n=1}^{\infty}(-1)^n$. Grouping the terms one way gives

$$(-1+1) + (-1+1) + (-1+1) + (-1+1) + \cdots = 0 + 0 + 0 + 0 + \cdots = 0,$$

whereas grouping in another yields

$$-1 + (1-1) + (1-1) + (1-1) + \cdots = -1 + 0 + 0 + 0 + \cdots = -1.$$

Manipulations that are legitimate in finite settings do not always extend to infinite settings. Deciding when they do and why they do not is one of the central themes of analysis.

2.2   The Limit of a Sequence

An understanding of infinite series depends heavily on a clear understanding of the theory of sequences. In fact, most of the concepts in analysis can be reduced to statements about the behavior of sequences. Thus, we will spend a significant amount of time investigating sequences before taking on infinite series.

Definition 2.2.1. A *sequence* is a function whose domain is $\mathbf{N}$.

This formal definition leads immediately to the familiar depiction of a sequence as an ordered list of real numbers. Given a function $f : \mathbf{N} \to \mathbf{R}$, $f(n)$ is just the nth term on the list. The notation for sequences reinforces this familiar understanding.

Example 2.2.2. Each of the following are common ways to describe a sequence.

(i) $(1, \frac{1}{2}, \frac{1}{3}, \frac{1}{4}, \cdots)$,

(ii) $(\frac{1+n}{n})_{n=1}^{\infty} = (\frac{2}{1}, \frac{3}{2}, \frac{4}{3}, \cdots)$,

(iii) (a_n), where $a_n = 2^n$ for each $n \in \mathbf{N}$,

(iv) (x_n), where $x_1 = 2$ and $x_{n+1} = \frac{x_n+1}{2}$.

On occasion, it will be more convenient to index a sequence beginning with $n = 0$ or $n = n_0$ for some natural number n_0 different from 1. These minor variations should cause no confusion. What is essential is that a sequence be an *infinite* list of real numbers. What happens at the beginning of such a list is of little importance in most cases. The business of analysis is concerned with the behavior of the infinite "tail" of a given sequence.

We now present what is arguably the most important definition in the book.

Definition 2.2.3 (Convergence of a Sequence). A sequence (a_n) *converges* to a real number a if, for every positive number ϵ, there exists an $N \in \mathbf{N}$ such that whenever $n \geq N$ it follows that $|a_n - a| < \epsilon$.

To indicate that (a_n) converges to a, we usually write either $\lim a_n = a$ or $(a_n) \to a$. The notation $\lim_{n \to \infty} a_n = a$ is also standard.

In an effort to decipher this complicated definition, it helps first to consider the ending phrase "$|a_n - a| < \epsilon$," and think about the points that satisfy an inequality of this type.

Definition 2.2.4. Given a real number $a \in \mathbf{R}$ and a positive number $\epsilon > 0$, the set

$$V_\epsilon(a) = \{x \in \mathbf{R} : |x - a| < \epsilon\}$$

is called the ϵ-*neighborhood of* a.

Notice that $V_\epsilon(a)$ consists of all of those points whose distance from a is less than ϵ. Said another way, $V_\epsilon(a)$ is an interval, centered at a, with radius ϵ.

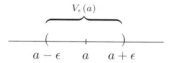

Recasting the definition of convergence in terms of ϵ-neighborhoods gives a more geometric impression of what is being described.

Definition 2.2.3B (Convergence of a Sequence: Topological Version). A sequence (a_n) converges to a if, given any ϵ-neighborhood $V_\epsilon(a)$ of a, there exists a point in the sequence after which all of the terms are in $V_\epsilon(a)$. In other words, every ϵ-neighborhood contains all but a finite number of the terms of (a_n).

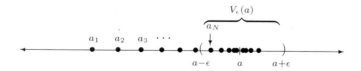

Definition 2.2.3 and Definition 2.2.3B say precisely the same thing; the natural number N in the original version of the definition is the point where the sequence (a_n) enters $V_\epsilon(a)$, never to leave. It should be apparent that *the value of N depends on the choice of ϵ*. The smaller the ϵ-neighborhood, the larger N may have to be.

Example 2.2.5. Consider the sequence (a_n), where $a_n = 1/\sqrt{n}$.

Our intuitive understanding of limits points confidently to the conclusion that

$$\lim \left(\frac{1}{\sqrt{n}} \right) = 0.$$

Before trying to prove this not too impressive fact, let's first explore the relationship between ϵ and N in the definition of convergence. For the moment, take ϵ to be $1/10$. This defines a sort of "target zone" for the terms in the sequence. By claiming that the limit of (a_n) is 0, we are saying that the terms in this sequence eventually get arbitrarily close to 0. How close? What do we mean by "eventually"? We have set $\epsilon = 1/10$ as our standard for closeness, which leads to the ϵ-neighborhood $(-1/10, 1/10)$ centered around the limit 0. How far out into the sequence must we look before the terms fall into this interval? The 100th term $a_{100} = 1/10$ puts us right on the boundary, and a little thought reveals that

$$\text{if} \quad n > 100, \quad \text{then} \quad a_n \in \left(-\frac{1}{10}, \frac{1}{10} \right).$$

Thus, for $\epsilon = 1/10$ we choose $N = 101$ (or anything larger) as our response.

Now, our choice of $\epsilon = 1/10$ was rather whimsical, and we can do this again, letting $\epsilon = 1/50$. In this case, our target neighborhood shrinks to $(-1/50, 1/50)$, and it is apparent that we must travel farther out into the sequence before a_n falls into this interval. How far? Essentially, we require that

$$\frac{1}{\sqrt{n}} < \frac{1}{50} \quad \text{which occurs as long as} \quad n > 50^2 = 2500.$$

Thus, $N = 2501$ is a suitable response to the challenge of $\epsilon = 1/50$.

It may seem as though this duel could continue forever, with different ϵ challenges being handed to us one after another, each one requiring a suitable value of N in response. In a sense, this is correct, except that the game is effectively over the instant we recognize a *rule* for how to choose N given an *arbitrary* $\epsilon > 0$. For this problem, the desired algorithm is implicit in the algebra carried out to compute the previous response of $N = 2501$. Whatever ϵ happens to be, we want

$$\frac{1}{\sqrt{n}} < \epsilon \quad \text{which is equivalent to insisting that} \quad n > \frac{1}{\epsilon^2}.$$

With this observation, we are ready to write the formal argument.

We claim that

$$\lim \left(\frac{1}{\sqrt{n}} \right) = 0.$$

Proof. Let $\epsilon > 0$ be an arbitrary positive number. Choose a natural number N satisfying

$$N > \frac{1}{\epsilon^2}.$$

We now verify that this choice of N has the desired property. Let $n \geq N$. Then,

$$n > \frac{1}{\epsilon^2} \quad \text{implies} \quad \frac{1}{\sqrt{n}} < \epsilon, \quad \text{and hence} \quad |a_n - 0| < \epsilon. \qquad \square$$

Quantifiers

The definition of convergence given earlier is the result of hundreds of years of refining the intuitive notion of limit into a mathematically rigorous statement. The logic involved is complicated and is intimately tied to the use of the quantifiers "for all" and "there exists." Learning to write a grammatically correct convergence proof goes hand in hand with a deep understanding of why the quantifiers appear in the order that they do.

The definition begins with the phrase,

"For all $\epsilon > 0$, there exists $N \in \mathbf{N}$ such that ..."

Looking back at our first example, we see that our formal proof begins with, "Let $\epsilon > 0$ be an arbitrary positive number." This is followed by a construction of N and then a demonstration that this choice of N has the desired property. This, in fact, is a basic outline for how every convergence proof should be presented.

TEMPLATE FOR A PROOF THAT $(x_n) \to x$:

- "Let $\epsilon > 0$ be arbitrary."

- Demonstrate a choice for $N \in \mathbf{N}$. This step usually requires the most work, almost all of which is done prior to actually writing the formal proof.

- Now, show that N actually works.

- "Assume $n \geq N$."

- With N well chosen, it should be possible to derive the inequality $|x_n - x| < \epsilon$.

Example 2.2.6. Show

$$\lim \left(\frac{n+1}{n} \right) = 1.$$

As mentioned, before attempting a formal proof, we first need to do some preliminary scratch work. In the first example, we experimented by assigning specific values to ϵ (and it is not a bad idea to do this again), but let us skip straight to the algebraic punch line. The last line of our proof should be that for suitably large values of n,

$$\left| \frac{n+1}{n} - 1 \right| < \epsilon.$$

Because
$$\left|\frac{n+1}{n} - 1\right| = \frac{1}{n},$$
this is equivalent to the inequality $1/n < \epsilon$ or $n > 1/\epsilon$. Thus, choosing N to be an integer greater than $1/\epsilon$ will suffice.

With the work of the proof done, all that remains is the formal writeup.

Proof. Let $\epsilon > 0$ be arbitrary. Choose $N \in \mathbf{N}$ with $N > 1/\epsilon$. To verify that this choice of N is appropriate, let $n \in \mathbf{N}$ satisfy $n \geq N$. Then, $n \geq N$ implies $n > 1/\epsilon$, which is the same as saying $1/n < \epsilon$. Finally, this means
$$\left|\frac{n+1}{n} - 1\right| < \epsilon,$$
as desired.   □

It is instructive to see what goes wrong in the previous example if we try to prove that our sequence converges to some limit other than 1.

Theorem 2.2.7 (Uniqueness of Limits). *The limit of a sequence, when it exists, must be unique.*

Proof. Exercise 2.2.6.   □

Divergence

Significant insight into the role of the quantifiers in the definition of convergence can be gained by studying an example of a sequence that does not have a limit.

Example 2.2.8. Consider the sequence
$$\left(1, -\frac{1}{2}, \frac{1}{3}, -\frac{1}{4}, \frac{1}{5}, -\frac{1}{5}, \frac{1}{5}, -\frac{1}{5}, \frac{1}{5}, -\frac{1}{5}, \frac{1}{5}, -\frac{1}{5}, \frac{1}{5}, -\frac{1}{5}, \cdots\right).$$

How can we argue that this sequence does not converge to zero? Looking at the first few terms, it seems the initial evidence actually supports such a conclusion. Given a challenge of $\epsilon = 1/2$, a little reflection reveals that after $N = 3$ all the terms fall into the neighborhood $(-1/2, 1/2)$. We could also handle $\epsilon = 1/4$. (What is the smallest possible N in this case?)

But the definition of convergence says "*For all* $\epsilon > 0 \ldots$," and it should be apparent that there is no response to a choice of $\epsilon = 1/10$, for instance. This leads us to an important observation about the logical negation of the definition of convergence of a sequence. To prove that a particular number x is *not* the limit of a sequence (x_n), we must produce a single value of ϵ for which no $N \in \mathbf{N}$ works. More generally speaking, the negation of a statement that begins "For all P, there exists Q..." is the statement, "For at least one P, no Q is possible..." For instance, how could we disprove the spurious claim that "At every college in the United States, there is a student who is at least seven feet tall"?

We have argued that the preceding sequence does not converge to 0. Let's argue against the claim that it converges to 1/5. Choosing $\epsilon = 1/10$ produces the neighborhood $(1/10, 3/10)$. Although the sequence continually revisits this neighborhood, there is no point at which it enters and never leaves as the definition requires. Thus, no N exists for $\epsilon = 1/10$, so the sequence does not converge to 1/5.

Of course, this sequence does not converge to any other real number, and it would be more satisfying to simply say that this sequence does not converge.

Definition 2.2.9. A sequence that does not converge is said to *diverge*.

Although it is not too difficult, we will postpone arguing for divergence in general until we develop a more economical divergence criterion later in Section 2.5.

Exercises

Exercise 2.2.1. What happens if we reverse the order of the quantifiers in Definition 2.2.3?

Definition: A sequence (x_n) *verconges* to x if *there exists* an $\epsilon > 0$ such that *for all* $N \in \mathbf{N}$ it is true that $n \geq N$ implies $|x_n - x| < \epsilon$.

Give an example of a vercongent sequence. Is there an example of a vercongent sequence that is divergent? Can a sequence verconge to two different values? What exactly is being described in this strange definition?

Exercise 2.2.2. Verify, using the definition of convergence of a sequence, that the following sequences converge to the proposed limit.

(a) $\lim \frac{2n+1}{5n+4} = \frac{2}{5}$.

(b) $\lim \frac{2n^2}{n^3+3} = 0$.

(c) $\lim \frac{\sin(n^2)}{\sqrt[3]{n}} = 0$.

Exercise 2.2.3. Describe what we would have to demonstrate in order to disprove each of the following statements.

(a) At every college in the United States, there is a student who is at least seven feet tall.

(b) For all colleges in the United States, there exists a professor who gives every student a grade of either A or B.

(c) There exists a college in the United States where every student is at least six feet tall.

Exercise 2.2.4. Give an example of each or state that the request is impossible. For any that are impossible, give a compelling argument for why that is the case.

(a) A sequence with an infinite number of ones that does not converge to one.

(b) A sequence with an infinite number of ones that converges to a limit not equal to one.

(c) A divergent sequence such that for every $n \in \mathbf{N}$ it is possible to find n consecutive ones somewhere in the sequence.

Exercise 2.2.5. Let $[[x]]$ be the greatest integer less than or equal to x. For example, $[[\pi]] = 3$ and $[[3]] = 3$. For each sequence, find $\lim a_n$ and verify it with the definition of convergence.

(a) $a_n = [[5/n]]$,

(b) $a_n = [[(12 + 4n)/3n]]$.

Reflecting on these examples, comment on the statement following Definition 2.2.3 that "the smaller the ϵ-neighborhood, the larger N may have to be."

Exercise 2.2.6. Prove Theorem 2.2.7. To get started, assume $(a_n) \to a$ and also that $(a_n) \to b$. Now argue $a = b$.

Exercise 2.2.7. Here are two useful definitions:

(i) A sequence (a_n) is *eventually* in a set $A \subseteq \mathbf{R}$ if there exists an $N \in \mathbf{N}$ such that $a_n \in A$ for all $n \geq N$.

(ii) A sequence (a_n) is *frequently* in a set $A \subseteq \mathbf{R}$ if, for every $N \in \mathbf{N}$, there exists an $n \geq N$ such that $a_n \in A$.

(a) Is the sequence $(-1)^n$ eventually or frequently in the set $\{1\}$?

(b) Which definition is stronger? Does frequently imply eventually or does eventually imply frequently?

(c) Give an alternate rephrasing of Definition 2.2.3B using either frequently or eventually. Which is the term we want?

(d) Suppose an infinite number of terms of a sequence (x_n) are equal to 2. Is (x_n) necessarily eventually in the interval $(1.9, 2.1)$? Is it frequently in $(1.9, 2.1)$?

Exercise 2.2.8. For some additional practice with nested quantifiers, consider the following invented definition:

Let's call a sequence (x_n) *zero-heavy* if there exists $M \in \mathbf{N}$ such that for all $N \in \mathbf{N}$ there exists n satisfying $N \leq n \leq N + M$ where $x_n = 0$.

(a) Is the sequence $(0, 1, 0, 1, 0, 1, \ldots)$ zero heavy?

(b) If a sequence is zero-heavy does it necessarily contain an infinite number of zeros? If not, provide a counterexample.

(c) If a sequence contains an infinite number of zeros, is it necessarily zero-heavy? If not, provide a counterexample.

(d) Form the logical negation of the above definition. That is, complete the sentence: A sequence is *not* zero-heavy if $\ldots$.

2.3 The Algebraic and Order Limit Theorems

The real purpose of creating a rigorous definition for convergence of a sequence is *not* to have a tool to verify computational statements such as $\lim 2n/(n+2) = 2$. Historically, a definition of the limit like Definition 2.2.3 came 150 years after the founders of calculus began working with intuitive notions of convergence. The point of having such a logically tight description of convergence is so that we can confidently *prove statements about convergent sequences in general.* We are ultimately trying to resolve arguments about what is and is not true regarding the behavior of limits with respect to the mathematical manipulations we intend to inflict on them.

As a first example, let us prove that convergent sequences are bounded. The term "bounded" has a rather familiar connotation but, like everything else, we need to be explicit about what it means in this context.

Definition 2.3.1. A sequence (x_n) is *bounded* if there exists a number $M > 0$ such that $|x_n| \leq M$ for all $n \in \mathbf{N}$.

Geometrically, this means that we can find an interval $[-M, M]$ that contains every term in the sequence (x_n).

Theorem 2.3.2. *Every convergent sequence is bounded.*

Proof. Assume (x_n) converges to a limit l. This means that given a particular value of ϵ, say $\epsilon = 1$, we know there must exist an $N \in \mathbf{N}$ such that if $n \geq N$, then x_n is in the interval $(l - 1, l + 1)$. Not knowing whether l is positive or negative, we can certainly conclude that

$$|x_n| < |l| + 1$$

for all $n \geq N$.

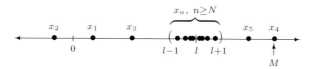

We still need to worry (slightly) about the terms in the sequence that come before the Nth term. Because there are only a finite number of these, we let

$$M = \max\{|x_1|, |x_2|, |x_3|, \ldots, |x_{N-1}|, |l| + 1\}.$$

It follows that $|x_n| \leq M$ for all $n \in \mathbf{N}$, as desired. $\qquad\square$

This chapter began with a demonstration of how applying familiar algebraic properties (commutativity of addition) to infinite objects (series) can lead to paradoxical results. These examples are meant to instill in us a sense of caution and justify the extreme care we are taking in drawing our conclusions. The following theorems illustrate that sequences behave extremely well with respect to the operations of addition, multiplication, division, and order.

Theorem 2.3.3 (Algebraic Limit Theorem). *Let* $\lim a_n = a$, *and* $\lim b_n = b$. *Then,*

   (i) $\lim(ca_n) = ca$, *for all* $c \in \mathbf{R}$;

   (ii) $\lim(a_n + b_n) = a + b$;

   (iii) $\lim(a_n b_n) = ab$;

   (iv) $\lim(a_n/b_n) = a/b$, *provided* $b \neq 0$.

Proof. (i) Consider the case where $c \neq 0$. We want to show that the sequence (ca_n) converges to ca, so the structure of the proof follows the template we described in Section 2.2. First, we let ϵ be some arbitrary positive number. Our goal is to find some point in the sequence (ca_n) after which we have

$$|ca_n - ca| < \epsilon.$$

Now,

$$|ca_n - ca| = |c||a_n - a|.$$

We are given that $(a_n) \to a$, so we know we can make $|a_n - a|$ as small as we like. In particular, we can choose an N such that

$$|a_n - a| < \frac{\epsilon}{|c|}$$

whenever $n \geq N$. To see that this N indeed works, observe that, for all $n \geq N$,

$$|ca_n - ca| = |c||a_n - a| < |c|\frac{\epsilon}{|c|} = \epsilon.$$

The case $c = 0$ reduces to showing that the constant sequence $(0, 0, 0, \ldots)$ converges to 0, which is easily verified.

Before continuing with parts (ii), (iii), and (iv), we should point out that the proof of (i), while somewhat short, is extremely typical for a convergence proof. Before embarking on a formal argument, it is a good idea to take an inventory of what we *want* to make less than ϵ, and what we are *given* can be made small for suitable choices of n. For the previous proof, we wanted to make $|ca_n - ca| < \epsilon$, and we were given $|a_n - a| <$ *anything we like* (for large values of n). Notice that in (i), and all of the ensuing arguments, the strategy each time is to bound the quantity we want to be less than ϵ, which in each case is

$$|(\text{terms of sequence}) - (\text{proposed limit})|,$$

with some algebraic combination of quantities over which we have control.

(ii) To prove this statement, we need to argue that the quantity

$$|(a_n + b_n) - (a + b)|$$

can be made less than an arbitrary ϵ using the assumptions that $|a_n - a|$ and $|b_n - b|$ can be made as small as we like for large n. The first step is to use the triangle inequality (Example 1.2.5) to say

$$|(a_n + b_n) - (a + b)| = |(a_n - a) + (b_n - b)| \leq |a_n - a| + |b_n - b|.$$

Again, we let $\epsilon > 0$ be arbitrary. The technique this time is to divide the ϵ between the two expressions on the right-hand side in the preceding inequality. Using the hypothesis that $(a_n) \to a$, we know there exists an N_1 such that

$$|a_n - a| < \frac{\epsilon}{2} \quad \text{whenever} \quad n \geq N_1.$$

Likewise, the assumption that $(b_n) \to b$ means that we can choose an N_2 so that

$$|b_n - b| < \frac{\epsilon}{2} \quad \text{whenever} \quad n \geq N_2.$$

The question now arises as to which of N_1 or N_2 we should take to be our choice of N. By choosing $N = \max\{N_1, N_2\}$, we ensure that if $n \geq N$, then $n \geq N_1$ and $n \geq N_2$. This allows us to conclude that

$$\begin{aligned}
|(a_n + b_n) - (a + b)| &\leq |a_n - a| + |b_n - b| \\
&< \frac{\epsilon}{2} + \frac{\epsilon}{2} = \epsilon
\end{aligned}$$

for all $n \geq N$, as desired.

(iii) To show that $(a_n b_n) \to ab$, we begin by observing that

$$\begin{aligned}
|a_n b_n - ab| &= |a_n b_n - ab_n + ab_n - ab| \\
&\leq |a_n b_n - ab_n| + |ab_n - ab| \\
&= |b_n||a_n - a| + |a||b_n - b|.
\end{aligned}$$

In the initial step, we subtracted and then added ab_n, which created an opportunity to use the triangle inequality. Essentially, we have broken up the distance from $a_n b_n$ to ab with a midway point and are using the sum of the two distances to overestimate the original distance. This clever trick will become a familiar technique in arguments to come.

Letting $\epsilon > 0$ be arbitrary, we again proceed with the strategy of making each piece in the preceding inequality less than $\epsilon/2$. For the piece on the right-hand side ($|a||b_n - b|$), if $a \neq 0$ we can choose N_1 so that

$$n \geq N_1 \quad \text{implies} \quad |b_n - b| < \frac{1}{|a|}\frac{\epsilon}{2}.$$

(The case when $a = 0$ is handled in Exercise 2.3.9.) Getting the term on the left-hand side ($|b_n||a_n - a|$) to be less than $\epsilon/2$ is complicated by the fact that we have a variable quantity $|b_n|$ to contend with as opposed to the constant $|a|$ we encountered in the right-hand term. The idea is to replace $|b_n|$ with a worst-case estimate. Using the fact that convergent sequences are bounded (Theorem 2.3.2), we know there exists a bound $M > 0$ satisfying $|b_n| \leq M$ for all $n \in \mathbf{N}$. Now, we can choose N_2 so that

$$|a_n - a| < \frac{1}{M}\frac{\epsilon}{2} \quad \text{whenever} \quad n \geq N_2.$$

To finish the argument, pick $N = \max\{N_1, N_2\}$, and observe that if $n \geq N$, then

$$
\begin{aligned}
|a_n b_n - ab| &\leq |a_n b_n - ab_n| + |ab_n - ab| \\
&= |b_n||a_n - a| + |a||b_n - b| \\
&\leq M|a_n - a| + |a||b_n - b| \\
&< M\left(\frac{\epsilon}{M2}\right) + |a|\left(\frac{\epsilon}{|a|2}\right) = \epsilon.
\end{aligned}
$$

(iv) This final statement will follow from (iii) if we can prove that

$$(b_n) \to b \quad \text{implies} \quad \left(\frac{1}{b_n}\right) \to \frac{1}{b}$$

whenever $b \neq 0$. We begin by observing that

$$\left|\frac{1}{b_n} - \frac{1}{b}\right| = \frac{|b - b_n|}{|b||b_n|}.$$

Because $(b_n) \to b$, we can make the preceding numerator as small as we like by choosing n large. The problem comes in that we need a worst-case estimate on the size of $1/(|b||b_n|)$. Because the b_n terms are in the denominator, we are no longer interested in an upper bound on $|b_n|$ but rather in an inequality of the form $|b_n| \geq \delta > 0$. This will then lead to a bound on the size of $1/(|b||b_n|)$.

The trick is to look far enough out into the sequence (b_n) so that the terms are closer to b than they are to 0. Consider the particular value $\epsilon_0 = |b|/2$. Because $(b_n) \to b$, there exists an N_1 such that $|b_n - b| < |b|/2$ for all $n \geq N_1$. This implies $|b_n| > |b|/2$.

Next, choose N_2 so that $n \geq N_2$ implies

$$|b_n - b| < \frac{\epsilon|b|^2}{2}.$$

Finally, if we let $N = \max\{N_1, N_2\}$, then $n \geq N$ implies

$$\left|\frac{1}{b_n} - \frac{1}{b}\right| = |b - b_n|\frac{1}{|b||b_n|} < \frac{\epsilon|b|^2}{2}\frac{1}{|b|\frac{|b|}{2}} = \epsilon. \qquad \square$$

Limits and Order

Although there are a few dangers to avoid (see Exercise 2.3.7), the Algebraic Limit Theorem verifies that the relationship between algebraic combinations of sequences and the limiting process is as trouble-free as we could hope for. Limits can be computed from the individual component sequences provided that each component limit exists. The limiting process is also well-behaved with respect to the order operation.

Theorem 2.3.4 (Order Limit Theorem). *Assume* $\lim a_n = a$ *and* $\lim b_n = b$.

(i) *If* $a_n \geq 0$ *for all* $n \in \mathbf{N}$, *then* $a \geq 0$.

(ii) *If* $a_n \leq b_n$ *for all* $n \in \mathbf{N}$, *then* $a \leq b$.

(iii) *If there exists* $c \in \mathbf{R}$ *for which* $c \leq b_n$ *for all* $n \in \mathbf{N}$, *then* $c \leq b$. *Similarly, if* $a_n \leq c$ *for all* $n \in \mathbf{N}$, *then* $a \leq c$.

Proof. (i) We will prove this by contradiction; thus, let's assume $a < 0$. The idea is to produce a term in the sequence (a_n) that is also less than zero. To do this, we consider the particular value $\epsilon = |a|$. The definition of convergence guarantees that we can find an N such that $|a_n - a| < |a|$ for all $n \geq N$. In particular, this would mean that $|a_N - a| < |a|$, which implies $a_N < 0$. This contradicts our hypothesis that $a_N \geq 0$. We therefore conclude that $a \geq 0$.

(ii) The Algebraic Limit Theorem ensures that the sequence $(b_n - a_n)$ converges to $b - a$. Because $b_n - a_n \geq 0$, we can apply part (i) to get that $b - a \geq 0$.

(iii) Take $a_n = c$ (or $b_n = c$) for all $n \in \mathbf{N}$, and apply (ii). $\qquad \square$

A word about the idea of "tails" is in order. Loosely speaking, limits and their properties do not depend at all on what happens at the beginning of the sequence but are strictly determined by what happens when n gets large. Changing the value of the first ten—or ten thousand—terms in a particular sequence has no effect on the limit. Theorem 2.3.4, part (i), for instance, assumes that $a_n \geq 0$ *for all* $n \in \mathbf{N}$. However, the hypothesis could be weakened by assuming only that there exists some point N_1 where $a_n \geq 0$ for all $n \geq N_1$. The theorem remains true, and in fact the same proof is valid with the provision that when N is chosen it be at least as large as N_1.

In the language of analysis, when a property (such as non-negativity) is not necessarily possessed by some finite number of initial terms but is possessed by all terms in the sequence after some point N, we say that the sequence *eventually* has this property. (See Exercise 2.2.7.) Theorem 2.3.4, part (i), could be restated, "Convergent sequences that are eventually nonnegative converge to nonnegative limits." Parts (ii) and (iii) have similar modifications, as will many other upcoming results.

Exercises

Exercise 2.3.1. Let $x_n \geq 0$ for all $n \in \mathbf{N}$.

  (a) If $(x_n) \to 0$, show that $(\sqrt{x_n}) \to 0$.

  (b) If $(x_n) \to x$, show that $(\sqrt{x_n}) \to \sqrt{x}$.

Exercise 2.3.2. Using only Definition 2.2.3, prove that if $(x_n) \to 2$, then

  (a) $\left(\frac{2x_n - 1}{3}\right) \to 1$;

  (b) $(1/x_n) \to 1/2$.

(For this exercise the Algebraic Limit Theorem is off-limits, so to speak.)

Exercise 2.3.3 (Squeeze Theorem). Show that if $x_n \leq y_n \leq z_n$ for all $n \in \mathbf{N}$, and if $\lim x_n = \lim z_n = l$, then $\lim y_n = l$ as well.

Exercise 2.3.4. Let $(a_n) \to 0$, and use the Algebraic Limit Theorem to compute each of the following limits (assuming the fractions are always defined):

  (a) $\lim \left(\frac{1 + 2a_n}{1 + 3a_n - 4a_n^2} \right)$

  (b) $\lim \left(\frac{(a_n + 2)^2 - 4}{a_n} \right)$

  (c) $\lim \left(\frac{\frac{2}{a_n} + 3}{\frac{1}{a_n} + 5} \right)$.

Exercise 2.3.5. Let (x_n) and (y_n) be given, and define (z_n) to be the "shuffled" sequence $(x_1, y_1, x_2, y_2, x_3, y_3, \ldots, x_n, y_n, \ldots)$. Prove that (z_n) is convergent if and only if (x_n) and (y_n) are both convergent with $\lim x_n = \lim y_n$.

Exercise 2.3.6. Consider the sequence given by $b_n = n - \sqrt{n^2 + 2n}$. Taking $(1/n) \to 0$ as given, and using both the Algebraic Limit Theorem and the result in Exercise 2.3.1, show $\lim b_n$ exists and find the value of the limit.

Exercise 2.3.7. Give an example of each of the following, or state that such a request is impossible by referencing the proper theorem(s):

  (a) sequences (x_n) and (y_n), which both diverge, but whose sum $(x_n + y_n)$ converges;

  (b) sequences (x_n) and (y_n), where (x_n) converges, (y_n) diverges, and $(x_n + y_n)$ converges;

(c) a convergent sequence (b_n) with $b_n \neq 0$ for all n such that $(1/b_n)$ diverges;

(d) an unbounded sequence (a_n) and a convergent sequence (b_n) with $(a_n - b_n)$ bounded;

(e) two sequences (a_n) and (b_n), where $(a_n b_n)$ and (a_n) converge but (b_n) does not.

Exercise 2.3.8. Let $(x_n) \to x$ and let $p(x)$ be a polynomial.

(a) Show $p(x_n) \to p(x)$.

(b) Find an example of a function $f(x)$ and a convergent sequence $(x_n) \to x$ where the sequence $f(x_n)$ converges, but not to $f(x)$.

Exercise 2.3.9.   (a) Let (a_n) be a bounded (not necessarily convergent) sequence, and assume $\lim b_n = 0$. Show that $\lim(a_n b_n) = 0$. Why are we not allowed to use the Algebraic Limit Theorem to prove this?

(b) Can we conclude anything about the convergence of $(a_n b_n)$ if we assume that (b_n) converges to some nonzero limit b?

(c) Use (a) to prove Theorem 2.3.3, part (iii), for the case when $a = 0$.

Exercise 2.3.10. Consider the following list of conjectures. Provide a short proof for those that are true and a counterexample for any that are false.

(a) If $\lim(a_n - b_n) = 0$, then $\lim a_n = \lim b_n$.

(b) If $(b_n) \to b$, then $|b_n| \to |b|$.

(c) If $(a_n) \to a$ and $(b_n - a_n) \to 0$, then $(b_n) \to a$.

(d) If $(a_n) \to 0$ and $|b_n - b| \le a_n$ for all $n \in \mathbf{N}$, then $(b_n) \to b$.

Exercise 2.3.11 (Cesaro Means).   (a) Show that if (x_n) is a convergent sequence, then the sequence given by the averages

$$y_n = \frac{x_1 + x_2 + \cdots + x_n}{n}$$

also converges to the same limit.

(b) Give an example to show that it is possible for the sequence (y_n) of averages to converge even if (x_n) does not.

Exercise 2.3.12. A typical task in analysis is to decipher whether a property possessed by every term in a convergent sequence is necessarily inherited by the limit. Assume $(a_n) \to a$, and determine the validity of each claim. Try to produce a counterexample for any that are false.

(a) If every a_n is an upper bound for a set B, then a is also an upper bound for B.

(b) If every a_n is in the complement of the interval $(0, 1)$, then a is also in the complement of $(0, 1)$.

(c) If every a_n is rational, then a is rational.

Exercise 2.3.13 (Iterated Limits). Given a doubly indexed array a_{mn} where $m, n \in \mathbf{N}$, what should $\lim_{m,n \to \infty} a_{mn}$ represent?

(a) Let $a_{mn} = m/(m + n)$ and compute the *iterated* limits

$$\lim_{n \to \infty} \left(\lim_{m \to \infty} a_{mn} \right) \quad \text{and} \quad \lim_{m \to \infty} \left(\lim_{n \to \infty} a_{mn} \right).$$

Define $\lim_{m,n \to \infty} a_{mn} = a$ to mean that for all $\epsilon > 0$ there exists an $N \in \mathbf{N}$ such that if both $m, n \geq N$, then $|a_{mn} - a| < \epsilon$.

(b) Let $a_{mn} = 1/(m+n)$. Does $\lim_{m,n \to \infty} a_{mn}$ exist in this case? Do the two iterated limits exist? How do these three values compare? Answer these same questions for $a_{mn} = mn/(m^2 + n^2)$.

(c) Produce an example where $\lim_{m,n \to \infty} a_{mn}$ exists but where neither iterated limit can be computed.

(d) Assume $\lim_{m,n \to \infty} a_{mn} = a$, and assume that for each fixed $m \in \mathbf{N}$, $\lim_{n \to \infty}(a_{mn}) \to b_m$. Show $\lim_{m \to \infty} b_m = a$.

(e) Prove that if $\lim_{m,n \to \infty} a_{mn}$ exists and the iterated limits both exist, then all three limits must be equal.

2.4   The Monotone Convergence Theorem and a First Look at Infinite Series

We showed in Theorem 2.3.2 that convergent sequences are bounded. The converse statement is certainly not true. It is not too difficult to produce an example of a bounded sequence that does not converge. On the other hand, if a bounded sequence is *monotone*, then in fact it does converge.

Definition 2.4.1. A sequence (a_n) is *increasing* if $a_n \leq a_{n+1}$ for all $n \in \mathbf{N}$ and *decreasing* if $a_n \geq a_{n+1}$ for all $n \in \mathbf{N}$. A sequence is *monotone* if it is either increasing or decreasing.

Theorem 2.4.2 (Monotone Convergence Theorem). *If a sequence is monotone and bounded, then it converges.*

Proof. Let (a_n) be monotone and bounded. To prove (a_n) converges using the definition of convergence, we are going to need a candidate for the limit. Let's assume the sequence is increasing (the decreasing case is handled similarly), and consider the *set* of points $\{a_n : n \in \mathbf{N}\}$. By assumption, this set is bounded, so we can let

$$s = \sup\{a_n : n \in \mathbf{N}\}.$$

It seems reasonable to claim that $\lim a_n = s$.

To prove this, let $\epsilon > 0$. Because s is the least upper bound for $\{a_n : n \in \mathbf{N}\}$, $s - \epsilon$ is not an upper bound, so there exists a point in the sequence a_N such that $s - \epsilon < a_N$. Now, the fact that (a_n) is increasing implies that if $n \geq N$, then $a_N \leq a_n$. Hence,

$$s - \epsilon < a_N \leq a_n \leq s < s + \epsilon,$$

which implies $|a_n - s| < \epsilon$, as desired. $\qquad\qquad\square$

The Monotone Convergence Theorem is extremely useful for the study of infinite series, largely because it asserts the convergence of a sequence without explicit mention of the actual limit. This is a good moment to do some preliminary investigations, so it is time to formalize the relationship between sequences and series.

Definition 2.4.3 (Convergence of a Series). Let (b_n) be a sequence. An *infinite series* is a formal expression of the form

$$\sum_{n=1}^{\infty} b_n = b_1 + b_2 + b_3 + b_4 + b_5 + \cdots .$$

We define the corresponding *sequence of partial sums* (s_m) by

$$s_m = b_1 + b_2 + b_3 + \cdots + b_m,$$

and say that the series $\sum_{n=1}^{\infty} b_n$ *converges* to B if the sequence (s_m) converges to B. In this case, we write $\sum_{n=1}^{\infty} b_n = B$.

Example 2.4.4. Consider

$$\sum_{n=1}^{\infty} \frac{1}{n^2}.$$

Because the terms in the sum are all positive, the sequence of partial sums given by

$$s_m = 1 + \frac{1}{4} + \frac{1}{9} + \cdots + \frac{1}{m^2}$$

is increasing. The question is whether or not we can find some upper bound on (s_m). To this end, observe

$$
\begin{aligned}
s_m &= 1 + \frac{1}{2 \cdot 2} + \frac{1}{3 \cdot 3} + \frac{1}{4 \cdot 4} + \cdots + \frac{1}{m^2} \\
&< 1 + \frac{1}{2 \cdot 1} + \frac{1}{3 \cdot 2} + \frac{1}{4 \cdot 3} + \cdots + \frac{1}{m(m-1)} \\
&= 1 + \left(1 - \frac{1}{2}\right) + \left(\frac{1}{2} - \frac{1}{3}\right) + \left(\frac{1}{3} - \frac{1}{4}\right) + \cdots + \left(\frac{1}{(m-1)} - \frac{1}{m}\right) \\
&= 1 + 1 - \frac{1}{m} \\
&< 2.
\end{aligned}
$$

Thus, 2 is an upper bound for the sequence of partial sums, so by the Monotone Convergence Theorem, $\sum_{n=1}^{\infty} 1/n^2$ converges to some (for the moment) unknown limit less than 2. (Finding the value of this limit is the subject of Sections 6.1 and 8.3.)

Example 2.4.5 (Harmonic Series). This time, consider the so-called *harmonic series*

$$
\sum_{n=1}^{\infty} \frac{1}{n}.
$$

Again, we have an increasing sequence of partial sums,

$$
s_m = 1 + \frac{1}{2} + \frac{1}{3} + \cdots + \frac{1}{m},
$$

that upon naive inspection appears as though it may be bounded. However, 2 is no longer an upper bound because

$$
s_4 = 1 + \frac{1}{2} + \left(\frac{1}{3} + \frac{1}{4}\right) > 1 + \frac{1}{2} + \left(\frac{1}{4} + \frac{1}{4}\right) = 2.
$$

A similar calculation shows that $s_8 > 2\frac{1}{2}$, and we can see that in general

$$
\begin{aligned}
s_{2^k} &= 1 + \frac{1}{2} + \left(\frac{1}{3} + \frac{1}{4}\right) + \left(\frac{1}{5} + \cdots + \frac{1}{8}\right) + \cdots + \left(\frac{1}{2^{k-1}+1} + \cdots + \frac{1}{2^k}\right) \\
&> 1 + \frac{1}{2} + \left(\frac{1}{4} + \frac{1}{4}\right) + \left(\frac{1}{8} + \cdots + \frac{1}{8}\right) + \cdots + \left(\frac{1}{2^k} + \cdots + \frac{1}{2^k}\right) \\
&= 1 + \frac{1}{2} + 2\left(\frac{1}{4}\right) + 4\left(\frac{1}{8}\right) + \cdots + 2^{k-1}\left(\frac{1}{2^k}\right) \\
&= 1 + \frac{1}{2} + \frac{1}{2} + \frac{1}{2} + \cdots + \frac{1}{2} \\
&= 1 + k\left(\frac{1}{2}\right),
\end{aligned}
$$

which is unbounded. Thus, despite the incredibly slow pace, the sequence of partial sums of $\sum_{n=1}^{\infty} 1/n$ eventually surpasses every number on the positive real line. Because convergent sequences are bounded, the harmonic series diverges.

The previous example is a special case of a general argument that can be used to determine the convergence or divergence of a large class of infinite series.

Theorem 2.4.6 (Cauchy Condensation Test). *Suppose (b_n) is decreasing and satisfies $b_n \geq 0$ for all $n \in \mathbf{N}$. Then, the series $\sum_{n=1}^{\infty} b_n$ converges if and only if the series*

$$\sum_{n=0}^{\infty} 2^n b_{2^n} = b_1 + 2b_2 + 4b_4 + 8b_8 + 16b_{16} + \cdots$$

converges.

Proof. First, assume that $\sum_{n=0}^{\infty} 2^n b_{2^n}$ converges. Theorem 2.3.2 guarantees that the partial sums

$$t_k = b_1 + 2b_2 + 4b_4 + \cdots + 2^k b_{2^k}$$

are bounded; that is, there exists an $M > 0$ such that $t_k \leq M$ for all $k \in \mathbf{N}$. We want to prove that $\sum_{n=1}^{\infty} b_n$ converges. Because $b_n \geq 0$, we know that the partial sums are increasing, so we only need to show that

$$s_m = b_1 + b_2 + b_3 + \cdots + b_m$$

is bounded.

Fix m and let k be large enough to ensure $m \leq 2^{k+1} - 1$. Then, $s_m \leq s_{2^{k+1}-1}$ and

$$
\begin{aligned}
s_{2^{k+1}-1} &= b_1 + (b_2 + b_3) + (b_4 + b_5 + b_6 + b_7) + \cdots + (b_{2^k} + \cdots + b_{2^{k+1}-1}) \\
&\leq b_1 + (b_2 + b_2) + (b_4 + b_4 + b_4 + b_4) + \cdots + (b_{2^k} + \cdots + b_{2^k}) \\
&= b_1 + 2b_2 + 4b_4 + \cdots + 2^k b_{2^k} = t_k.
\end{aligned}
$$

Thus, $s_m \leq t_k \leq M$, and the sequence (s_m) is bounded. By the Monotone Convergence Theorem, we can conclude that $\sum_{n=1}^{\infty} b_n$ converges.

The proof that $\sum_{n=0}^{\infty} 2^n b_{2^n}$ diverges implies $\sum_{n=1}^{\infty} b_n$ diverges is similar to Example 2.4.5. The details are requested in Exercise 2.4.9.   ☐

Corollary 2.4.7. *The series $\sum_{n=1}^{\infty} 1/n^p$ converges if and only if $p > 1$.*

A rigorous argument for this corollary requires a few basic facts about geometric series. The proof is requested in Exercise 2.7.5 at the end of Section 2.7 where geometric series are discussed.

Exercises

Exercise 2.4.1.   (a) Prove that the sequence defined by $x_1 = 3$ and

$$x_{n+1} = \frac{1}{4 - x_n}$$

converges.

(b) Now that we know $\lim x_n$ exists, explain why $\lim x_{n+1}$ must also exist and equal the same value.

(c) Take the limit of each side of the recursive equation in part (a) to explicitly compute $\lim x_n$.

Exercise 2.4.2.   (a) Consider the recursively defined sequence $y_1 = 1$,

$$y_{n+1} = 3 - y_n,$$

and set $y = \lim y_n$. Because (y_n) and (y_{n+1}) have the same limit, taking the limit across the recursive equation gives $y = 3 - y$. Solving for y, we conclude $\lim y_n = 3/2$.

What is wrong with this argument?

(b) This time set $y_1 = 1$ and $y_{n+1} = 3 - \frac{1}{y_n}$. Can the strategy in (a) be applied to compute the limit of this sequence?

Exercise 2.4.3.   (a) Show that

$$\sqrt{2}, \sqrt{2 + \sqrt{2}}, \sqrt{2 + \sqrt{2 + \sqrt{2}}}, \ldots$$

converges and find the limit.

(b) Does the sequence

$$\sqrt{2}, \sqrt{2\sqrt{2}}, \sqrt{2\sqrt{2\sqrt{2}}}, \ldots$$

converge? If so, find the limit.

Exercise 2.4.4.   (a) In Section 1.4 we used the Axiom of Completeness (AoC) to prove the Archimedean Property of $\mathbf{R}$ (Theorem 1.4.2). Show that the Monotone Convergence Theorem can also be used to prove the Archimedean Property without making any use of AoC.

(b) Use the Monotone Convergence Theorem to supply a proof for the Nested Interval Property (Theorem 1.4.1) that doesn't make use of AoC.

These two results suggest that we could have used the Monotone Convergence Theorem in place of AoC as our starting axiom for building a proper theory of the real numbers.

Exercise 2.4.5 (Calculating Square Roots). Let $x_1 = 2$, and define

$$x_{n+1} = \frac{1}{2}\left(x_n + \frac{2}{x_n}\right).$$

(a) Show that x_n^2 is always greater than or equal to 2, and then use this to prove that $x_n - x_{n+1} \geq 0$. Conclude that $\lim x_n = \sqrt{2}$.

(b) Modify the sequence (x_n) so that it converges to $\sqrt{c}$.

Exercise 2.4.6 (Arithmetic–Geometric Mean). (a) Explain why $\sqrt{xy} \leq (x+y)/2$ for any two positive real numbers x and y. (The geometric mean is always less than the arithmetic mean.)

(b) Now let $0 \leq x_1 \leq y_1$ and define

$$x_{n+1} = \sqrt{x_n y_n} \quad \text{and} \quad y_{n+1} = \frac{x_n + y_n}{2}.$$

Show $\lim x_n$ and $\lim y_n$ both exist and are equal.

Exercise 2.4.7 (Limit Superior). Let (a_n) be a bounded sequence.

(a) Prove that the sequence defined by $y_n = \sup\{a_k : k \geq n\}$ converges.

(b) The *limit superior* of (a_n), or $\limsup a_n$, is defined by

$$\limsup a_n = \lim y_n,$$

where y_n is the sequence from part (a) of this exercise. Provide a reasonable definition for $\liminf a_n$ and briefly explain why it always exists for any bounded sequence.

(c) Prove that $\liminf a_n \leq \limsup a_n$ for every bounded sequence, and give an example of a sequence for which the inequality is strict.

(d) Show that $\liminf a_n = \limsup a_n$ if and only if $\lim a_n$ exists. In this case, all three share the same value.

Exercise 2.4.8. For each series, find an explicit formula for the sequence of partial sums and determine if the series converges.

$$\text{(a)} \sum_{n=1}^{\infty} \frac{1}{2^n} \qquad \text{(b)} \sum_{n=1}^{\infty} \frac{1}{n(n+1)} \qquad \text{(c)} \sum_{n=1}^{\infty} \log\left(\frac{n+1}{n}\right)$$

(In (c), $\log(x)$ refers to the natural logarithm function from calculus.)

Exercise 2.4.9. Complete the proof of Theorem 2.4.6 by showing that if the series $\sum_{n=0}^{\infty} 2^n b_{2^n}$ diverges, then so does $\sum_{n=1}^{\infty} b_n$. Example 2.4.5 may be a useful reference.

Exercise 2.4.10 (Infinite Products). A close relative of infinite series is the *infinite product*

$$\prod_{n=1}^{\infty} b_n = b_1 b_2 b_3 \cdots$$

which is understood in terms of its sequence of *partial products*

$$p_m = \prod_{n=1}^{m} b_n = b_1 b_2 b_3 \cdots b_m.$$

Consider the special class of infinite products of the form

$$\prod_{n=1}^{\infty}(1 + a_n) = (1 + a_1)(1 + a_2)(1 + a_3)\cdots, \qquad \text{where } a_n \geq 0.$$

(a) Find an explicit formula for the sequence of partial products in the case where $a_n = 1/n$ and decide whether the sequence converges. Write out the first few terms in the sequence of partial products in the case where $a_n = 1/n^2$ and make a conjecture about the convergence of this sequence.

(b) Show, in general, that the sequence of partial products converges if and only if $\sum_{n=1}^{\infty} a_n$ converges. (The inequality $1 + x \leq 3^x$ for positive x will be useful in one direction.)

2.5   Subsequences and the Bolzano–Weierstrass Theorem

In Example 2.4.5, we showed that the sequence of partial sums (s_m) of the harmonic series does not converge by focusing our attention on a particular *subsequence* (s_{2^k}) of the original sequence. For the moment, we will put the topic of infinite series aside and more fully develop the important concept of subsequences.

Definition 2.5.1. Let (a_n) be a sequence of real numbers, and let $n_1 < n_2 < n_3 < n_4 < n_5 < \dots$ be an increasing sequence of natural numbers. Then the sequence

$$(a_{n_1}, a_{n_2}, a_{n_3}, a_{n_4}, a_{n_5}, \dots)$$

is called a *subsequence* of (a_n) and is denoted by (a_{n_k}), where $k \in \mathbf{N}$ indexes the subsequence.

Notice that the order of the terms in a subsequence is the same as in the original sequence, and repetitions are not allowed. Thus if

$$(a_n) = \left(1, \frac{1}{2}, \frac{1}{3}, \frac{1}{4}, \frac{1}{5}, \frac{1}{6}, \cdots\right),$$

then

$$\left(\frac{1}{2}, \frac{1}{4}, \frac{1}{6}, \frac{1}{8}, \cdots\right) \qquad \text{and} \qquad \left(\frac{1}{10}, \frac{1}{100}, \frac{1}{1000}, \frac{1}{10000}, \cdots\right)$$

are examples of legitimate subsequences, whereas

$$\left(\frac{1}{10}, \frac{1}{5}, \frac{1}{100}, \frac{1}{50}, \frac{1}{1000}, \frac{1}{500}, \cdots\right) \qquad \text{and} \qquad \left(1, 1, \frac{1}{3}, \frac{1}{3}, \frac{1}{5}, \frac{1}{5}, \cdots\right)$$

are not.

Theorem 2.5.2. *Subsequences of a convergent sequence converge to the same limit as the original sequence.*

Proof. Assume $(a_n) \to a$, and let (a_{n_k}) be a subsequence. Given $\epsilon > 0$, there exists N such that $|a_n - a| < \epsilon$ whenever $n \geq N$. Because $n_k \geq k$ for all k, the same N will suffice for the subsequence; that is, $|a_{n_k} - a| < \epsilon$ whenever $k \geq N$. $\qquad\square$

This not too surprising result has several somewhat surprising applications. It is the key ingredient for understanding when infinite sums are associative (Exercise 2.5.3). We can also use it in the following clever way to compute values of some familiar limits.

Example 2.5.3. Let $0 < b < 1$. Because

$$b > b^2 > b^3 > b^4 > \cdots > 0,$$

the sequence (b^n) is decreasing and bounded below. The Monotone Convergence Theorem allows us to conclude that (b^n) converges to some l satisfying $b > l \geq 0$. To compute l, notice that (b^{2n}) is a subsequence, so $(b^{2n}) \to l$ by Theorem 2.5.2. But $b^{2n} = b^n \cdot b^n$, so by the Algebraic Limit Theorem, $(b^{2n}) \to l \cdot l = l^2$. Because limits are unique (Theorem 2.2.7), $l^2 = l$, and thus $l = 0$.

Without much trouble (Exercise 2.5.7), we can generalize this example to conclude $(b^n) \to 0$ if and only if $-1 < b < 1$.

Example 2.5.4 (Divergence Criterion). Theorem 2.5.2 is also useful for providing economical proofs for divergence. In Example 2.2.8, we were quite sure that

$$\left(1, -\frac{1}{2}, \frac{1}{3}, -\frac{1}{4}, \frac{1}{5}, -\frac{1}{5}, \frac{1}{5}, -\frac{1}{5}, \frac{1}{5}, -\frac{1}{5}, \frac{1}{5}, -\frac{1}{5}, \cdots\right)$$

did not converge to any proposed limit. Notice that

$$\left(\frac{1}{5}, \frac{1}{5}, \frac{1}{5}, \frac{1}{5}, \frac{1}{5}, \cdots\right)$$

is a subsequence that converges to $1/5$. Also,

$$\left(-\frac{1}{5}, -\frac{1}{5}, -\frac{1}{5}, -\frac{1}{5}, -\frac{1}{5}, \cdots\right)$$

is a different subsequence of the original sequence that converges to $-1/5$. Because we have two subsequences converging to two different limits, we can rigorously conclude that the original sequence diverges.

The Bolzano–Weierstrass Theorem

In the previous example, it was rather easy to spot a convergent subsequence (or two) hiding in the original sequence. For *bounded* sequences, it turns out that it is always possible to find at least one such convergent subsequence.

Theorem 2.5.5 (Bolzano–Weierstrass Theorem). *Every bounded sequence contains a convergent subsequence.*

Proof. Let (a_n) be a bounded sequence so that there exists $M > 0$ satisfying $|a_n| \leq M$ for all $n \in \mathbf{N}$. Bisect the closed interval $[-M, M]$ into the two closed intervals $[-M, 0]$ and $[0, M]$. (The midpoint is included in both halves.) Now, it must be that at least one of these closed intervals contains an infinite number of the terms in the sequence (a_n). Select a half for which this is the case and label that interval as I_1. Then, let a_{n_1} be some term in the sequence (a_n) satisfying $a_{n_1} \in I_1$.

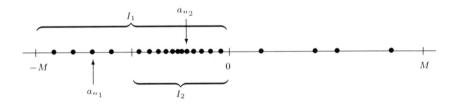

Next, we bisect I_1 into closed intervals of equal length, and let I_2 be a half that again contains an infinite number of terms of the original sequence. Because there are an infinite number of terms from (a_n) to choose from, we can select an a_{n_2} from the original sequence with $n_2 > n_1$ and $a_{n_2} \in I_2$. In general, we construct the closed interval I_k by taking a half of I_{k-1} containing an infinite number of terms of (a_n) and then select $n_k > n_{k-1} > \cdots > n_2 > n_1$ so that $a_{n_k} \in I_k$.

We want to argue that (a_{n_k}) is a convergent subsequence, but we need a candidate for the limit. The sets

$$I_1 \supseteq I_2 \supseteq I_3 \supseteq \cdots$$

form a nested sequence of closed intervals, and by the Nested Interval Property there exists at least one point $x \in \mathbf{R}$ contained in every I_k. This provides us with the candidate we were looking for. It just remains to show that $(a_{n_k}) \to x$.

Let $\epsilon > 0$. By construction, the length of I_k is $M(1/2)^{k-1}$ which converges to zero. (This follows from Example 2.5.3 and the Algebraic Limit Theorem.) Choose N so that $k \geq N$ implies that the length of I_k is less than ϵ. Because x and a_{n_k} are both in I_k, it follows that $|a_{n_k} - x| < \epsilon$. $\qquad\square$

Exercises

Exercise 2.5.1. Give an example of each of the following, or argue that such a request is impossible.

(a) A sequence that has a subsequence that is bounded but contains no subsequence that converges.

(b) A sequence that does not contain 0 or 1 as a term but contains subsequences converging to each of these values.

(c) A sequence that contains subsequences converging to every point in the infinite set $\{1, 1/2, 1/3, 1/4, 1/5, \ldots\}$.

(d) A sequence that contains subsequences converging to every point in the infinite set $\{1, 1/2, 1/3, 1/4, 1/5, \ldots\}$, and no subsequences converging to points outside of this set.

Exercise 2.5.2. Decide whether the following propositions are true or false, providing a short justification for each conclusion.

(a) If every proper subsequence of (x_n) converges, then (x_n) converges as well.

(b) If (x_n) contains a divergent subsequence, then (x_n) diverges.

(c) If (x_n) is bounded and diverges, then there exist two subsequences of (x_n) that converge to different limits.

(d) If (x_n) is monotone and contains a convergent subsequence, then (x_n) converges.

Exercise 2.5.3. (a) Prove that if an infinite series converges, then the associative property holds. Assume $a_1 + a_2 + a_3 + a_4 + a_5 + \cdots$ converges to a limit L (i.e., the sequence of partial sums $(s_n) \to L$). Show that any regrouping of the terms

$$(a_1 + a_2 + \cdots + a_{n_1}) + (a_{n_1+1} + \cdots + a_{n_2}) + (a_{n_2+1} + \cdots + a_{n_3}) + \cdots$$

leads to a series that also converges to L.

(b) Compare this result to the example discussed at the end of Section 2.1 where infinite addition was shown not to be associative. Why doesn't our proof in (a) apply to this example?

Exercise 2.5.4. The Bolzano–Weierstrass Theorem is extremely important, and so is the strategy employed in the proof. To gain some more experience with this technique, assume the Nested Interval Property is true and use it to provide a proof of the Axiom of Completeness. To prevent the argument from being circular, assume also that $(1/2^n) \to 0$. (Why precisely is this last assumption needed to avoid circularity?)

Exercise 2.5.5. Assume (a_n) is a bounded sequence with the property that every convergent subsequence of (a_n) converges to the same limit $a \in \mathbf{R}$. Show that (a_n) must converge to a.

Exercise 2.5.6. Use a similar strategy to the one in Example 2.5.3 to show $\lim b^{1/n}$ exists for all $b \geq 0$ and find the value of the limit. (The results in Exercise 2.3.1 may be assumed.)

Exercise 2.5.7. Extend the result proved in Example 2.5.3 to the case $|b| < 1$: that is, show $\lim(b^n) = 0$ if and only if $-1 < b < 1$.

Exercise 2.5.8. Another way to prove the Bolzano–Weierstrass Theorem is to show that every sequence contains a monotone subsequence. A useful device in this endeavor is the notion of a *peak term*. Given a sequence (x_n), a particular term x_m is a peak term if no later term in the sequence exceeds it; i.e., if $x_m \geq x_n$ for all $n \geq m$.

    (a) Find examples of sequences with zero, one, and two peak terms. Find an example of a sequence with infinitely many peak terms that is not monotone.

    (b) Show that every sequence contains a monotone subsequence and explain how this furnishes a new proof of the Bolzano–Weierstrass Theorem.

Exercise 2.5.9. Let (a_n) be a bounded sequence, and define the set

$$S = \{x \in \mathbf{R} : x < a_n \text{ for infinitely many terms } a_n\}.$$

Show that there exists a subsequence (a_{n_k}) converging to $s = \sup S$. (This is a direct proof of the Bolzano–Weierstrass Theorem using the Axiom of Completeness.)

2.6 The Cauchy Criterion

The following definition bears a striking resemblance to the definition of convergence for a sequence.

Definition 2.6.1. A sequence (a_n) is called a *Cauchy sequence* if, for every $\epsilon > 0$, there exists an $N \in \mathbf{N}$ such that whenever $m, n \geq N$ it follows that $|a_n - a_m| < \epsilon$.

To make the comparison easier, let's restate the definition of convergence.

Definition 2.2.3. A sequence (a_n) *converges* to a real number a if, for every $\epsilon > 0$, there exists an $N \in \mathbf{N}$ such that whenever $n \geq N$ it follows that $|a_n - a| < \epsilon$.

    As we have discussed, the definition of convergence asserts that, given an arbitrary positive ϵ, it is possible to find a point in the sequence after which the terms of the sequence are all closer to the limit a than the given ϵ. On the other hand, a sequence is a Cauchy sequence if, for every ϵ, there is a point in the sequence after which the terms are all closer *to each other* than the given ϵ. To spoil the surprise, we will argue in this section that in fact these two definitions are equivalent: Convergent sequences are Cauchy sequences, and Cauchy sequences converge. The significance of the definition of a Cauchy sequence is that there is no mention of a limit. This is somewhat like the situation with the Monotone Convergence Theorem in that we will have another way of proving that sequences converge without having any explicit knowledge of what the limit might be.

Theorem 2.6.2. *Every convergent sequence is a Cauchy sequence.*

Proof. Assume (x_n) converges to x. To prove that (x_n) is Cauchy, we must find a point in the sequence after which we have $|x_n - x_m| < \epsilon$. This can be done using an application of the triangle inequality. The details are requested in Exercise 2.6.1. $\qquad\square$

The converse is a bit more difficult to prove, mainly because, in order to prove that a sequence converges, we must have a proposed limit for the sequence to approach. We have been in this situation before in the proofs of the Monotone Convergence Theorem and the Bolzano–Weierstrass Theorem. Our strategy here will be to use the Bolzano–Weierstrass Theorem. This is the reason for the next lemma. (Compare this with Theorem 2.3.2.)

Lemma 2.6.3. *Cauchy sequences are bounded.*

Proof. Given $\epsilon = 1$, there exists an N such that $|x_m - x_n| < 1$ for all $m, n \geq N$. Thus, we must have $|x_n| < |x_N| + 1$ for all $n \geq N$. It follows that

$$M = \max\{|x_1|, |x_2|, |x_3|, \ldots, |x_{N-1}|, |x_N| + 1\}$$

is a bound for the sequence (x_n). $\qquad\square$

Theorem 2.6.4 (Cauchy Criterion). *A sequence converges if and only if it is a Cauchy sequence.*

Proof. ($\Rightarrow$) This direction is Theorem 2.6.2.

($\Leftarrow$) For this direction, we start with a Cauchy sequence (x_n). Lemma 2.6.3 guarantees that (x_n) is bounded, so we may use the Bolzano–Weierstrass Theorem to produce a convergent subsequence (x_{n_k}). Set

$$x = \lim x_{n_k}.$$

The idea is to show that the original sequence (x_n) converges to this same limit. Once again, we will use a triangle inequality argument. We know the terms in the subsequence are getting close to the limit x, and the assumption that (x_n) is Cauchy implies the terms in the "tail" of the sequence are close to each other. Thus, we want to make each of these distances less than half of the prescribed ϵ.

Let $\epsilon > 0$. Because (x_n) is Cauchy, there exists N such that

$$|x_n - x_m| < \frac{\epsilon}{2}$$

whenever $m, n \geq N$. Now, we also know that $(x_{n_k}) \to x$, so choose a term in this subsequence, call it x_{n_K}, with $n_K \geq N$ and

$$|x_{n_K} - x| < \frac{\epsilon}{2}.$$

To see that N has the desired property (for the original sequence (x_n)), observe that if $n \geq N$, then

$$
\begin{aligned}
|x_n - x| &= |x_n - x_{n_K} + x_{n_K} - x| \\
&\leq |x_n - x_{n_K}| + |x_{n_K} - x| \\
&< \frac{\epsilon}{2} + \frac{\epsilon}{2} = \epsilon.
\end{aligned}
$$

$\square$

The Cauchy Criterion is named after the French mathematician Augustin Louis Cauchy. Cauchy is a major figure in the history of many branches of mathematics—number theory and the theory of finite groups, to name a few— but he is most widely recognized for his enormous contributions in analysis, especially complex analysis. He is deservedly credited with inventing the ϵ-based definition of limits we use today, although it is probably better to view him as a pioneer of analysis in the sense that his work did not attain the level of refinement that modern mathematicians have come to expect. The Cauchy Criterion, for instance, was devised and used by Cauchy to study infinite series, but he never actually proved it in both directions. The fact that there were gaps in Cauchy's work should not diminish his brilliance in any way. The issues of the day were both difficult and subtle, and Cauchy was far and away the most influential in laying the groundwork for modern standards of rigor. Karl Weierstrass played a major role in sharpening Cauchy's arguments. We will hear a good deal more from Weierstrass, most notably in Chapter 6 when we take up uniform convergence. Bernhard Bolzano was working in Prague and was writing and thinking about many of these same issues surrounding limits and continuity. Because his work was not widely available to the rest of the mathematical community, his historical reputation never achieved the distinction that his impressive accomplishments would seem to merit.

Completeness Revisited

In the first chapter, we established the Axiom of Completeness (AoC) to be the assertion that nonempty sets bounded above have least upper bounds. We then used this axiom as the crucial step in the proof of the Nested Interval Property (NIP). In this chapter, AoC was the central step in the Monotone Convergence Theorem (MCT), and NIP was the key to proving the Bolzano–Weierstrass

Theorem (BW). Finally, we needed BW in our proof of the Cauchy Criterion (CC) for convergent sequences. The list of implications then looks like

$$\text{AoC} \Rightarrow \left\{ \begin{array}{l} \text{NIP} \quad \Rightarrow \quad \text{BW} \quad \Rightarrow \quad \text{CC.} \\ \text{MCT.} \end{array} \right.$$

But this one-directional list is not the whole story. Recall that in our original discussions about completeness, the fundamental problem was that the rational numbers contained "gaps." The reason for moving from the rational numbers to the real numbers to do analysis is so that when we encounter a sequence that looks as if it is converging to some number—say $\sqrt{2}$—then we can be assured that there is indeed a number there that we can call the limit. The assertion that "nonempty sets bounded above have least upper bounds" is simply one way to mathematically articulate our insistence that there be no "holes" in our ordered field, but it is not the only way. Instead, we could have taken MCT to be our defining axiom and used it to prove NIP and the existence of least upper bounds. This is the content of Exercise 2.4.4.

How about NIP? Could this property serve as a starting point for a proper axiomatic treatment of the real numbers? Almost. In Exercise 2.5.4 we showed that NIP implies AoC, but to prevent the argument from making implicit use of AoC we needed an extra assumption that is equivalent to the Archimedean Property (Theorem 1.4.2). This extra hypothesis is unavoidable. Whereas AoC and MCT can both be used to prove that $\mathbf{N}$ is not a bounded subset of $\mathbf{R}$, there is no way to prove this same fact starting from NIP. The upshot is that NIP is a perfectly reasonable candidate to use as the fundamental axiom of the real numbers provided that we also include the Archimedean Property as a second unproven assumption.

In fact, if we assume the Archimedean Property holds, then AoC, NIP, MCT, BW, and CC are equivalent in the sense that once we take any one of them to be true, it is possible to derive the other four. However, because we have an example of an ordered field that is not complete—namely, the set of rational numbers—we know it is impossible to prove any of them using only the field and order properties. Just how we decide which should be the axiom and which then become theorems depends largely on preference and context, and in the end is not especially significant. What is important is that we understand all of these results as belonging to the same family, each asserting the completeness of $\mathbf{R}$ in its own particular language.

One loose end in this conversation is the curious and somewhat unpredictable relationship of the Archimedean Property to these other results. As we have mentioned, the Archimedean Property follows as a consequence of AoC as well as MCT, but not from NIP. Starting from BW, it is possible to prove MCT and thus also the Archimedean Property. On the other hand, the Cauchy Criterion is like NIP in that it cannot be used on its own to prove the Archimedean Property.[1]

---

[1] A thorough account of the logical dependence between these various results can be found in [23].

Exercises

Exercise 2.6.1. Supply a proof for Theorem 2.6.2.

Exercise 2.6.2. Give an example of each of the following, or argue that such a request is impossible.

(a) A Cauchy sequence that is not monotone.

(b) A Cauchy sequence with an unbounded subsequence.

(c) A divergent monotone sequence with a Cauchy subsequence.

(d) An unbounded sequence containing a subsequence that is Cauchy.

Exercise 2.6.3. If (x_n) and (y_n) are Cauchy sequences, then one easy way to prove that $(x_n + y_n)$ is Cauchy is to use the Cauchy Criterion. By Theorem 2.6.4, (x_n) and (y_n) must be convergent, and the Algebraic Limit Theorem then implies $(x_n + y_n)$ is convergent and hence Cauchy.

(a) Give a direct argument that $(x_n + y_n)$ is a Cauchy sequence that does not use the Cauchy Criterion or the Algebraic Limit Theorem.

(b) Do the same for the product $(x_n y_n)$.

Exercise 2.6.4. Let (a_n) and (b_n) be Cauchy sequences. Decide whether each of the following sequences is a Cauchy sequence, justifying each conclusion.

(a) $c_n = |a_n - b_n|$

(b) $c_n = (-1)^n a_n$

(c) $c_n = [[a_n]]$, where $[[x]]$ refers to the greatest integer less than or equal to x.

Exercise 2.6.5. Consider the following (invented) definition: A sequence (s_n) is *pseudo-Cauchy* if, for all $\epsilon > 0$, there exists an N such that if $n \geq N$, then $|s_{n+1} - s_n| < \epsilon$.
  Decide which one of the following two propositions is actually true. Supply a proof for the valid statement and a counterexample for the other.

(i) Pseudo-Cauchy sequences are bounded.

(ii) If (x_n) and (y_n) are pseudo-Cauchy, then $(x_n + y_n)$ is pseudo-Cauchy as well.

Exercise 2.6.6. Let's call a sequence (a_n) *quasi-increasing* if for all $\epsilon > 0$ there exists an N such that whenever $n > m \geq N$ it follows that $a_n > a_m - \epsilon$.

(a) Give an example of a sequence that is quasi-increasing but not monotone or eventually monotone.

(b) Give an example of a quasi-increasing sequence that is divergent and not monotone or eventually monotone.

(c) Is there an analogue of the Monotone Convergence Theorem for quasi-increasing sequences? Give an example of a bounded, quasi-increasing sequence that doesn't converge, or prove that no such sequence exists.

Exercise 2.6.7. Exercises 2.4.4 and 2.5.4 establish the equivalence of the Axiom of Completeness and the Monotone Convergence Theorem. They also show the Nested Interval Property is equivalent to these other two in the presence of the Archimedean Property.

(a) Assume the Bolzano–Weierstrass Theorem is true and use it to construct a proof of the Monotone Convergence Theorem without making any appeal to the Archimedean Property. This shows that BW, AoC, and MCT are all equivalent.

(b) Use the Cauchy Criterion to prove the Bolzano–Weierstrass Theorem, and find the point in the argument where the Archimedean Property is implicitly required. This establishes the final link in the equivalence of the five characterizations of completeness discussed at the end of Section 2.6.

(c) How do we know it is impossible to prove the Axiom of Completeness starting from the Archimedean Property?

2.7   Properties of Infinite Series

Given an infinite series $\sum_{k=1}^{\infty} a_k$, it is important to keep a clear distinction between

(i) the sequence of *terms*: $(a_1, a_2, a_3, \ldots)$ and

(ii) the sequence of *partial sums*: $(s_1, s_2, s_3, \ldots)$, where $s_n = a_1 + a_2 + \cdots + a_n$.

The convergence of the series $\sum_{k=1}^{\infty} a_k$ is defined in terms of the sequence (s_n). Specifically, the statement

$$\sum_{k=1}^{\infty} a_k = A \quad \text{means that} \quad \lim s_n = A.$$

It is for this reason that we can immediately translate many of our results from the study of sequences into statements about the behavior of infinite series.

Theorem 2.7.1 (Algebraic Limit Theorem for Series). *If $\sum_{k=1}^{\infty} a_k = A$ and $\sum_{k=1}^{\infty} b_k = B$, then*

(i) $\sum_{k=1}^{\infty} c a_k = cA$ *for all $c \in \mathbf{R}$ and*

(ii) $\sum_{k=1}^{\infty} (a_k + b_k) = A + B$.

Proof. (i) In order to show that $\sum_{k=1}^{\infty} ca_k = cA$, we must argue that the sequence of partial sums

$$t_m = ca_1 + ca_2 + ca_3 + \cdots + ca_m$$

converges to cA. But we are given that $\sum_{k=1}^{\infty} a_k$ converges to A, meaning that the partial sums

$$s_m = a_1 + a_2 + a_3 + \cdots + a_m$$

converge to A. Because $t_m = cs_m$, applying the Algebraic Limit Theorem for sequences (Theorem 2.3.3) yields $(t_m) \to cA$, as desired.

The proof of part (ii) is analogous and is left as an unofficial exercise. $\square$

One way to summarize Theorem 2.7.1 (i) is to say that infinite addition still satisfies the distributive property. Part (ii) verifies that series can be added in the usual way. Missing from this theorem is any statement about the *product* of two infinite series. At the heart of this question is the issue of commutativity, which requires a more delicate analysis and so is postponed until Section 2.8.

Theorem 2.7.2 (Cauchy Criterion for Series). *The series $\sum_{k=1}^{\infty} a_k$ converges if and only if, given $\epsilon > 0$, there exists an $N \in \mathbf{N}$ such that whenever $n > m \geq N$ it follows that*

$$|a_{m+1} + a_{m+2} + \cdots + a_n| < \epsilon.$$

Proof. Observe that

$$|s_n - s_m| = |a_{m+1} + a_{m+2} + \cdots + a_n|$$

and apply the Cauchy Criterion for sequences. $\square$

The Cauchy Criterion leads to economical proofs of several basic facts about series.

Theorem 2.7.3. *If the series $\sum_{k=1}^{\infty} a_k$ converges, then $(a_k) \to 0$.*

Proof. Consider the special case $n = m + 1$ in the Cauchy Criterion for Series.
                                                                            $\square$

Every statement of this result should be accompanied with a reminder to look at the harmonic series (Example 2.4.5) to erase any misconception that the converse statement is true. Knowing (a_k) tends to 0 does not imply that the series converges.

Theorem 2.7.4 (Comparison Test). *Assume (a_k) and (b_k) are sequences satisfying $0 \leq a_k \leq b_k$ for all $k \in \mathbf{N}$.*

(i) *If $\sum_{k=1}^{\infty} b_k$ converges, then $\sum_{k=1}^{\infty} a_k$ converges.*

(ii) *If $\sum_{k=1}^{\infty} a_k$ diverges, then $\sum_{k=1}^{\infty} b_k$ diverges.*

Proof. Both statements follow immediately from the Cauchy Criterion for Series and the observation that

$$|a_{m+1} + a_{m+2} + \cdots + a_n| \leq |b_{m+1} + b_{m+2} + \cdots + b_n|.$$

Alternate proofs using the Monotone Convergence Theorem are requested in the exercises. □

This is a good point to remind ourselves again that statements about convergence of sequences and series are immune to changes in some finite number of initial terms. In the Comparison Test, the requirement that $0 \leq a_k \leq b_k$ does not really need to hold for *all* $k \in \mathbf{N}$ but just needs to be *eventually* true. A weaker, but sufficient, hypothesis would be to assume that there exists some point $M \in \mathbf{N}$ such that the inequality $a_k \leq b_k$ is true for all $k \geq M$.

The Comparison Test is used to deduce the convergence or divergence of one series based on the behavior of another. Thus, for this test to be of any great use, we need a catalog of series we can use as measuring sticks. In Section 2.4, we proved the Cauchy Condensation Test, which led to the general statement that the series $\sum_{n=1}^{\infty} 1/n^p$ converges if and only if $p > 1$.

The next example summarizes the situation for another important class of series.

Example 2.7.5 (Geometric Series). A series is called *geometric* if it is of the form

$$\sum_{k=0}^{\infty} ar^k = a + ar + ar^2 + ar^3 + \cdots.$$

If $r = 1$ and $a \neq 0$, the series evidently diverges. For $r \neq 1$, the algebraic identity

$$(1 - r)(1 + r + r^2 + r^3 + \cdots + r^{m-1}) = 1 - r^m$$

enables us to rewrite the partial sum

$$s_m = a + ar + ar^2 + ar^3 + \cdots + ar^{m-1} = \frac{a(1 - r^m)}{1 - r}.$$

Now the Algebraic Limit Theorem (for sequences) and Example 2.5.3 justify the conclusion

$$\sum_{k=0}^{\infty} ar^k = \frac{a}{1 - r}$$

if and only if $|r| < 1$.

Although the Comparison Test requires that the terms of the series be positive, it is often used in conjunction with the next theorem to handle series that contain some negative terms.

Theorem 2.7.6 (Absolute Convergence Test). *If the series $\sum_{n=1}^{\infty} |a_n|$ converges, then $\sum_{n=1}^{\infty} a_n$ converges as well.*

Proof. This proof makes use of both the necessity (the "if" direction) and the sufficiency (the "only if" direction) of the Cauchy Criterion for Series. Because $\sum_{n=1}^{\infty} |a_n|$ converges, we know that, given an $\epsilon > 0$, there exists an $N \in \mathbf{N}$ such that

$$|a_{m+1}| + |a_{m+2}| + \cdots + |a_n| < \epsilon$$

for all $n > m \geq N$. By the triangle inequality,

$$|a_{m+1} + a_{m+2} + \cdots + a_n| \leq |a_{m+1}| + |a_{m+2}| + \cdots + |a_n|,$$

so the sufficiency of the Cauchy Criterion guarantees that $\sum_{n=1}^{\infty} a_n$ also converges. $\square$

The converse of this theorem is false. In the opening discussion of this chapter, we considered the *alternating harmonic series*

$$1 - \frac{1}{2} + \frac{1}{3} - \frac{1}{4} + \frac{1}{5} - \frac{1}{6} + \cdots .$$

Taking absolute values of the terms gives us the harmonic series $\sum_{n=1}^{\infty} 1/n$, which we have seen diverges. However, it is not too difficult to prove that with the alternating negative signs the series indeed converges. This is a special case of the Alternating Series Test.

Theorem 2.7.7 (Alternating Series Test). *Let (a_n) be a sequence satisfying,*

(i) $a_1 \geq a_2 \geq a_3 \geq \cdots \geq a_n \geq a_{n+1} \geq \cdots$ *and*

(ii) $(a_n) \to 0$.

Then, the alternating series $\sum_{n=1}^{\infty} (-1)^{n+1} a_n$ converges.

Proof. A consequence of conditions (i) and (ii) is that $a_n \geq 0$. Several proofs of this theorem are outlined in Exercise 2.7.1. $\square$

Definition 2.7.8. If $\sum_{n=1}^{\infty} |a_n|$ converges, then we say that the original series $\sum_{n=1}^{\infty} a_n$ *converges absolutely*. If, on the other hand, the series $\sum_{n=1}^{\infty} a_n$ converges but the series of absolute values $\sum_{n=1}^{\infty} |a_n|$ does not converge, then we say that the original series $\sum_{n=1}^{\infty} a_n$ *converges conditionally*.

In terms of this newly defined jargon, we have shown that

$$\sum_{n=1}^{\infty} \frac{(-1)^{n+1}}{n}$$

converges conditionally, whereas

$$\sum_{n=1}^{\infty} \frac{(-1)^{n+1}}{n^2}, \quad \sum_{n=1}^{\infty} \frac{1}{2^n} \quad \text{and} \quad \sum_{n=1}^{\infty} \frac{(-1)^{n+1}}{2^n}$$

converge absolutely. In particular, any convergent series with (all but finitely many) positive terms must converge absolutely.

The Alternating Series Test is the most accessible test for conditional convergence, but several others are explored in the exercises. In particular, Abel's Test, outlined in Exercise 2.7.13, will prove useful in our investigations of power series in Chapter 6.

Rearrangements

Informally speaking, a rearrangement of a series is obtained by permuting the terms in the sum into some other order. It is important that all of the original terms eventually appear in the new ordering and that no term gets repeated. In an earlier discussion from Section 2.1, we formed a rearrangement of the alternating harmonic series by taking two positive terms for each negative term:

$$1 + \frac{1}{3} - \frac{1}{2} + \frac{1}{5} + \frac{1}{7} - \frac{1}{4} + \cdots .$$

There are clearly an infinite number of rearrangements of any sum; however, it is helpful to see why neither

$$1 + \frac{1}{2} - \frac{1}{3} + \frac{1}{4} + \frac{1}{5} - \frac{1}{6} + \cdots$$

nor

$$1 + \frac{1}{3} - \frac{1}{4} + \frac{1}{5} + \frac{1}{7} - \frac{1}{8} + \frac{1}{9} + \frac{1}{11} - \frac{1}{12} + \cdots$$

is considered a rearrangement of the original alternating harmonic series.

Definition 2.7.9. Let $\sum_{k=1}^{\infty} a_k$ be a series. A series $\sum_{k=1}^{\infty} b_k$ is called a *rearrangement* of $\sum_{k=1}^{\infty} a_k$ if there exists a one-to-one, onto function $f : \mathbf{N} \to \mathbf{N}$ such that $b_{f(k)} = a_k$ for all $k \in \mathbf{N}$.

We now have all the tools and notation in place to resolve an issue raised at the beginning of the chapter. In Section 2.1, we constructed a particular rearrangement of the alternating harmonic series that converges to a limit different from that of the original series. This happens because the convergence is *conditional*.

Theorem 2.7.10. *If a series converges absolutely, then any rearrangement of this series converges to the same limit.*

Proof. Assume $\sum_{k=1}^{\infty} a_k$ converges absolutely to A, and let $\sum_{k=1}^{\infty} b_k$ be a rearrangement of $\sum_{k=1}^{\infty} a_k$. Let's use

$$s_n = \sum_{k=1}^{n} a_k = a_1 + a_2 + \cdots + a_n$$

for the partial sums of the original series and use

$$t_m = \sum_{k=1}^{m} b_k = b_1 + b_2 + \cdots + b_m$$

for the partial sums of the rearranged series. Thus we want to show that $(t_m) \to A$.

Let $\epsilon > 0$. By hypothesis, $(s_n) \to A$, so choose N_1 such that

$$|s_n - A| < \frac{\epsilon}{2}$$

for all $n \geq N_1$. Because the convergence is absolute, we can choose N_2 so that

$$\sum_{k=m+1}^{n} |a_k| < \frac{\epsilon}{2}$$

for all $n > m \geq N_2$. Now, take $N = \max\{N_1, N_2\}$. We know that the finite set of terms $\{a_1, a_2, a_3, \ldots, a_N\}$ must all appear in the rearranged series, and we want to move far enough out in the series $\sum_{n=1}^{\infty} b_n$ so that we have included all of these terms. Thus, choose

$$M = \max\{f(k) : 1 \leq k \leq N\}.$$

It should now be evident that if $m \geq M$, then $(t_m - s_N)$ consists of a finite set of terms, the absolute values of which appear in the tail $\sum_{k=N+1}^{\infty} |a_k|$. Our choice of N_2 earlier then guarantees $|t_m - s_N| < \epsilon/2$, and so

$$
\begin{aligned}
|t_m - A| &= |t_m - s_N + s_N - A| \\
&\leq |t_m - s_N| + |s_N - A| \\
&< \frac{\epsilon}{2} + \frac{\epsilon}{2} = \epsilon
\end{aligned}
$$

whenever $m \geq M$. $\square$

Exercises

Exercise 2.7.1. Proving the Alternating Series Test (Theorem 2.7.7) amounts to showing that the sequence of partial sums

$$s_n = a_1 - a_2 + a_3 - \cdots \pm a_n$$

converges. (The opening example in Section 2.1 includes a typical illustration of (s_n).) Different characterizations of completeness lead to different proofs.

(a) Prove the Alternating Series Test by showing that (s_n) is a Cauchy sequence.

(b) Supply another proof for this result using the Nested Interval Property (Theorem 1.4.1).

(c) Consider the subsequences (s_{2n}) and (s_{2n+1}), and show how the Monotone Convergence Theorem leads to a third proof for the Alternating Series Test.

Exercise 2.7.2. Decide whether each of the following series converges or diverges:

(a) $\sum_{n=1}^{\infty} \frac{1}{2^n + n}$ (b) $\sum_{n=1}^{\infty} \frac{\sin(n)}{n^2}$

(c) $1 - \frac{3}{4} + \frac{4}{6} - \frac{5}{8} + \frac{6}{10} - \frac{7}{12} + \cdots$

(d) $1 + \frac{1}{2} - \frac{1}{3} + \frac{1}{4} + \frac{1}{5} - \frac{1}{6} + \frac{1}{7} + \frac{1}{8} - \frac{1}{9} + \cdots$

(e) $1 - \frac{1}{2^2} + \frac{1}{3} - \frac{1}{4^2} + \frac{1}{5} - \frac{1}{6^2} + \frac{1}{7} - \frac{1}{8^2} + \cdots$

Exercise 2.7.3.   (a) Provide the details for the proof of the Comparison Test (Theorem 2.7.4) using the Cauchy Criterion for Series.

(b) Give another proof for the Comparison Test, this time using the Monotone Convergence Theorem.

Exercise 2.7.4. Give an example of each or explain why the request is impossible referencing the proper theorem(s).

(a) Two series $\sum x_n$ and $\sum y_n$ that both diverge but where $\sum x_n y_n$ converges.

(b) A convergent series $\sum x_n$ and a bounded sequence (y_n) such that $\sum x_n y_n$ diverges.

(c) Two sequences (x_n) and (y_n) where $\sum x_n$ and $\sum (x_n + y_n)$ both converge but $\sum y_n$ diverges.

(d) A sequence (x_n) satisfying $0 \le x_n \le 1/n$ where $\sum (-1)^n x_n$ diverges.

Exercise 2.7.5. Now that we have proved the basic facts about geometric series, supply a proof for Corollary 2.4.7.

Exercise 2.7.6. Let's say that a series *subverges* if the sequence of partial sums contains a subsequence that converges. Consider this (invented) definition for a moment, and then decide which of the following statements are valid propositions about subvergent series:

(a) If (a_n) is bounded, then $\sum a_n$ subverges.

(b) All convergent series are subvergent.

(c) If $\sum |a_n|$ subverges, then $\sum a_n$ subverges as well.

(d) If $\sum a_n$ subverges, then (a_n) has a convergent subsequence.

Exercise 2.7.7.   (a) Show that if $a_n > 0$ and $\lim(n a_n) = l$ with $l \ne 0$, then the series $\sum a_n$ diverges.

(b) Assume $a_n > 0$ and $\lim(n^2 a_n)$ exists. Show that $\sum a_n$ converges.

Exercise 2.7.8. Consider each of the following propositions. Provide short proofs for those that are true and counterexamples for any that are not.

(a) If $\sum a_n$ converges absolutely, then $\sum a_n^2$ also converges absolutely.

(b) If $\sum a_n$ converges and (b_n) converges, then $\sum a_n b_n$ converges.

(c) If $\sum a_n$ converges conditionally, then $\sum n^2 a_n$ diverges.

Exercise 2.7.9 (Ratio Test). Given a series $\sum_{n=1}^{\infty} a_n$ with $a_n \neq 0$, the Ratio Test states that if (a_n) satisfies

$$\lim \left| \frac{a_{n+1}}{a_n} \right| = r < 1,$$

then the series converges absolutely.

(a) Let r' satisfy $r < r' < 1$. Explain why there exists an N such that $n \geq N$ implies $|a_{n+1}| \leq |a_n| r'$.

(b) Why does $|a_N| \sum (r')^n$ converge?

(c) Now, show that $\sum |a_n|$ converges, and conclude that $\sum a_n$ converges.

Exercise 2.7.10 (Infinite Products). Review Exercise 2.4.10 about infinite products and then answer the following questions:

(a) Does $\frac{2}{1} \cdot \frac{3}{2} \cdot \frac{5}{4} \cdot \frac{9}{8} \cdot \frac{17}{16} \cdots$ converge?

(b) The infinite product $\frac{1}{2} \cdot \frac{3}{4} \cdot \frac{5}{6} \cdot \frac{7}{8} \cdot \frac{9}{10} \cdots$ certainly converges. (Why?) Does it converge to zero?

(c) In 1655, John Wallis famously derived the formula

$$\left(\frac{2 \cdot 2}{1 \cdot 3} \right) \left(\frac{4 \cdot 4}{3 \cdot 5} \right) \left(\frac{6 \cdot 6}{5 \cdot 7} \right) \left(\frac{8 \cdot 8}{7 \cdot 9} \right) \cdots = \frac{\pi}{2}.$$

Show that the left side of this identity at least converges to something. (A complete proof of this result is taken up in Section 8.3.)

Exercise 2.7.11. Find examples of two series $\sum a_n$ and $\sum b_n$ both of which diverge but for which $\sum \min\{a_n, b_n\}$ converges. To make it more challenging, produce examples where (a_n) and (b_n) are strictly positive and decreasing.

Exercise 2.7.12 (Summation-by-parts). Let (x_n) and (y_n) be sequences, let $s_n = x_1 + x_2 + \cdots + x_n$ and set $s_0 = 0$. Use the observation that $x_j = s_j - s_{j-1}$ to verify the formula

$$\sum_{j=m}^{n} x_j y_j = s_n y_{n+1} - s_{m-1} y_m + \sum_{j=m}^{n} s_j (y_j - y_{j+1}).$$

Exercise 2.7.13 (Abel's Test). Abel's Test for convergence states that if the series $\sum_{k=1}^{\infty} x_k$ converges, and if (y_k) is a sequence satisfying

$$y_1 \geq y_2 \geq y_3 \geq \cdots \geq 0,$$

then the series $\sum_{k=1}^{\infty} x_k y_k$ converges.

(a) Use Exercise 2.7.12 to show that

$$\sum_{k=1}^{n} x_k y_k = s_n y_{n+1} + \sum_{k=1}^{n} s_k (y_k - y_{k+1}),$$

where $s_n = x_1 + x_2 + \cdots + x_n$.

(b) Use the Comparison Test to argue that $\sum_{k=1}^{\infty} s_k (y_k - y_{k+1})$ converges absolutely, and show how this leads directly to a proof of Abel's Test.

Exercise 2.7.14 (Dirichlet's Test). Dirichlet's Test for convergence states that if the partial sums of $\sum_{k=1}^{\infty} x_k$ are bounded (but not necessarily convergent), and if (y_k) is a sequence satisfying $y_1 \geq y_2 \geq y_3 \geq \cdots \geq 0$ with $\lim y_k = 0$, then the series $\sum_{k=1}^{\infty} x_k y_k$ converges.

(a) Point out how the hypothesis of Dirichlet's Test differs from that of Abel's Test in Exercise 2.7.13, but show that essentially the same strategy can be used to provide a proof.

(b) Show how the Alternating Series Test (Theorem 2.7.7) can be derived as a special case of Dirichlet's Test.

2.8 Double Summations and Products of Infinite Series

Given a doubly indexed array of real numbers $\{a_{ij} : i, j \in \mathbf{N}\}$, we discovered in Section 2.1 that there is a dangerous ambiguity in how we might define $\sum_{i,j=1}^{\infty} a_{ij}$. Performing the sum over first one of the variables and then the other is referred to as an *iterated* summation. In our specific example, summing the rows first and then taking the sum of these totals produced a different result than first computing the sum of each column and adding these sums together. In short,

$$\sum_{j=1}^{\infty} \sum_{i=1}^{\infty} a_{ij} \neq \sum_{i=1}^{\infty} \sum_{j=1}^{\infty} a_{ij}.$$

There are still other ways to reasonably define $\sum_{i,j=1}^{\infty} a_{ij}$. One natural idea is to calculate a kind of partial sum by adding together finite numbers of terms in larger and larger "rectangles" in the array; that is, for $m, n \in \mathbf{N}$, set

(1)
$$s_{mn} = \sum_{i=1}^{m} \sum_{j=1}^{n} a_{ij}.$$

The order of the sum here is irrelevant because the sum is finite. Of particular interest to our discussion are the sums s_{nn} (sums over "squares"), which form a legitimate sequence indexed by n and thus can be subjected to our arsenal of theorems and definitions. If the sequence (s_{nn}) converges, for instance, we might wish to define

$$\sum_{i,j=1}^{\infty} a_{ij} = \lim_{n \to \infty} s_{nn}.$$

Exercise 2.8.1. Using the particular array (a_{ij}) from Section 2.1, compute $\lim_{n \to \infty} s_{nn}$. How does this value compare to the two iterated values for the sum already computed?

There is a deep similarity between the issue of how to define a double summation and the topic of rearrangements discussed at the end of Section 2.7. Both relate to the commutativity of addition in an infinite setting. For rearrangements, the resolution came with the added hypothesis of *absolute* convergence, and it is not surprising that the same remedy applies for double summations. Under the assumption of absolute convergence, each of the methods discussed for computing the value of a double sum yields the same result.

Exercise 2.8.2. Show that if the iterated series

$$\sum_{i=1}^{\infty} \sum_{j=1}^{\infty} |a_{ij}|$$

converges (meaning that for each fixed $i \in \mathbf{N}$ the series $\sum_{j=1}^{\infty} |a_{ij}|$ converges to some real number b_i, and the series $\sum_{i=1}^{\infty} b_i$ converges as well), then the iterated series

$$\sum_{i=1}^{\infty} \sum_{j=1}^{\infty} a_{ij}$$

converges.

Theorem 2.8.1. *Let $\{a_{ij} : i, j \in \mathbf{N}\}$ be a doubly indexed array of real numbers. If*

$$\sum_{i=1}^{\infty} \sum_{j=1}^{\infty} |a_{ij}|$$

converges, then both $\sum_{i=1}^{\infty} \sum_{j=1}^{\infty} a_{ij}$ and $\sum_{j=1}^{\infty} \sum_{i=1}^{\infty} a_{ij}$ converge to the same value. Moreover,

$$\lim_{n \to \infty} s_{nn} = \sum_{i=1}^{\infty} \sum_{j=1}^{\infty} a_{ij} = \sum_{j=1}^{\infty} \sum_{i=1}^{\infty} a_{ij},$$

where $s_{nn} = \sum_{i=1}^{n} \sum_{j=1}^{n} a_{ij}$.

Proof. In the same way that we defined the rectangular partial sums s_{mn} above in equation (1), define

$$t_{mn} = \sum_{i=1}^{m} \sum_{j=1}^{n} |a_{ij}|.$$

Exercise 2.8.3.   (a) Prove that (t_{nn}) converges.

(b) Now, use the fact that (t_{nn}) is a Cauchy sequence to argue that (s_{nn}) converges.

We can now set

$$S = \lim_{n \to \infty} s_{nn}.$$

In order to prove the theorem, we must show that the two iterated sums converge to this same limit. We will first show that

$$S = \sum_{i=1}^{\infty} \sum_{j=1}^{\infty} a_{ij}.$$

Because $\{t_{mn} : m, n \in \mathbf{N}\}$ is bounded above, we can let

$$B = \sup\{t_{mn} : m, n \in \mathbf{N}\}.$$

Exercise 2.8.4.   (a) Let $\epsilon > 0$ be arbitrary and argue that there exists an $N_1 \in \mathbf{N}$ such that $m, n \geq N_1$ implies $B - \frac{\epsilon}{2} < t_{mn} \leq B$.

(b) Now, show that there exists an N such that

$$|s_{mn} - S| < \epsilon$$

for all $m, n \geq N$.

For the moment, consider $m \in \mathbf{N}$ to be fixed and write s_{mn} as

$$s_{mn} = \sum_{j=1}^{n} a_{1j} + \sum_{j=1}^{n} a_{2j} + \cdots + \sum_{j=1}^{n} a_{mj}.$$

Our hypothesis guarantees that for each fixed row i, the series $\sum_{j=1}^{\infty} a_{ij}$ converges absolutely to some real number r_i.

Exercise 2.8.5.   (a) Show that for all $m \geq N$

$$|(r_1 + r_2 + \cdots + r_m) - S| \leq \epsilon.$$

Conclude that the iterated sum $\sum_{i=1}^{\infty} \sum_{j=1}^{\infty} a_{ij}$ converges to S.

(b) Finish the proof by showing that the other iterated sum, $\sum_{j=1}^{\infty} \sum_{i=1}^{\infty} a_{ij}$, converges to S as well. Notice that the same argument can be used once it is established that, for each fixed column j, the sum $\sum_{i=1}^{\infty} a_{ij}$ converges to some real number c_j.   □

One final common way of computing a double summation is to sum along diagonals where $i + j$ equals a constant. Given a doubly indexed array $\{a_{ij} : i, j \in \mathbf{N}\}$, let

$$d_2 = a_{11}, \quad d_3 = a_{12} + a_{21}, \quad d_4 = a_{13} + a_{22} + a_{31},$$

and in general set

$$d_k = a_{1,k-1} + a_{2,k-2} + \cdots + a_{k-1,1}.$$

Then, $\sum_{k=2}^{\infty} d_k$ represents another reasonable way of summing over every a_{ij} in the array.

Exercise 2.8.6.   (a) Assuming the hypothesis—and hence the conclusion—of Theorem 2.8.1, show that $\sum_{k=2}^{\infty} d_k$ converges absolutely.

(b) Imitate the strategy in the proof of Theorem 2.8.1 to show that $\sum_{k=2}^{\infty} d_k$ converges to $S = \lim_{n \to \infty} s_{nn}$.

Products of Series

Conspicuously missing from the Algebraic Limit Theorem for Series (Theorem 2.7.1) is any statement about the product of two convergent series. One way to formally carry out the algebra on such a product is to write

$$\left(\sum_{i=1}^{\infty} a_i \right) \left(\sum_{j=1}^{\infty} b_j \right) = (a_1 + a_2 + a_3 + \cdots)(b_1 + b_2 + b_3 + \cdots)$$

$$= a_1 b_1 + (a_1 b_2 + a_2 b_1) + (a_3 b_1 + a_2 b_2 + a_1 b_3) + \cdots$$

$$= \sum_{k=2}^{\infty} d_k,$$

where

$$d_k = a_1 b_{k-1} + a_2 b_{k-2} + \cdots + a_{k-1} b_1.$$

This particular form of the product, examined earlier in Exercise 2.8.6, is called the *Cauchy product* of two series. Although there is something algebraically natural about writing the product in this form, it may very well be that computing the value of the sum is more easily done via one or the other iterated summation. The question remains, then, as to how the value of the Cauchy product—if it exists—is related to these other values of the double sum. If the two series being multiplied converge absolutely, it is not too difficult to prove that the sum may be computed in whatever way is most convenient.

Exercise 2.8.7. Assume that $\sum_{i=1}^{\infty} a_i$ converges absolutely to A, and $\sum_{j=1}^{\infty} b_j$ converges absolutely to B.

(a) Show that the iterated sum $\sum_{i=1}^{\infty} \sum_{j=1}^{\infty} |a_i b_j|$ converges so that we may apply Theorem 2.8.1.

(b) Let $s_{nn} = \sum_{i=1}^{n} \sum_{j=1}^{n} a_i b_j$, and prove that $\lim_{n \to \infty} s_{nn} = AB$. Conclude that

$$\sum_{i=1}^{\infty} \sum_{j=1}^{\infty} a_i b_j = \sum_{j=1}^{\infty} \sum_{i=1}^{\infty} a_i b_j = \sum_{k=2}^{\infty} d_k = AB,$$

where, as before, $d_k = a_1 b_{k-1} + a_2 b_{k-2} + \cdots + a_{k-1} b_1$.

2.9   Epilogue

Theorems 2.7.10 and 2.8.1 make it clear that absolute convergence is an extremely desirable quality to have when manipulating series. On the other hand, the situation for conditionally convergent series is delightfully patholog-ical. In the case of rearrangements, not only are they no longer guaranteed to converge to the same limit, but in fact if $\sum_{n=1}^{\infty} a_n$ converges conditionally, then for *any* $r \in \mathbf{R}$ there exists a rearrangement of $\sum_{n=1}^{\infty} a_n$ that converges to r. To see why, let's look again at the alternating harmonic series

$$\sum_{n=1}^{\infty} \frac{(-1)^{n+1}}{n} = 1 - \frac{1}{2} + \frac{1}{3} - \frac{1}{4} + \frac{1}{5} - \frac{1}{6} + \cdots .$$

The negative terms taken alone form the series $\sum_{n=1}^{\infty} (-1)/2n$. The partial sums of this series are precisely $-1/2$ the partial sums of the harmonic series, and so march off (at half speed) to negative infinity. A similar argument shows that the sum of positive terms $\sum_{n=1}^{\infty} 1/(2n-1)$ also diverges to infinity. It is not too difficult to argue that this situation is *always* the case for conditionally convergent series. Now, let r be some proposed limit, which, for the sake of this argument, we take to be positive. The idea is to take as many positive terms as necessary to form the first partial sum greater than r. We then add negative terms until the partial sum falls below r, at which point we switch back to positive terms. The fact that there is no bound on the sums of either the positive terms or the negative terms allows this process to continue indefinitely. The fact that the terms themselves tend to zero is enough to guarantee that the partial sums, when constructed in this manner, indeed converge to r as they oscillate around this target value.

Perhaps the best way to summarize the situation is to say that the hypothe-sis of absolute convergence essentially allows us to treat infinite sums as though they were finite sums. This assessment extends to double sums as well, although there are a few subtleties to address. In the case of products, we showed in Ex-ercise 2.8.7 that the Cauchy product of two absolutely convergent infinite series converges to the product of the two factors, but in fact the same conclusion follows if we only have absolute convergence in one of the two original series. In the notation of Exercise 2.8.7, if $\sum a_n$ converges absolutely to A, and if $\sum b_n$ converges (perhaps conditionally) to B, then the Cauchy product $\sum d_k = AB$.

On the other hand, if both $\sum a_n$ and $\sum b_n$ converge conditionally, then it is possible for the Cauchy product to diverge. Squaring $\sum (-1)^n/\sqrt{n}$ provides an example of this phenomenon. Of course, it is also possible to find $\sum a_n = A$ conditionally and $\sum b_n = B$ conditionally whose Cauchy product $\sum d_k$ converges. If this is the case, then the convergence is to the right value, namely $\sum d_k = AB$. A proof of this last fact will be offered in Chapter 6 (Exercise 6.5.9), where we undertake the study of *power series*. Here is the connection. A power series has the form $a_0 + a_1 x + a_2 x^2 + \cdots$. If we multiply two power series together as though they were polynomials, then when we collect common powers of x the result is

$$
\begin{aligned}
&(a_0 + a_1 x + a_2 x^2 + \cdots)(b_0 + b_1 x + b_2 x^2 + \cdots) \\
=\quad & a_0 b_0 + (a_0 b_1 + a_1 b_0)x + (a_0 b_2 + a_1 b_1 + a_2 b_0)x^2 + \cdots \\
=\quad & d_0 + d_1 x + d_2 x^2 + \cdots,
\end{aligned}
$$

which is the Cauchy product of $\sum a_n x^n$ and $\sum b_n x^n$. (The index starts with $n = 0$ rather than $n = 1$.) Upcoming results about the good behavior of power series will lead to a proof that convergent Cauchy products sum to the proper value. In the other direction, Exercise 2.8.7 will be useful in establishing a theorem about the product of two power series.

Chapter 3

Basic Topology of R

3.1 Discussion: The Cantor Set

What follows is a fascinating mathematical construction, due to Georg Cantor, which is extremely useful for extending the horizons of our intuition about the nature of subsets of the real line. Cantor's name has already appeared in the first chapter in our discussion of uncountable sets. Indeed, Cantor's proof that **R** is uncountable occupies another spot on the short list of the most significant contributions toward understanding the mathematical infinite. In the words of the mathematician David Hilbert, "No one shall expel us from the paradise that Cantor has created for us."

Let C_0 be the closed interval $[0, 1]$, and define C_1 to be the set that results when the open middle third is removed; that is,

$$C_1 = C_0 \setminus \left(\frac{1}{3}, \frac{2}{3}\right) = \left[0, \frac{1}{3}\right] \cup \left[\frac{2}{3}, 1\right].$$

Now, construct C_2 in a similar way by removing the open middle third of each of the two components of C_1:

$$C_2 = \left(\left[0, \frac{1}{9}\right] \cup \left[\frac{2}{9}, \frac{1}{3}\right]\right) \cup \left(\left[\frac{2}{3}, \frac{7}{9}\right] \cup \left[\frac{8}{9}, 1\right]\right).$$

If we continue this process inductively, then for each $n = 0, 1, 2, \ldots$ we get a set C_n consisting of 2^n closed intervals each having length $1/3^n$. Finally, we define the *Cantor set* C (Fig. 3.1) to be the intersection

$$C = \bigcap_{n=0}^{\infty} C_n.$$

© Springer Science+Business Media New York 2015
S. Abbott, *Understanding Analysis*, Undergraduate Texts in Mathematics, DOI 10.1007/978-1-4939-2712-8_3

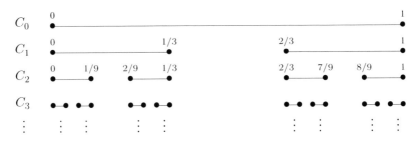

Figure 3.1: Defining the Cantor set; $C = \bigcap_{n=0}^{\infty} C_n$.

It may be useful to understand C as the remainder of the interval $[0,1]$ after the iterative process of removing open middle thirds is taken to infinity:

$$C = [0,1] \setminus \left[\left(\frac{1}{3}, \frac{2}{3} \right) \cup \left(\frac{1}{9}, \frac{2}{9} \right) \cup \left(\frac{7}{9}, \frac{8}{9} \right) \cup \cdots \right].$$

There is some initial doubt whether anything remains at all, but notice that because we are always removing open middle thirds, then for every $n \in \mathbf{N}$, $0 \in C_n$ and hence $0 \in C$. The same argument shows $1 \in C$. In fact, if y is the endpoint of some closed interval of some particular set C_n, then it is also an endpoint of one of the intervals of C_{n+1}. Because, at each stage, endpoints are never removed, it follows that $y \in C_n$ for all n. Thus, C at least contains the endpoints of all of the intervals that make up each of the sets C_n.

Is there anything else? Is C countable? Does C contain any intervals? Any irrational numbers? These are difficult questions at the moment. All of the endpoints mentioned earlier are rational numbers (they have the form $m/3^n$), which means that if it is true that C consists of only these endpoints, then C would be a subset of $\mathbf{Q}$ and hence countable. We shall see about this. There is some strong evidence that not much is left in C if we consider the total length of the intervals removed. To form C_1, an open interval of length $1/3$ was taken out. In the second step, we removed two intervals of length $1/9$, and to construct C_n we removed 2^{n-1} middle thirds of length $1/3^n$. There is some logic, then, to defining the "length" of C to be 1 minus the total

$$\frac{1}{3} + 2 \left(\frac{1}{9} \right) + 4 \left(\frac{1}{27} \right) + \cdots + 2^{n-1} \left(\frac{1}{3^n} \right) + \cdots = \frac{\frac{1}{3}}{1 - \frac{2}{3}} = 1.$$

The Cantor set has *zero length*.

To this point, the information we have collected suggests a mental picture of C as a relatively small, thin set. For these reasons, the set C is often referred to as Cantor "dust." But there are some strong counterarguments that imply a very different picture. First, C is actually *uncountable*, with cardinality equal to the cardinality of **R**. One slightly intuitive but convincing way to see this is to create a 1–1 correspondence between C and sequences of the form $(a_n)_{n=1}^{\infty}$, where $a_n = 0$ or 1. For each $c \in C$, set $a_1 = 0$ if c falls in the left-hand component

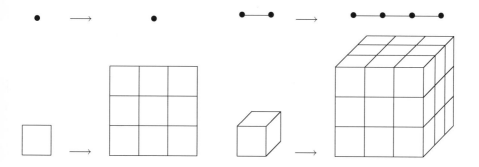

Figure 3.2: Magnifying sets by a factor of 3.

of C_1 and set $a_1 = 1$ if c falls in the right-hand component. Having established where in C_1 the point c is located, there are now two possible components of C_2 that might contain c. This time, we set $a_2 = 0$ or 1 depending on whether c falls in the left or right half of these two components of C_2. Continuing in this way, we come to see that every element $c \in C$ yields a sequence $(a_1, a_2, a_3, \ldots)$ of zeros and ones that acts as a set of directions for how to locate c within C. Likewise, every such sequence corresponds to a point in the Cantor set. Because the set of sequences of zeros and ones is uncountable (Exercise 1.6.4), we must conclude that C is uncountable as well.

What does this imply? In the first place, because the endpoints of the approximating sets C_n form a countable set, we are forced to accept the fact that not only are there other points in C but there are uncountably many of them. From the point of view of *cardinality*, C is quite large—as large as **R**, in fact. This should be contrasted with the fact that from the point of view of *length*, C measures the same size as a single point. We conclude this discussion with a demonstration that from the point of view of *dimension*, C strangely falls somewhere in between.

There is a sensible agreement that a point has dimension zero, a line segment has dimension one, a square has dimension two, and a cube has dimension three. Without attempting a formal definition of dimension (of which there are several), we can nevertheless get a sense of how one might be defined by observing how the dimension affects the result of magnifying each particular set by a factor of 3 (Fig. 3.2). (The reason for the choice of 3 will become clear when we turn our attention back to the Cantor set). A single point undergoes no change at all, whereas a line segment triples in length. For the square, magnifying each length by a factor of 3 results in a larger square that contains 9 copies of the original square. Finally, the magnified cube yields a cube that contains 27 copies of the original cube within its volume. Notice that, in each case, to compute the "size" of the new set, the dimension appears as the exponent of the magnification factor.

	dim	×3	new copies
point	0	→	$1 = 3^0$
segment	1	→	$3 = 3^1$
square	2	→	$9 = 3^2$
cube	3	→	$27 = 3^3$
C	x	→	$2 = 3^x$

Figure 3.3: Dimension of C; $2 = 3^x \Rightarrow x = \log 2 / \log 3$.

Now, apply this transformation to the Cantor set. The set $C_0 = [0, 1]$ becomes the interval $[0, 3]$. Deleting the middle third leaves $[0, 1] \cup [2, 3]$, which is where we started in the original construction except that we now stand to produce an additional copy of C in the interval $[2, 3]$. Magnifying the Cantor set by a factor of 3 yields *two* copies of the original set. Thus, if x is the dimension of C, then x should satisfy $2 = 3^x$, or $x = \log 2 / \log 3 \approx .631$ (Fig. 3.3).

The notion of a noninteger or fractional dimension is the impetus behind the term "fractal," coined in 1975 by Benoit Mandlebrot to describe a class of sets whose intricate structures have much in common with the Cantor set. Cantor's construction, however, is over a hundred years old and for us represents an invaluable testing ground for the upcoming theorems and conjectures about the often elusive nature of subsets of the real line.

3.2 Open and Closed Sets

Given $a \in \mathbf{R}$ and $\epsilon > 0$, recall that the ϵ-*neighborhood* of a is the set

$$V_\epsilon(a) = \{x \in \mathbf{R} : |x - a| < \epsilon\}.$$

In other words, $V_\epsilon(a)$ is the open interval $(a - \epsilon, a + \epsilon)$, centered at a with radius ϵ.

Definition 3.2.1. A set $O \subseteq \mathbf{R}$ is *open* if for all points $a \in O$ there exists an ϵ-neighborhood $V_\epsilon(a) \subseteq O$.

Example 3.2.2.   (i) Perhaps the simplest example of an open set is **R** itself. Given an arbitrary element $a \in \mathbf{R}$, we are free to pick any ϵ-neighborhood we like and it will always be true that $V_\epsilon(a) \subseteq \mathbf{R}$. It is also the case that the logical structure of Definition 3.2.1 requires us to classify the empty set $\emptyset$ as an open subset of the real line.

(ii) For a more useful collection of examples, consider the open interval

$$(c, d) = \{x \in \mathbf{R} : c < x < d\}.$$

To see that (c, d) is open in the sense just defined, let $x \in (c, d)$ be arbitrary. If we take $\epsilon = \min\{x - c, d - x\}$, then it follows that $V_\epsilon(x) \subseteq (c, d)$. It is important to see where this argument breaks down if the interval includes either one of its endpoints.

The union of open intervals is another example of an open set. This observation leads to the next result.

Theorem 3.2.3. (i) *The union of an arbitrary collection of open sets is open.*

(ii) *The intersection of a finite collection of open sets is open.*

Proof. To prove (i), we let $\{O_\lambda : \lambda \in \Lambda\}$ be a collection of open sets and let $O = \bigcup_{\lambda \in \Lambda} O_\lambda$. Let a be an arbitrary element of O. In order to show that O is open, Definition 3.2.1 insists that we produce an ϵ-neighborhood of a completely contained in O. But $a \in O$ implies that a is an element of at least one particular $O_{\lambda'}$. Because we are assuming $O_{\lambda'}$ is open, we can use Definition 3.2.1 to assert that there exists $V_\epsilon(a) \subseteq O_{\lambda'}$. The fact that $O_{\lambda'} \subseteq O$ allows us to conclude that $V_\epsilon(a) \subseteq O$. This completes the proof of (i).

For (ii), let $\{O_1, O_2, \ldots, O_N\}$ be a finite collection of open sets. Now, if $a \in \bigcap_{k=1}^{N} O_k$, then a is an element of each of the open sets. By the definition of an open set, we know that, for each $1 \leq k \leq N$, there exists $V_{\epsilon_k}(a) \subseteq O_k$. We are in search of a *single* ϵ-neighborhood of a that is contained in every O_k, so the trick is to take the smallest one. Letting $\epsilon = \min\{\epsilon_1, \epsilon_2, \ldots, \epsilon_N\}$, it follows that $V_\epsilon(a) \subseteq V_{\epsilon_k}(a)$ for all k, and hence $V_\epsilon(a) \subseteq \bigcap_{k=1}^{N} O_k$, as desired. $\square$

Closed Sets

Definition 3.2.4. A point x is a *limit point* of a set A if every ϵ-neighborhood $V_\epsilon(x)$ of x intersects the set A at some point other than x.

Limit points are also often referred to as "cluster points" or "accumulation points," but the phrase "x is a limit point of A" has the advantage of explicitly reminding us that x is quite literally the limit of a sequence in A.

Theorem 3.2.5. *A point x is a limit point of a set A if and only if $x = \lim a_n$ for some sequence (a_n) contained in A satisfying $a_n \neq x$ for all $n \in \mathbf{N}$.*

Proof. ($\Rightarrow$) Assume x is a limit point of A. In order to produce a sequence (a_n) converging to x, we are going to consider the particular ϵ-neighborhoods obtained using $\epsilon = 1/n$. By Definition 3.2.4, every neighborhood of x intersects A in some point other than x. This means that, for each $n \in \mathbf{N}$, we are justified in picking a point

$$a_n \in V_{1/n}(x) \cap A$$

with the stipulation that $a_n \neq x$. It should not be too difficult to see why $(a_n) \to x$. Given an arbitrary $\epsilon > 0$, choose N such that $1/N < \epsilon$. It follows that $|a_n - x| < \epsilon$ for all $n \geq N$.

($\Leftarrow$) For the reverse implication we assume $\lim a_n = x$ where $a_n \in A$ but $a_n \neq x$, and let $V_\epsilon(x)$ be an arbitrary ϵ-neighborhood. The definition of convergence assures us that there exists a term a_N in the sequence satisfying $a_N \in V_\epsilon(x)$, and the proof is complete. $\qquad\qquad\qquad\qquad\qquad\qquad\qquad\qquad\qquad\qquad\qquad\quad\square$

The restriction that $a_n \neq x$ in Theorem 3.2.5 deserves a comment. Given a point $a \in A$, it is always the case that a is the limit of a sequence in A if we are allowed to consider the constant sequence $(a, a, a, \dots)$. There will be occasions where we will want to avoid this somewhat uninteresting situation, so it is important to have a vocabulary that can distinguish limit points of a set from *isolated points*.

Definition 3.2.6. A point $a \in A$ is an *isolated point of A* if it is not a limit point of A.

As a word of caution, we need to be a little careful about how we understand the relationship between these concepts. Whereas an isolated point is always an element of the relevant set A, it is quite possible for a limit point of A not to belong to A. As an example, consider the endpoint of an open interval. This situation is the subject of the next important definition.

Definition 3.2.7. A set $F \subseteq \mathbf{R}$ is *closed* if it contains its limit points.

The adjective "closed" appears in several other mathematical contexts and is usually employed to mean that an operation on the elements of a given set does not take us out of the set. In linear algebra, for example, a vector space is a set that is "closed" under addition and scalar multiplication. In analysis, the operation we are concerned with is the limiting operation. Topologically speaking, a closed set is one where convergent sequences within the set have limits that are also in the set.

Theorem 3.2.8. *A set $F \subseteq \mathbf{R}$ is closed if and only if every Cauchy sequence contained in F has a limit that is also an element of F.*

Proof. Exercise 3.2.5. $\qquad\qquad\qquad\qquad\qquad\qquad\qquad\qquad\qquad\qquad\qquad\qquad\quad\square$

Example 3.2.9. (i) Consider

$$A = \left\{ \frac{1}{n} : n \in \mathbf{N} \right\}.$$

Let's show that each point of A is isolated. Given $1/n \in A$, choose $\epsilon = 1/n - 1/(n+1)$. Then,

$$V_\epsilon(1/n) \cap A = \left\{ \frac{1}{n} \right\}.$$

It follows from Definition 3.2.4 that $1/n$ is not a limit point and so is isolated. Although all of the points of A are isolated, the set does have one limit point, namely 0. This is because every neighborhood centered at zero, no matter how small, is going to contain points of A. Because $0 \notin A$, A is not closed. The set $F = A \cup \{0\}$ is an example of a closed set and is called the *closure of A*. (The closure of a set is discussed in a moment.)

(ii) Let's prove that a closed interval

$$[c, d] = \{x \in \mathbf{R} : c \le x \le d\}$$

is a closed set using Definition 3.2.7. If x is a limit point of $[c, d]$, then by Theorem 3.2.5 there exists $(x_n) \subseteq [c, d]$ with $(x_n) \to x$. We need to prove that $x \in [c, d]$.

The key to this argument is contained in the Order Limit Theorem (Theorem 2.3.4), which summarizes the relationship between inequalities and the limiting process. Because $c \le x_n \le d$, it follows from Theorem 2.3.4 (iii) that $c \le x \le d$ as well. Thus, $[c, d]$ is closed.

(iii) Consider the set $\mathbf{Q} \subseteq \mathbf{R}$ of rational numbers. An extremely important property of $\mathbf{Q}$ is that its set of limit points is actually *all of $\mathbf{R}$*. To see why this is so, recall Theorem 1.4.3 from Chapter 1, which is referred to as the density property of $\mathbf{Q}$ in $\mathbf{R}$.

Let $y \in \mathbf{R}$ be arbitrary, and consider any neighborhood $V_\epsilon(y) = (y - \epsilon, y + \epsilon)$. Theorem 1.4.3 allows us to conclude that there exists a rational number $r \ne y$ that falls in this neighborhood. Thus, y is a limit point of $\mathbf{Q}$.

The density property of $\mathbf{Q}$ can now be reformulated in the following way.

Theorem 3.2.10 (Density of Q in R). *For every $y \in \mathbf{R}$, there exists a sequence of rational numbers that converges to y.*

Proof. Combine the preceding discussion with Theorem 3.2.5. $\qquad \square$

The same argument can also be used to show that every real number is the limit of a sequence of irrational numbers. Although interesting, part of the allure of the rational numbers is that, in addition to being dense in $\mathbf{R}$, they are countable. As we will see, this tangible aspect of $\mathbf{Q}$ makes it an extremely useful set, both for proving theorems and for producing interesting counterexamples.

Closure

Definition 3.2.11. Given a set $A \subseteq \mathbf{R}$, let L be the set of all limit points of A. The *closure* of A is defined to be $\overline{A} = A \cup L$.

In Example 3.2.9 (i), we saw that if $A = \{1/n : n \in \mathbf{N}\}$, then the closure of A is $\overline{A} = A \cup \{0\}$. Example 3.2.9 (iii) verifies that $\overline{\mathbf{Q}} = \mathbf{R}$. If A is an open interval (a, b), then $\overline{A} = [a, b]$. If A is a closed interval, then $\overline{A} = A$. It is not for lack of imagination that in each of these examples $\overline{A}$ is always a closed set.

Theorem 3.2.12. *For any $A \subseteq \mathbf{R}$, the closure $\overline{A}$ is a closed set and is the smallest closed set containing A.*

Proof. If L is the set of limit points of A, then it is immediately clear that $\overline{A}$ contains the limit points of A. There is still something more to prove, however, because taking the union of L with A could potentially produce some new limit points of $\overline{A}$. In Exercise 3.2.7, we outline the argument that this does not happen.

Now, *any* closed set containing A must contain L as well. This shows that $\overline{A} = A \cup L$ is the smallest closed set containing A. $\square$

Complements

The mathematical notions of open and closed are not antonyms the way they are in standard English. If a set is not open, that does not imply it must be closed. Many sets such as the half-open interval $(c, d] = \{x \in \mathbf{R} : c < x \le d\}$ are neither open nor closed. The sets $\mathbf{R}$ and $\emptyset$ are both simultaneously open and closed although, thankfully, these are the only ones with this disorienting property (Exercise 3.2.13). There is, however, an important relationship between open and closed sets. Recall that the *complement* of a set $A \subseteq \mathbf{R}$ is defined to be the set

$$A^c = \{x \in \mathbf{R} : x \notin A\}.$$

Theorem 3.2.13. *A set O is open if and only if O^c is closed. Likewise, a set F is closed if and only if F^c is open.*

Proof. Given an open set $O \subseteq \mathbf{R}$, let's first prove that O^c is a closed set. To prove O^c is closed, we need to show that it contains all of its limit points. If x is a limit point of O^c, then *every* neighborhood of x contains some point of O^c. But that is enough to conclude that x cannot be in the open set O because $x \in O$ would imply that there exists a neighborhood $V_\epsilon(x) \subseteq O$. Thus, $x \in O^c$, as desired.

For the converse statement, we assume O^c is closed and argue that O is open. Thus, given an arbitrary point $x \in O$, we must produce an ϵ-neighborhood $V_\epsilon(x) \subseteq O$. Because O^c is closed, we can be sure that x is *not* a limit point of O^c. Looking at the definition of limit point, we see that this implies that there must be some neighborhood $V_\epsilon(x)$ of x that does not intersect the set O^c. But this means $V_\epsilon(x) \subseteq O$, which is precisely what we needed to show.

The second statement in Theorem 3.2.13 follows quickly from the first using the observation that $(E^c)^c = E$ for any set $E \subseteq \mathbf{R}$. $\square$

The last theorem of this section should be compared to Theorem 3.2.3.

Theorem 3.2.14.   (i) *The union of a finite collection of closed sets is closed.*

(ii) *The intersection of an arbitrary collection of closed sets is closed.*

Proof. De Morgan's Laws state that for any collection of sets $\{E_\lambda : \lambda \in \Lambda\}$ it is true that

$$\left(\bigcup_{\lambda \in \Lambda} E_\lambda\right)^c = \bigcap_{\lambda \in \Lambda} E_\lambda^c \quad \text{and} \quad \left(\bigcap_{\lambda \in \Lambda} E_\lambda\right)^c = \bigcup_{\lambda \in \Lambda} E_\lambda^c.$$

The result follows directly from these statements and Theorem 3.2.3. The details are requested in Exercise 3.2.9. $\square$

Exercises

Exercise 3.2.1. (a) Where in the proof of Theorem 3.2.3 part (ii) does the assumption that the collection of open sets be *finite* get used?

(b) Give an example of a countable collection of open sets $\{O_1, O_2, O_3, \ldots\}$ whose intersection $\bigcap_{n=1}^{\infty} O_n$ is closed, not empty and not all of $\mathbf{R}$.

Exercise 3.2.2. Let

$$A = \left\{(-1)^n + \frac{2}{n} : n = 1, 2, 3, \ldots\right\} \quad \text{and} \quad B = \{x \in \mathbf{Q} : 0 < x < 1\}.$$

Answer the following questions for each set:

(a) What are the limit points?

(b) Is the set open? Closed?

(c) Does the set contain any isolated points?

(d) Find the closure of the set.

Exercise 3.2.3. Decide whether the following sets are open, closed, or neither. If a set is not open, find a point in the set for which there is no ϵ-neighborhood contained in the set. If a set is not closed, find a limit point that is not contained in the set.

(a) $\mathbf{Q}$.

(b) $\mathbf{N}$.

(c) $\{x \in \mathbf{R} : x \neq 0\}$.

(d) $\{1 + 1/4 + 1/9 + \cdots + 1/n^2 : n \in \mathbf{N}\}$.

(e) $\{1 + 1/2 + 1/3 + \cdots + 1/n : n \in \mathbf{N}\}$.

Exercise 3.2.4. Let A be nonempty and bounded above so that $s = \sup A$ exists.

(a) Show that $s \in \overline{A}$.

(b) Can an open set contain its supremum?

Exercise 3.2.5. Prove Theorem 3.2.8.

Exercise 3.2.6. Decide whether the following statements are true or false. Provide counterexamples for those that are false, and supply proofs for those that are true.

(a) An open set that contains every rational number must necessarily be all of **R**.

(b) The Nested Interval Property remains true if the term "closed interval" is replaced by "closed set."

(c) Every nonempty open set contains a rational number.

(d) Every bounded infinite closed set contains a rational number.

(e) The Cantor set is closed.

Exercise 3.2.7. Given $A \subseteq \mathbf{R}$, let L be the set of all limit points of A.

(a) Show that the set L is closed.

(b) Argue that if x is a limit point of $A \cup L$, then x is a limit point of A. Use this observation to furnish a proof for Theorem 3.2.12.

Exercise 3.2.8. Assume A is an open set and B is a closed set. Determine if the following sets are definitely open, definitely closed, both, or neither.

(a) $\overline{A \cup B}$

(b) $A \backslash B = \{x \in A : x \notin B\}$

(c) $(A^c \cup B)^c$

(d) $(A \cap B) \cup (A^c \cap B)$

(e) $\overline{A}^c \cap \overline{A^c}$

Exercise 3.2.9 (De Morgan's Laws). A proof for De Morgan's Laws in the case of two sets is outlined in Exercise 1.2.5. The general argument is similar.

(a) Given a collection of sets $\{E_\lambda : \lambda \in \Lambda\}$, show that

$$\left(\bigcup_{\lambda \in \Lambda} E_\lambda \right)^c = \bigcap_{\lambda \in \Lambda} E_\lambda^c \quad \text{and} \quad \left(\bigcap_{\lambda \in \Lambda} E_\lambda \right)^c = \bigcup_{\lambda \in \Lambda} E_\lambda^c.$$

(b) Now, provide the details for the proof of Theorem 3.2.14.

Exercise 3.2.10. Only one of the following three descriptions can be realized. Provide an example that illustrates the viable description, and explain why the other two cannot exist.

(i) A countable set contained in $[0,1]$ with no limit points.

(ii) A countable set contained in $[0,1]$ with no isolated points.

(iii) A set with an uncountable number of isolated points.

Exercise 3.2.11. (a) Prove that $\overline{A \cup B} = \overline{A} \cup \overline{B}$.

(b) Does this result about closures extend to infinite unions of sets?

Exercise 3.2.12. Let A be an uncountable set and let B be the set of real numbers that divides A into two uncountable sets; that is, $s \in B$ if both $\{x : x \in A \text{ and } x < s\}$ and $\{x : x \in A \text{ and } x > s\}$ are uncountable. Show B is nonempty and open.

Exercise 3.2.13. Prove that the only sets that are both open and closed are $\mathbf{R}$ and the empty set $\emptyset$.

Exercise 3.2.14. A dual notion to the closure of a set is the interior of a set. The *interior* of E is denoted E° and is defined as

$$E^\circ = \{x \in E : \text{there exists } V_\epsilon(x) \subseteq E\}.$$

Results about closures and interiors possess a useful symmetry.

(a) Show that E is closed if and only if $\overline{E} = E$. Show that E is open if and only if $E^\circ = E$.

(b) Show that $\overline{E}^c = (E^c)^\circ$, and similarly that $(E^\circ)^c = \overline{E^c}$.

Exercise 3.2.15. A set A is called an F_σ set if it can be written as the countable union of closed sets. A set B is called a G_δ set if it can be written as the countable intersection of open sets.

(a) Show that a closed interval $[a, b]$ is a G_δ set.

(b) Show that the half-open interval $(a, b]$ is both a G_δ and an F_σ set.

(c) Show that $\mathbf{Q}$ is an F_σ set, and the set of irrationals $\mathbf{I}$ forms a G_δ set. (We will see in Section 3.5 that $\mathbf{Q}$ is *not* a G_δ set, nor is $\mathbf{I}$ an F_σ set.)

3.3   Compact Sets

The central challenge in analysis is to exploit the power of the mathematical infinite—via limits, series, derivatives, integrals, etc.—without falling victim to erroneous logic or faulty intuition. A major tool for maintaining a rigorous footing in this endeavor is the concept of compact sets. In ways that will become clear, especially in our upcoming study of continuous functions, employing compact sets in a proof often has the effect of bringing a finite quality to the argument, thereby making it much more tractable.

Definition 3.3.1 (Compactness). A set $K \subseteq \mathbf{R}$ is *compact* if every sequence in K has a subsequence that converges to a limit that is also in K.

Example 3.3.2. The most basic example of a compact set is a closed interval. To see this, notice that if (a_n) is contained in an interval $[c, d]$, then the Bolzano–Weierstrass Theorem guarantees that we can find a convergent subsequence (a_{n_k}). Because a closed interval is a closed set (Example 3.2.9, (ii)), we know that the limit of this subsequence is also in $[c, d]$.

What are the properties of closed intervals that we used in the preceding argument? The Bolzano–Weierstrass Theorem requires boundedness, and we used the fact that closed sets contain their limit points. As we are about to see, these two properties completely characterize compact sets in $\mathbf{R}$. The term "bounded" has thus far only been used to describe sequences (Definition 2.3.1), but an analogous statement can easily be made about sets.

Definition 3.3.3. A set $A \subseteq \mathbf{R}$ is *bounded* if there exists $M > 0$ such that $|a| \leq M$ for all $a \in A$.

Theorem 3.3.4 (Characterization of Compactness in R). *A set $K \subseteq \mathbf{R}$ is compact if and only if it is closed and bounded.*

Proof. Let K be compact. We will first prove that K must be bounded, so assume, for contradiction, that K is not a bounded set. The idea is to produce a sequence in K that marches off to infinity in such a way that it cannot have a convergent subsequence as the definition of compact requires. To do this, notice that because K is not bounded there must exist an element $x_1 \in K$ satisfying $|x_1| > 1$. Likewise, there must exist $x_2 \in K$ with $|x_2| > 2$, and in general, given any $n \in \mathbf{N}$, we can produce $x_n \in K$ such that $|x_n| > n$.

Now, because K is assumed to be compact, (x_n) should have a convergent subsequence (x_{n_k}). But the elements of the subsequence must satisfy $|x_{n_k}| > n_k$, and consequently (x_{n_k}) is unbounded. Because convergent sequences are bounded (Theorem 2.3.2), we have a contradiction. Thus, K must at least be a bounded set.

Next, we will show that K is also closed. To see that K contains its limit points, we let $x = \lim x_n$, where (x_n) is contained in K and argue that x must be in K as well. By Definition 3.3.1, the sequence (x_n) has a convergent subsequence (x_{n_k}), and by Theorem 2.5.2, we know (x_{n_k}) converges to the same limit x. Finally, Definition 3.3.1 requires that $x \in K$. This proves that K is closed.

The proof of the converse statement is requested in Exercise 3.3.3. $\qquad\square$

There may be a temptation to consider closed intervals as being a kind of standard archetype for compact sets, but this is misleading. The structure of compact sets can be much more intricate and interesting. For instance, one implication of Theorem 3.3.4 is that the Cantor set is compact. It is more useful to think of compact sets as generalizations of closed intervals. Whenever a fact involving closed intervals is true, it is often the case that the same result holds when we replace "closed interval" with "compact set." As an example, let's experiment with the Nested Interval Property proved in the first chapter.

Theorem 3.3.5 (Nested Compact Set Property). *If*

$$K_1 \supseteq K_2 \supseteq K_3 \supseteq K_4 \supseteq \cdots$$

is a nested sequence of nonempty compact sets, then the intersection $\bigcap_{n=1}^{\infty} K_n$ is not empty.

Proof. In order to take advantage of the compactness of each K_n, we are going to produce a sequence that is eventually in each of these sets. Thus, for each $n \in \mathbf{N}$, pick a point $x_n \in K_n$. Because the compact sets are nested, it follows that the sequence (x_n) is contained in K_1. By Definition 3.3.1, (x_n) has a convergent subsequence (x_{n_k}) whose limit $x = \lim x_{n_k}$ is an element of K_1.

In fact, x is an element of *every* K_n for essentially the same reason. Given a particular $n_0 \in \mathbf{N}$, the terms in the sequence (x_n) are contained in K_{n_0} as long as $n \geq n_0$. Ignoring the finite number of terms for which $n_k < n_0$, the same subsequence (x_{n_k}) is then also contained in K_{n_0}. The conclusion is that $x = \lim x_{n_k}$ is an element of K_{n_0}. Because n_0 was arbitrary, $x \in \bigcap_{n=1}^{\infty} K_n$ and the proof is complete. $\qquad\square$

Open Covers

Defining compactness for sets in $\mathbf{R}$ is reminiscent of the situation we encountered with completeness in that there are a number of equivalent ways to describe this phenomenon. We demonstrated the equivalence of two such characterizations in Theorem 3.3.4. What this theorem implies is that we could have decided to *define* compact sets to be sets that are closed and bounded, and then *proved* that sequences contained in compact sets have convergent subsequences with limits in the set. There are some larger issues involved in deciding what the definition should be, but what is important at this moment is that we be versatile enough to use whatever description of compactness is most appropriate for a given situation.

Although Theorem 3.3.4 is sufficient for most of our purposes, there is a third important characterization of compactness, equivalent to the two others, which is described in terms of *open covers* and *finite subcovers*.

Definition 3.3.6. Let $A \subseteq \mathbf{R}$. An *open cover* for A is a (possibly infinite) collection of open sets $\{O_\lambda : \lambda \in \Lambda\}$ whose union contains the set A; that is, $A \subseteq \bigcup_{\lambda \in \Lambda} O_\lambda$. Given an open cover for A, a *finite subcover* is a finite subcollection of open sets from the original open cover whose union still manages to completely contain A.

Example 3.3.7. Consider the open interval $(0, 1)$. For each point $x \in (0, 1)$, let O_x be the open interval $(x/2, 1)$. Taken together, the infinite collection $\{O_x : x \in (0, 1)\}$ forms an open cover for the open interval $(0, 1)$. Notice, however, that it is impossible to find a finite subcover. Given any proposed finite subcollection

$$\{O_{x_1}, O_{x_2}, \dots, O_{x_n}\},$$

set $x' = \min\{x_1, x_2, \dots, x_n\}$ and observe that any real number y satisfying $0 < y \le x'/2$ is not contained in the union $\bigcup_{i=1}^{n} O_{x_i}$.

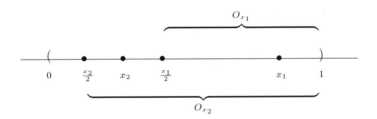

Now, consider a similar cover for the closed interval $[0, 1]$. For $x \in (0, 1)$, the sets $O_x = (x/2, 1)$ do a fine job covering $(0, 1)$, but in order to have an open cover of the closed interval $[0, 1]$, we must also cover the endpoints. To remedy this, we could fix $\epsilon > 0$, and let $O_0 = (-\epsilon, \epsilon)$ and $O_1 = (1 - \epsilon, 1 + \epsilon)$. Then, the collection

$$\{O_0, O_1, O_x : x \in (0, 1)\}$$

is an open cover for $[0, 1]$. But this time, notice there is a finite subcover. Because of the addition of the set O_0, we can choose x' so that $x'/2 < \epsilon$. It follows that $\{O_0, O_{x'}, O_1\}$ is a finite subcover for the closed interval $[0, 1]$.

Theorem 3.3.8 (Heine–Borel Theorem). *Let K be a subset of $\mathbf{R}$. All of the following statements are equivalent in the sense that any one of them implies the two others:*

(i) *K is compact.*

(ii) *K is closed and bounded.*

(iii) *Every open cover for K has a finite subcover.*

Proof. The equivalence of (i) and (ii) is the content of Theorem 3.3.4. What remains is to show that (iii) is equivalent to (i) and (ii). Let's first assume (iii), and prove that it implies (ii) (and thus (i) as well).

To show that K is bounded, we construct an open cover for K by defining O_x to be an open interval of radius 1 around each point $x \in K$. In the language of neighborhoods, $O_x = V_1(x)$. The open cover $\{O_x : x \in K\}$ then must have a finite subcover $\{O_{x_1}, O_{x_2}, \ldots, O_{x_n}\}$. Because K is contained in a finite union of bounded sets, K must itself be bounded.

The proof that K is closed is more delicate, and we argue it by contradiction. Let (y_n) be a Cauchy sequence contained in K with $\lim y_n = y$. To show that K is closed, we must demonstrate that $y \in K$, so assume for contradiction that this is not the case. If $y \notin K$, then every $x \in K$ is some positive distance away from y. We now construct an open cover by taking O_x to be an interval of radius $|x - y|/2$ around each point x in K. Because we are assuming (iii), the resulting open cover $\{O_x : x \in K\}$ must have a finite subcover $\{O_{x_1}, O_{x_2}, \ldots, O_{x_n}\}$. The contradiction arises when we realize that, in the spirit of Example 3.3.7, this finite subcover cannot contain all of the elements of the sequence (y_n). To make this explicit, set

$$\epsilon_0 = \min\left\{ \frac{|x_i - y|}{2} : 1 \le i \le n \right\}.$$

Because $(y_n) \to y$, we can certainly find a term y_N satisfying $|y_N - y| < \epsilon_0$. But such a y_N must necessarily be excluded from each O_{x_i}, meaning that

$$y_N \notin \bigcup_{i=1}^{n} O_{x_i}.$$

Thus our supposed subcover does not actually cover all of K. This contradiction implies that $y \in K$, and hence K is closed and bounded.

The proof that (ii) implies (iii) is outlined in Exercise 3.3.9. To be historically accurate, it is this particular implication that is most appropriately referred to as the Heine–Borel Theorem. □

Exercises

Exercise 3.3.1. Show that if K is compact and nonempty, then $\sup K$ and $\inf K$ both exist and are elements of K.

Exercise 3.3.2. Decide which of the following sets are compact. For those that are not compact, show how Definition 3.3.1 breaks down. In other words, give an example of a sequence contained in the given set that does not possess a subsequence converging to a limit in the set.

(a) $\mathbf{N}$.

(b) $\mathbf{Q} \cap [0, 1]$.

(c) The Cantor set.

(d) $\{1 + 1/2^2 + 1/3^2 + \cdots + 1/n^2 : n \in \mathbf{N}\}$.

(e) $\{1, 1/2, 2/3, 3/4, 4/5, \ldots\}$.

Exercise 3.3.3. Prove the converse of Theorem 3.3.4 by showing that if a set $K \subseteq \mathbf{R}$ is closed and bounded, then it is compact.

Exercise 3.3.4. Assume K is compact and F is closed. Decide if the following sets are definitely compact, definitely closed, both, or neither.

(a) $K \cap F$

(b) $\overline{F^c \cup K^c}$

(c) $K \backslash F = \{x \in K : x \notin F\}$

(d) $\overline{K \cap F^c}$

Exercise 3.3.5. Decide whether the following propositions are true or false. If the claim is valid, supply a short proof, and if the claim is false, provide a counterexample.

(a) The arbitrary intersection of compact sets is compact.

(b) The arbitrary union of compact sets is compact.

(c) Let A be arbitrary, and let K be compact. Then, the intersection $A \cap K$ is compact.

(d) If $F_1 \supseteq F_2 \supseteq F_3 \supseteq F_4 \supseteq \cdots$ is a nested sequence of nonempty closed sets, then the intersection $\bigcap_{n=1}^{\infty} F_n \neq \emptyset$.

Exercise 3.3.6. This exercise is meant to illustrate the point made in the opening paragraph to Section 3.3. Verify that the following three statements are true if every blank is filled in with the word "finite." Which are true if every blank is filled in with the word "compact"? Which are true if every blank is filled in with the word "closed"?

(a) Every _____ set has a maximum.

(b) If A and B are _____, then $A + B = \{a + b : a \in A, b \in B\}$ is also _____.

(c) If $\{A_n : n \in \mathbf{N}\}$ is a collection of _____ sets with the property that every finite subcollection has a nonempty intersection, then $\bigcap_{n=1}^{\infty} A_n$ is nonempty as well.

Exercise 3.3.7. As some more evidence of the surprising nature of the Cantor set, follow these steps to show that the sum $C + C = \{x + y : x, y \in C\}$ is equal to the closed interval $[0, 2]$. (Keep in mind that C has zero length and contains no intervals.)

Because $C \subseteq [0, 1]$, $C + C \subseteq [0, 2]$, so we only need to prove the reverse inclusion $[0, 2] \subseteq \{x + y : x, y \in C\}$. Thus, given $s \in [0, 2]$, we must find two elements $x, y \in C$ satisfying $x + y = s$.

(a) Show that there exist $x_1, y_1 \in C_1$ for which $x_1 + y_1 = s$. Show in general that, for an arbitrary $n \in \mathbf{N}$, we can always find $x_n, y_n \in C_n$ for which $x_n + y_n = s$.

(b) Keeping in mind that the sequences (x_n) and (y_n) do not necessarily converge, show how they can nevertheless be used to produce the desired x and y in C satisfying $x + y = s$.

Exercise 3.3.8. Let K and L be nonempty compact sets, and define

$$d = \inf\{|x - y| : x \in K \text{ and } y \in L\}.$$

This turns out to be a reasonable definition for the *distance* between K and L.

(a) If K and L are disjoint, show $d > 0$ and that $d = |x_0 - y_0|$ for some $x_0 \in K$ and $y_0 \in L$.

(b) Show that it's possible to have $d = 0$ if we assume only that the disjoint sets K and L are closed.

Exercise 3.3.9. Follow these steps to prove the final implication in Theorem 3.3.8.

Assume K satisfies (i) and (ii), and let $\{O_\lambda : \lambda \in \Lambda\}$ be an open cover for K. For contradiction, let's assume that no finite subcover exists. Let I_0 be a closed interval containing K.

(a) Show that there exists a nested sequence of closed intervals $I_0 \supseteq I_1 \supseteq I_2 \supseteq \cdots$ with the property that, for each n, $I_n \cap K$ cannot be finitely covered and $\lim |I_n| = 0$.

(b) Argue that there exists an $x \in K$ such that $x \in I_n$ for all n.

(c) Because $x \in K$, there must exist an open set O_{λ_0} from the original collection that contains x as an element. Explain how this leads to the desired contradiction.

Exercise 3.3.10. Here is an alternate proof to the one given in Exercise 3.3.9 for the final implication in the Heine–Borel Theorem.

Consider the special case where K is a closed interval. Let $\{O_\lambda : \lambda \in \Lambda\}$ be an open cover for $[a, b]$ and define S to be the set of all $x \in [a, b]$ such that $[a, x]$ has a finite subcover from $\{O_\lambda : \lambda \in \Lambda\}$.

(a) Argue that S is nonempty and bounded, and thus $s = \sup S$ exists.

(b) Now show $s = b$, which implies $[a, b]$ has a finite subcover.

(c) Finally, prove the theorem for an arbitrary closed and bounded set K.

Exercise 3.3.11. Consider each of the sets listed in Exercise 3.3.2. For each one that is not compact, find an open cover for which there is no finite subcover.

Exercise 3.3.12. Using the concept of open covers (and explicitly avoiding the Bolzano–Weierstrass Theorem), prove that every bounded infinite set has a limit point.

Exercise 3.3.13. Let's call a set *clompact* if it has the property that every *closed* cover (i.e., a cover consisting of closed sets) admits a finite subcover. Describe all of the clompact subsets of **R**.

3.4   Perfect Sets and Connected Sets

One of the underlying goals of topology is to strip away all of the extraneous information that comes with our intuitive picture of the real numbers and isolate just those properties that are responsible for the phenomenon we are studying. For example, we were quick to observe that any closed interval is a compact set. The content of Theorem 3.3.4, however, is that the compactness of a closed interval has nothing to do with the fact that the set is an *interval* but is a consequence of the set being bounded and closed. In Chapter 1, we argued that the set of real numbers between 0 and 1 is an uncountable set. This turns out to be the case for any nonempty closed set that does not contain isolated points.

Perfect Sets

Definition 3.4.1. A set $P \subseteq \mathbf{R}$ is *perfect* if it is closed and contains no isolated points.

Closed intervals (other than the singleton sets $[a, a]$) serve as the most obvious class of perfect sets, but there are more interesting examples.

Example 3.4.2 (Cantor Set). It is not too hard to see that the Cantor set is perfect. In Section 3.1, we defined the Cantor set as the intersection

$$C = \bigcap_{n=0}^{\infty} C_n,$$

where each C_n is a finite union of closed intervals. By Theorem 3.2.14, each C_n is closed, and by the same theorem, C is closed as well. It remains to show that no point in C is isolated.

Let $x \in C$ be arbitrary. To convince ourselves that x is not isolated, we must construct a sequence (x_n) of points in C, different from x, that converges to x. From our earlier discussion, we know that C at least contains the endpoints of the intervals that make up each C_n. In Exercise 3.4.3, we sketch the argument that these are all that is needed to construct (x_n).

One argument for the uncountability of the Cantor set was presented in Section 3.1. Another, perhaps more satisfying, argument for the same conclusion can be obtained from the next theorem.

Theorem 3.4.3. *A nonempty perfect set is uncountable.*

Proof. If P is perfect and nonempty, then it must be infinite because otherwise it would consist only of isolated points. Let's assume, for contradiction, that P is countable. Thus, we can write

$$P = \{x_1, x_2, x_3, \ldots\},$$

where *every* element of P appears on this list. The idea is to construct a sequence of nested compact sets K_n, all contained in P, with the property that

$x_1 \notin K_2$, $x_2 \notin K_3$, $x_3 \notin K_4$, ... Some care must be taken to ensure that each K_n is nonempty, for then we can use Theorem 3.3.5 to produce an

$$x \in \bigcap_{n=1}^{\infty} K_n \subseteq P$$

that *cannot* be on the list $\{x_1, x_2, x_3, \ldots\}$.

Let I_1 be a closed interval that contains x_1 in its interior (i.e., x_1 is not an endpoint of I_1). Now, x_1 is not isolated, so there exists some other point $y_2 \in P$ that is also in the interior of I_1. Construct a closed interval I_2, centered on y_2, so that $I_2 \subseteq I_1$ but $x_1 \notin I_2$. More explicitly, if $I_1 = [a, b]$, let

$$\epsilon = \min\{y_2 - a, b - y_2, |x_1 - y_2|\}.$$

Then, the interval $I_2 = [y_2 - \epsilon/2, y_2 + \epsilon/2]$ has the desired properties.

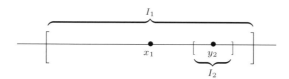

This process can be continued. Because $y_2 \in P$ is not isolated, there must exist another point $y_3 \in P$ in the interior of I_2, and we may insist that $y_3 \neq x_2$. Now, construct I_3 centered on y_3 and small enough so that $x_2 \notin I_3$ and $I_3 \subseteq I_2$. Observe that $I_3 \cap P \neq \emptyset$ because this intersection contains at least y_3.

If we carry out this construction inductively, the result is a sequence of closed intervals I_n satisfying

(i) $I_{n+1} \subseteq I_n$,

(ii) $x_n \notin I_{n+1}$, and

(iii) $I_n \cap P \neq \emptyset$.

To finish the proof, we let $K_n = I_n \cap P$. For each $n \in \mathbf{N}$, we have that K_n is closed because it is the intersection of closed sets, and bounded because it is contained in the bounded set I_n. Hence, K_n is compact. By construction, K_n is not empty and $K_{n+1} \subseteq K_n$. Thus, we can employ the Nested Compact Set Property (Theorem 3.3.5) to conclude that the intersection

$$\bigcap_{n=1}^{\infty} K_n \neq \emptyset.$$

But each K_n is a subset of P, and the fact that $x_n \notin I_{n+1}$ leads to the conclusion that $\bigcap_{n=1}^{\infty} K_n = \emptyset$, which is the sought-after contradiction. $\square$

Connected Sets

Although the two open intervals $(1, 2)$ and $(2, 5)$ have the limit point $x = 2$ in common, there is still some space between them in the sense that no limit point of one of these intervals is actually contained in the other. Said another way, the closure of $(1, 2)$ (see Definition 3.2.11) is disjoint from $(2, 5)$, and the closure of $(2, 5)$ does not intersect $(1, 2)$. Notice that this same observation cannot be made about $(1, 2]$ and $(2, 5)$, even though these latter sets are disjoint.

Definition 3.4.4. Two nonempty sets $A, B \subseteq \mathbf{R}$ are *separated* if $\overline{A} \cap B$ and $A \cap \overline{B}$ are both empty. A set $E \subseteq \mathbf{R}$ is *disconnected* if it can be written as $E = A \cup B$, where A and B are nonempty separated sets.

A set that is not disconnected is called a *connected* set.

Example 3.4.5.   (i) If we let $A = (1, 2)$ and $B = (2, 5)$, then it is not difficult to verify that $E = (1, 2) \cup (2, 5)$ is disconnected. Notice that the sets $C = (1, 2]$ and $D = (2, 5)$ are not separated because $C \cap \overline{D} = \{2\}$ is not empty. This should be comforting. The union $C \cup D$ is equal to the interval $(1, 5)$, which better not qualify as a disconnected set. We will prove in a moment that every interval is a connected subset of **R** and vice versa.

(ii) Let's show that the set of rational numbers is disconnected. If we let

$$A = \mathbf{Q} \cap (-\infty, \sqrt{2}) \quad \text{and} \quad B = \mathbf{Q} \cap (\sqrt{2}, \infty),$$

then we certainly have $\mathbf{Q} = A \cup B$. The fact that $A \subseteq (-\infty, \sqrt{2})$ implies (by the Order Limit Theorem) that any limit point of A will necessarily fall in $(-\infty, \sqrt{2}]$. Because this is disjoint from B, we get $\overline{A} \cap B = \emptyset$. We can similarly show that $A \cap \overline{B} = \emptyset$, which implies that A and B are separated.

The definition of connected is stated as the negation of disconnected, but a little care with the logical negation of the quantifiers in Definition 3.4.4 results in a positive characterization of connectedness. Essentially, a set E is connected if, no matter how it is partitioned into two nonempty disjoint sets, it is always possible to show that at least one of the sets contains a limit point of the other.

Theorem 3.4.6. *A set $E \subseteq \mathbf{R}$ is connected if and only if, for all nonempty disjoint sets A and B satisfying $E = A \cup B$, there always exists a convergent sequence $(x_n) \to x$ with (x_n) contained in one of A or B, and x an element of the other.*

Proof. Exercise 3.4.6.   □

The concept of connectedness is more relevant when working with subsets of the plane and other higher-dimensional spaces. This is because, in **R**, the connected sets coincide precisely with the collection of intervals (with the understanding that unbounded intervals such as $(-\infty, 3)$ and $[0, \infty)$ are included).

Theorem 3.4.7. *A set $E \subseteq \mathbf{R}$ is connected if and only if whenever $a < c < b$ with $a, b \in E$, it follows that $c \in E$ as well.*

Proof. Assume E is connected, and let $a, b \in E$ and $a < c < b$. Set

$$A = (-\infty, c) \cap E \quad \text{and} \quad B = (c, \infty) \cap E.$$

Because $a \in A$ and $b \in B$, neither set is empty and, just as in Example 3.4.5 (ii), neither set contains a limit point of the other. If $E = A \cup B$, then we would have that E is disconnected, which it is not. It must then be that $A \cup B$ is missing some element of E, and c is the only possibility. Thus, $c \in E$.

Conversely, assume that E is an interval in the sense that whenever $a, b \in E$ satisfy $a < c < b$ for some c, then $c \in E$. Our intent is to use the characterization of connected sets in Theorem 3.4.6, so let $E = A \cup B$, where A and B are nonempty and disjoint. We need to show that one of these sets contains a limit point of the other. Pick $a_0 \in A$ and $b_0 \in B$, and, for the sake of the argument, assume $a_0 < b_0$. Because E is itself an interval, the interval $I_0 = [a_0, b_0]$ is contained in E. Now, bisect I_0 into two equal halves. The midpoint of I_0 must either be in A or B, and so choose $I_1 = [a_1, b_1]$ to be the half that allows us to have $a_1 \in A$ and $b_1 \in B$. Continuing this process yields a sequence of nested intervals $I_n = [a_n, b_n]$, where $a_n \in A$, $b_n \in B$, and the length $(b_n - a_n) \to 0$. The remainder of this argument should feel familiar. By the Nested Interval Property, there exists an

$$x \in \bigcap_{n=0}^{\infty} I_n,$$

and it is straightforward to show that the sequences of endpoints each satisfy $\lim a_n = x$ and $\lim b_n = x$. But now $x \in E$ must belong to either A or B, thus making it a limit point of the other. This completes the argument. $\square$

Exercises

Exercise 3.4.1. If P is a perfect set and K is compact, is the intersection $P \cap K$ always compact? Always perfect?

Exercise 3.4.2. Does there exist a perfect set consisting of only rational numbers?

Exercise 3.4.3. Review the portion of the proof given in Example 3.4.2 and follow these steps to complete the argument.

(a) Because $x \in C_1$, argue that there exists an $x_1 \in C \cap C_1$ with $x_1 \neq x$ satisfying $|x - x_1| \leq 1/3$.

(b) Finish the proof by showing that for each $n \in \mathbf{N}$, there exists $x_n \in C \cap C_n$, different from x, satisfying $|x - x_n| \leq 1/3^n$.

Exercise 3.4.4. Repeat the Cantor construction from Section 3.1 starting with the interval $[0, 1]$. This time, however, remove the open middle *fourth* from each component.

(a) Is the resulting set compact? Perfect?

(b) Using the algorithms from Section 3.1, compute the length and dimension of this Cantor-like set.

Exercise 3.4.5. Let A and B be nonempty subsets of **R**. Show that if there exist disjoint open sets U and V with $A \subseteq U$ and $B \subseteq V$, then A and B are separated.

Exercise 3.4.6. Prove Theorem 3.4.6.

Exercise 3.4.7. A set E is *totally disconnected* if, given any two distinct points $x, y \in E$, there exist separated sets A and B with $x \in A$, $y \in B$, and $E = A \cup B$.

(a) Show that **Q** is totally disconnected.

(b) Is the set of irrational numbers totally disconnected?

Exercise 3.4.8. Follow these steps to show that the Cantor set is totally disconnected in the sense described in Exercise 3.4.7.
Let $C = \bigcap_{n=0}^{\infty} C_n$, as defined in Section 3.1.

(a) Given $x, y \in C$, with $x < y$, set $\epsilon = y - x$. For each $n = 0, 1, 2, \ldots$, the set C_n consists of a finite number of closed intervals. Explain why there must exist an N large enough so that it is impossible for x and y both to belong to the same closed interval of C_N.

(b) Show that C is totally disconnected.

Exercise 3.4.9. Let $\{r_1, r_2, r_3, \ldots\}$ be an enumeration of the rational numbers, and for each $n \in \mathbf{N}$ set $\epsilon_n = 1/2^n$. Define $O = \bigcup_{n=1}^{\infty} V_{\epsilon_n}(r_n)$, and let $F = O^c$.

(a) Argue that F is a closed, nonempty set consisting only of irrational numbers.

(b) Does F contain any nonempty open intervals? Is F totally disconnected? (See Exercise 3.4.7 for the definition.)

(c) Is it possible to know whether F is perfect? If not, can we modify this construction to produce a nonempty perfect set of irrational numbers?

3.5   Baire's Theorem

The nature of the real line can be deceptively elusive. The closer we look, the more intricate and enigmatic **R** becomes, and the more we are reminded to proceed carefully (i.e., axiomatically) with all of our conclusions about properties of subsets of **R**. The structure of open sets is fairly straightforward. Every open set is either a finite or countable union of open intervals. Standing in opposition to this tidy description of all open sets is the Cantor set. The Cantor set is a closed, uncountable set that contains no intervals of any kind. Thus, no such characterization of closed sets should be anticipated.

Recall that the arbitrary union of open sets is always an open set. Likewise, the arbitrary intersection of closed sets is closed. By taking unions of closed sets or intersections of open sets, however, it is possible to obtain a new selection of subsets of $\mathbf{R}$.

Definition 3.5.1. A set $A \subseteq \mathbf{R}$ is called an F_σ *set* if it can be written as the countable union of closed sets. A set $B \subseteq \mathbf{R}$ is called a G_δ *set* if it can be written as the countable intersection of open sets.

Exercise 3.5.1. Argue that a set A is a G_δ set if and only if its complement is an F_σ set.

Exercise 3.5.2. Replace each _____ with the word *finite* or *countable*, depending on which is more appropriate.

(a) The _____ union of F_σ sets is an F_σ set.

(b) The _____ intersection of F_σ sets is an F_σ set.

(c) The _____ union of G_δ sets is a G_δ set.

(d) The _____ intersection of G_δ sets is a G_δ set.

Exercise 3.5.3. (This exercise has already appeared as Exercise 3.2.15.)

(a) Show that a closed interval $[a, b]$ is a G_δ set.

(b) Show that the half-open interval $(a, b]$ is both a G_δ and an F_σ set.

(c) Show that $\mathbf{Q}$ is an F_σ set, and the set of irrationals $\mathbf{I}$ forms a G_δ set.

It is not readily obvious that the class F_σ does not include every subset of $\mathbf{R}$, but we are now ready to argue that $\mathbf{I}$ is not an F_σ set (and consequently $\mathbf{Q}$ is not a G_δ set). This will follow from a theorem due to René Louis Baire (1874–1932).

Recall that a set $G \subseteq \mathbf{R}$ is *dense* in $\mathbf{R}$ if, given any two real numbers $a < b$, it is possible to find a point $x \in G$ with $a < x < b$.

Theorem 3.5.2. *If $\{G_1, G_2, G_3, \ldots\}$ is a countable collection of dense, open sets, then the intersection $\bigcap_{n=1}^{\infty} G_n$ is not empty.*

Proof. Before embarking on the proof, notice that we have seen a conclusion like this before. Theorem 3.3.5 asserts that a nested sequence of compact sets has a nontrivial intersection. In this theorem, we are dealing with dense, open sets, but as it turns out, we are going to use Theorem 3.3.5—and actually, just the Nested Interval Property—as the crucial step in the argument.

Exercise 3.5.4. Starting with $n = 1$, inductively construct a nested sequence of *closed* intervals $I_1 \supseteq I_2 \supseteq I_3 \supseteq \cdots$ satisfying $I_n \subseteq G_n$. Give special attention to the issue of the endpoints of each I_n. Show how this leads to a proof of the theorem. □

Exercise 3.5.5. Show that it is impossible to write

$$\mathbf{R} = \bigcup_{n=1}^{\infty} F_n,$$

where for each $n \in \mathbf{N}$, F_n is a closed set containing no nonempty open intervals.

Exercise 3.5.6. Show how the previous exercise implies that the set **I** of irrationals cannot be an F_σ set, and **Q** cannot be a G_δ set.

Exercise 3.5.7. Using Exercise 3.5.6 and versions of the statements in Exercise 3.5.2, construct a set that is neither in F_σ nor in G_δ.

Nowhere-Dense Sets

We have encountered several equivalent ways to assert that a particular set G is dense in **R**. In Section 3.2, we observed that G is dense in **R** if and only if every point of **R** is a limit point of G. Because the closure of any set is obtained by taking the union of the set and its limit points, we have that

$$G \text{ is dense in } \mathbf{R} \text{ if and only if } \overline{G} = \mathbf{R}.$$

The set **Q** is dense in **R**; the set **Z** is clearly not. In fact, in the jargon of analysis, **Z** is nowhere-dense in **R**.

Definition 3.5.3. A set E is *nowhere-dense* if $\overline{E}$ contains no nonempty open intervals.

Exercise 3.5.8. Show that a set E is nowhere-dense in **R** if and only if the complement of $\overline{E}$ is dense in **R**.

Exercise 3.5.9. Decide whether the following sets are dense in **R**, nowhere-dense in **R**, or somewhere in between.

(a) $A = \mathbf{Q} \cap [0, 5]$.

(b) $B = \{1/n : n \in \mathbf{N}\}$.

(c) the set of irrationals.

(d) the Cantor set.

We can now restate Theorem 3.5.2 in a slightly more general form.

Theorem 3.5.4 (Baire's Theorem). *The set of real numbers* **R** *cannot be written as the countable union of nowhere-dense sets.*

Proof. For contradiction, assume that $E_1, E_2, E_3, \ldots$ are each nowhere-dense and satisfy $\mathbf{R} = \bigcup_{n=1}^{\infty} E_n$.

Exercise 3.5.10. Finish the proof by finding a contradiction to the results in this section.   □

3.6 Epilogue

Baire's Theorem is yet another statement about the size of **R**. We have already encountered several ways to describe the sizes of infinite sets. In terms of cardinality, countable sets are relatively small whereas uncountable sets are large. We also briefly discussed the concept of "length," or "measure," in Section 3.1. Baire's Theorem offers a third perspective. From this point of view, nowhere-dense sets are considered to be "thin" sets. Any set that is the countable union—i.e., a not very large union—of these small sets is called a "meager" set or a set of "first category." A set that is not of first category is of "second category." Intuitively, sets of the second category are the "fat" subsets. The Baire Category Theorem, as it is often called, states that **R** is of second category.

There is a significance to the Baire Category Theorem that is difficult to appreciate at the moment because we are only seeing a special case of this result. The real numbers are an example of a *complete metric space.* Metric spaces are discussed in some detail in Section 8.2, but here is the basic idea. Given a set of mathematical objects such as real numbers, points in the plane or continuous functions defined on [0,1], a "metric" is a rule that assigns a "distance" between two elements in the set. In **R**, we have been using $|x-y|$ as the distance between the real numbers x and y. The point is that if we can create a satisfactory notion of "distance" on these other spaces (we will need the triangle inequality to hold, for instance), then the concepts of convergence, Cauchy sequences, and open sets, for example, can be naturally transferred over. A complete metric space is any set with a suitably defined metric in which Cauchy sequences have limits. We have spent a good deal of time discussing the fact that **R** is a complete metric space whereas **Q** is not.

The Baire Category Theorem in its more general form states that *any* complete metric space must be too large to be the countable union of nowhere-dense subsets. One particularly interesting example of a complete metric space is the set of continuous functions defined on the interval $[0, 1]$. (The distance between two functions f and g in this space is defined to be $\sup |f(x) - g(x)|$, where $x \in [0, 1]$.) Now, in this space we will see that the collection of continuous functions that are differentiable at even one point *can* be written as the countable union of nowhere-dense sets. Thus, a fascinating consequence of Baire's Theorem in this setting is that *most continuous functions do not have derivatives at any point.* Chapter 5 concludes with a construction of one such function. This odd situation mirrors the roles of **Q** and **I** as subsets of **R**. Just as the familiar rational numbers constitute a minute proportion of the real line, the differentiable functions of calculus are exceedingly atypical of continuous functions in general.

Chapter 4

Functional Limits and Continuity

4.1 Discussion: Examples of Dirichlet and Thomae

Although it is a common practice in calculus courses to discuss continuity before differentiation, historically mathematicians' attention to the concept of continuity came long after the derivative was in wide use. Pierre de Fermat (1601–1665) was using tangent lines to solve optimization problems as early as 1629. On the other hand, it was not until around 1820 that Cauchy, Bolzano, Weierstrass, and others began to characterize continuity in terms more rigorous than prevailing intuitive notions such as "unbroken curves" or "functions which have no jumps or gaps." The basic reason for this two-hundred year waiting period lies in the fact that, for most of this time, the very notion of *function* did not really permit discontinuities. Functions were entities such as polynomials, sines, and cosines, always smooth and continuous over their relevant domains. The gradual liberation of the term function to its modern understanding—a rule associating a unique output with a given input—was simultaneous with 19th century investigations into the behavior of infinite series. Extensions of the power of calculus were intimately tied to the ability to represent a function $f(x)$ as a limit of polynomials (called a *power series*) or as a limit of sums of sines and cosines (called a *trigonometric* or *Fourier* series). A typical question for Cauchy and his contemporaries was whether the continuity of the limiting polynomials or trigonometric functions necessarily implied that the limit f would also be continuous.

Sequences and series of functions are the topics of Chapter 6. What is relevant at this moment is that we realize why the issue of finding a rigorous

© Springer Science+Business Media New York 2015
S. Abbott, *Understanding Analysis*, Undergraduate Texts in Mathematics, DOI 10.1007/978-1-4939-2712-8_4

Figure 4.1: Dirichlet's Function, $g(x)$.

definition for continuity finally made its way to the fore. Any significant progress on the question of whether the limit of continuous functions is continuous (for Cauchy and for us) necessarily depends on a definition of continuity that does not rely on imprecise notions such as "no holes" or "gaps." With a mathematically unambiguous definition for the limit of a sequence in hand, we are well on our way toward a rigorous understanding of continuity.

Given a function f with domain $A \subseteq \mathbf{R}$, we want to define continuity at a point $c \in A$ to mean that if $x \in A$ is chosen *near* c, then $f(x)$ will be *near* $f(c)$. Symbolically, we will say f is continuous at c if

$$\lim_{x \to c} f(x) = f(c).$$

The problem is that, at present, we only have a definition for the limit of a sequence, and it is not entirely clear what is meant by $\lim_{x \to c} f(x)$. The subtleties that arise as we try to fashion such a definition are well-illustrated via a family of examples, all based on an idea of the prominent German mathematician, Peter Lejeune Dirichlet. Dirichlet's idea was to define a function g in a piecewise manner based on whether or not the input variable x is rational or irrational. Specifically, let

$$g(x) = \begin{cases} 1 & \text{if } x \in \mathbf{Q} \\ 0 & \text{if } x \notin \mathbf{Q}. \end{cases}$$

The intricate way that $\mathbf{Q}$ and $\mathbf{I}$ fit inside of $\mathbf{R}$ makes an accurate graph of g technically impossible to draw, but Figure 4.1 illustrates the basic idea.

Does it make sense to attach a value to the expression $\lim_{x \to 1/2} g(x)$? One idea is to consider a sequence $(x_n) \to 1/2$. Using our notion of the limit of a sequence, we might try to define $\lim_{x \to 1/2} g(x)$ as simply the limit of the sequence $g(x_n)$. But notice that this limit depends on how the sequence (x_n) is chosen. If each x_n is rational, then

$$\lim_{n \to \infty} g(x_n) = 1.$$

On the other hand, if x_n is irrational for each n, then

$$\lim_{n \to \infty} g(x_n) = 0.$$

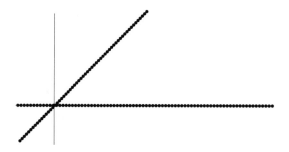

Figure 4.2: Modified Dirichlet Function, $h(x)$**.**

This unacceptable situation demands that we work harder on our definition of functional limits. Generally speaking, we want the value of $\lim_{x \to c} g(x)$ to be independent of how we approach c. In this particular case, the definition of a functional limit that we agree on should lead to the conclusion that

$$\lim_{x \to 1/2} g(x) \quad \text{does not exist.}$$

Postponing the search for formal definitions for the moment, we should nonetheless realize that Dirichlet's function is not continuous at $c = 1/2$. In fact, the real significance of this function is that there is nothing unique about the point $c = 1/2$. Because both $\mathbf{Q}$ and $\mathbf{I}$ (the set of irrationals) are dense in the real line, it follows that for any $z \in \mathbf{R}$ we can find sequences $(x_n) \subseteq \mathbf{Q}$ and $(y_n) \subseteq \mathbf{I}$ such that

$$\lim x_n = \lim y_n = z.$$

(See Example 3.2.9 (iii).) Because

$$\lim g(x_n) \neq \lim g(y_n),$$

the same line of reasoning reveals that $g(x)$ is not continuous at z. In the jargon of analysis, Dirichlet's function is a *nowhere-continuous* function on $\mathbf{R}$.

What happens if we adjust the definition of $g(x)$ in the following way? Define a new function h (Fig. 4.2) on $\mathbf{R}$ by setting

$$h(x) = \begin{cases} x & \text{if } x \in \mathbf{Q} \\ 0 & \text{if } x \notin \mathbf{Q}. \end{cases}$$

If we take c different from zero, then just as before we can construct sequences $(x_n) \to c$ of rationals and $(y_n) \to c$ of irrationals so that

$$\lim h(x_n) = c \quad \text{and} \quad \lim h(y_n) = 0.$$

Thus, h is not continuous at every point $c \neq 0$.

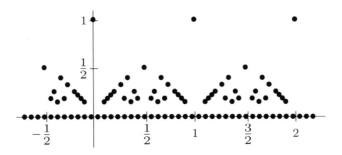

Figure 4.3: Thomae's Function, $t(x)$.

If $c = 0$, however, then these two limits are both equal to $h(0) = 0$. In fact, it appears as though no matter how we construct a sequence (z_n) converging to zero, it will always be the case that $\lim h(z_n) = 0$. This observation goes to the heart of what we want functional limits to entail. To assert that

$$\lim_{x \to c} h(x) = L$$

should imply that

$$h(z_n) \to L \quad \text{for all sequences} \quad (z_n) \to c.$$

For reasons not yet apparent, it is beneficial to fashion the definition for functional limits in terms of neighborhoods constructed around c and L. We will quickly see, however, that this topological formulation is equivalent to the sequential characterization we have arrived at here.

To this point, we have been discussing continuity of a function at a particular point in its domain. This is a significant departure from thinking of continuous functions as curves that can be drawn without lifting the pen from the paper, and it leads to some fascinating questions. In 1875, K.J. Thomae discovered the function

$$t(x) = \begin{cases} 1 & \text{if } x = 0 \\ 1/n & \text{if } x = m/n \in \mathbf{Q}\backslash\{0\} \text{ is in lowest terms with } n > 0 \\ 0 & \text{if } x \notin \mathbf{Q}. \end{cases}$$

If $c \in \mathbf{Q}$, then $t(c) > 0$. Because the set of irrationals is dense in $\mathbf{R}$, we can find a sequence (y_n) in $\mathbf{I}$ converging to c. The result is that

$$\lim t(y_n) = 0 \neq t(c),$$

and Thomae's function (Fig. 4.3) fails to be continuous at any rational point.

The twist comes when we try this argument on some irrational point in the domain such as $c = \sqrt{2}$. All irrational values get mapped to zero by t, so the natural thing would be to consider a sequence (x_n) of rational numbers that converges to $\sqrt{2}$. Now, $\sqrt{2} \approx 1.414213\ldots$, so a good start on a particular sequence of rational approximations for $\sqrt{2}$ might be

$$\left(1, \frac{14}{10}, \frac{141}{100}, \frac{1414}{1000}, \frac{14142}{10000}, \frac{141421}{100000}, \ldots\right).$$

But notice that the denominators of these fractions are getting larger. In this case, the sequence $t(x_n)$ begins,

$$\left(1, \frac{1}{5}, \frac{1}{100}, \frac{1}{500}, \frac{1}{5000}, \frac{1}{100000}, \ldots\right)$$

and is fast approaching $0 = t(\sqrt{2})$. We will see that this always happens. The closer a rational number is chosen to a fixed irrational number, the larger its denominator must necessarily be. As a consequence, Thomae's function has the bizarre property of being continuous at every irrational point on $\mathbf{R}$ and discontinuous at every rational point.

Is there an example of a function with the opposite property? In other words, does there exist a function defined on all of $\mathbf{R}$ that is continuous on $\mathbf{Q}$ but fails to be continuous on $\mathbf{I}$? Can the set of discontinuities of a particular function be arbitrary? If we are given some set $A \subseteq \mathbf{R}$, is it always possible to find a function that is continuous only on the set A^c? In each of the examples in this section, the functions were defined to have erratic oscillations around points in the domain. What conclusions can we draw if we restrict our attention to functions that are somewhat less volatile? One such class is the set of so-called *monotone* functions, which are either increasing or decreasing on a given domain. What might we be able to say about the set of discontinuities of a monotone function on $\mathbf{R}$?

4.2 Functional Limits

Consider a function $f : A \to \mathbf{R}$. Recall that a limit point c of A is a point with the property that every ϵ-neighborhood $V_\epsilon(c)$ intersects A in some point other than c. Equivalently, c is a limit point of A if and only if $c = \lim x_n$ for some sequence $(x_n) \subseteq A$ with $x_n \neq c$. It is important to remember that limit points of A do not necessarily belong to the set A unless A is closed.

If c is a limit point of the domain of f, then, intuitively, the statement

$$\lim_{x \to c} f(x) = L$$

is intended to convey that values of $f(x)$ get arbitrarily close to L as x is chosen closer and closer to c. The issue of what happens when $x = c$ is irrelevant from the point of view of functional limits. In fact, c need not even be in the domain of f.

The structure of the definition of functional limits follows the "challenge–response" pattern established in the definition for the limit of a sequence. Recall that given a sequence (a_n), the assertion $\lim a_n = L$ implies that for every

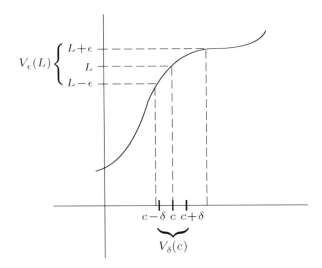

Figure 4.4: Definition of Functional Limit.

ϵ-neighborhood $V_\epsilon(L)$ centered at L, there is a point in the sequence—call it a_N—after which all of the terms a_n fall in $V_\epsilon(L)$. Each ϵ-neighborhood represents a particular challenge, and each N is the respective response. For functional limit statements such as $\lim_{x \to c} f(x) = L$, the challenges are still made in the form of an arbitrary ϵ-neighborhood around L, but the response this time is a δ-neighborhood centered at c.

Definition 4.2.1 (Functional Limit). Let $f : A \to \mathbf{R}$, and let c be a limit point of the domain A. We say that $\lim_{x \to c} f(x) = L$ provided that, for all $\epsilon > 0$, there exists a $\delta > 0$ such that whenever $0 < |x - c| < \delta$ (and $x \in A$) it follows that $|f(x) - L| < \epsilon$.

This is often referred to as the "ϵ–δ version" of the definition for functional limits. Recall that the statement

$$|f(x) - L| < \epsilon \text{ is equivalent to } f(x) \in V_\epsilon(L).$$

Likewise, the statement

$$|x - c| < \delta \text{ is satisfied if and only if } x \in V_\delta(c).$$

The additional restriction $0 < |x - c|$ is just an economical way of saying $x \neq c$. Recasting Definition 4.2.1 in terms of neighborhoods—just as we did for the definition of convergence of a sequence in Section 2.2—amounts to little more than a change of notation, but it does help emphasize the geometrical nature of what is happening (Fig. 4.4).

Definition 4.2.1B (Functional Limit: Topological Version). Let c be a limit point of the domain of $f : A \to \mathbf{R}$. We say $\lim_{x \to c} f(x) = L$ provided that, for every ϵ-neighborhood $V_\epsilon(L)$ of L, there exists a δ-neighborhood $V_\delta(c)$ around c with the property that for all $x \in V_\delta(c)$ different from c (with $x \in A$) it follows that $f(x) \in V_\epsilon(L)$.

The parenthetical reminder "$(x \in A)$" present in both versions of the definition is included to ensure that x is an allowable input for the function in question. When no confusion is likely, we may omit this reminder with the understanding that the appearance of $f(x)$ carries with it the implicit assumption that x is in the domain of f. On a related note, there is no reason to discuss functional limits at isolated points of the domain. Thus, functional limits will only be considered as x tends toward a limit point of the function's domain.

Example 4.2.2. (i) To familiarize ourselves with Definition 4.2.1, let's prove that if $f(x) = 3x + 1$, then

$$\lim_{x \to 2} f(x) = 7.$$

Let $\epsilon > 0$. Definition 4.2.1 requires that we produce a $\delta > 0$ so that $0 < |x - 2| < \delta$ leads to the conclusion $|f(x) - 7| < \epsilon$. Notice that

$$|f(x) - 7| = |(3x + 1) - 7| = |3x - 6| = 3|x - 2|.$$

Thus, if we choose $\delta = \epsilon/3$, then $0 < |x - 2| < \delta$ implies $|f(x) - 7| < 3(\epsilon/3) = \epsilon$.

(ii) Let's show

$$\lim_{x \to 2} g(x) = 4,$$

where $g(x) = x^2$. Given an arbitrary $\epsilon > 0$, our goal this time is to make $|g(x) - 4| < \epsilon$ by restricting $|x - 2|$ to be smaller than some carefully chosen δ. As in the previous problem, a little algebra reveals

$$|g(x) - 4| = |x^2 - 4| = |x + 2||x - 2|.$$

We can make $|x - 2|$ as small as we like, but we need an upper bound on $|x+2|$ in order to know how small to choose δ. The presence of the variable x causes some initial confusion, but keep in mind that we are discussing the limit as x approaches 2. If we agree that our δ-neighborhood around $c = 2$ must have radius no bigger than $\delta = 1$, then we get the upper bound $|x + 2| \leq |3 + 2| = 5$ for all $x \in V_\delta(c)$.

Now, choose $\delta = \min\{1, \epsilon/5\}$. If $0 < |x - 2| < \delta$, then it follows that

$$|x^2 - 4| = |x + 2||x - 2| < (5)\frac{\epsilon}{5} = \epsilon,$$

and the limit is proved.

Sequential Criterion for Functional Limits

We worked very hard in Chapter 2 to derive an impressive list of properties enjoyed by sequential limits. In particular, the Algebraic Limit Theorem (Theorem 2.3.3) and the Order Limit Theorem (Theorem 2.3.4) proved invaluable in a large number of the arguments that followed. Not surprisingly, we are going to need analogous statements for functional limits. Although it is not difficult to generate independent proofs for these statements, all of them will follow quite naturally from their sequential analogs once we derive the sequential criterion for functional limits motivated in the opening discussion of this chapter.

Theorem 4.2.3 (Sequential Criterion for Functional Limits). *Given a function $f : A \to \mathbf{R}$ and a limit point c of A, the following two statements are equivalent:*

(i) $\lim_{x \to c} f(x) = L$.

(ii) *For all sequences $(x_n) \subseteq A$ satisfying $x_n \neq c$ and $(x_n) \to c$, it follows that $f(x_n) \to L$.*

Proof. ($\Rightarrow$) Let's first assume that $\lim_{x \to c} f(x) = L$. To prove (ii), we consider an arbitrary sequence (x_n), which converges to c and satisfies $x_n \neq c$. Our goal is to show that the image sequence $f(x_n)$ converges to L. This is most easily seen using the topological formulation of the definition.

Let $\epsilon > 0$. Because we are assuming (i), Definition 4.2.1B implies that there exists $V_\delta(c)$ with the property that all $x \in V_\delta(c)$ different from c satisfy $f(x) \in V_\epsilon(L)$. All we need to do then is argue that our particular sequence (x_n) is eventually in $V_\delta(c)$. But we are assuming that $(x_n) \to c$. This implies that there exists a point x_N after which $x_n \in V_\delta(c)$. It follows that $n \geq N$ implies $f(x_n) \in V_\epsilon(L)$, as desired.

($\Leftarrow$) For this implication we give a contrapositive proof, which is essentially a proof by contradiction. Thus, we assume that statement (ii) is true, and carefully negate statement (i). To say that

$$\lim_{x \to c} f(x) \neq L$$

means that there exists at least one particular $\epsilon_0 > 0$ for which no δ is a suitable response. In other words, no matter what $\delta > 0$ we try, there will always be at least one point

$$x \in V_\delta(c) \quad \text{with} \quad x \neq c \quad \text{for which} \quad f(x) \notin V_{\epsilon_0}(L).$$

Now consider $\delta_n = 1/n$. From the preceding discussion, it follows that for each $n \in \mathbf{N}$ we may pick an $x_n \in V_{\delta_n}(c)$ with $x_n \neq c$ and $f(x_n) \notin V_{\epsilon_0}(L)$. But now notice that the result of this is a sequence $(x_n) \to c$ with $x_n \neq c$, where the image sequence $f(x_n)$ certainly does *not* converge to L.

Because this contradicts (ii), which we are assuming is true for this argument, we may conclude that (i) must also hold. $\square$

Theorem 4.2.3 has several useful corollaries. In addition to the previously advertised benefit of granting us some short proofs of statements about how functional limits interact with algebraic combinations of functions, we also get an economical way of establishing that certain limits do not exist.

Corollary 4.2.4 (Algebraic Limit Theorem for Functional Limits). *Let f and g be functions defined on a domain $A \subseteq \mathbf{R}$, and assume $\lim_{x \to c} f(x) = L$ and $\lim_{x \to c} g(x) = M$ for some limit point c of A. Then,*

(i) $\lim_{x \to c} k f(x) = kL$ *for all $k \in \mathbf{R}$,*

(ii) $\lim_{x \to c} [f(x) + g(x)] = L + M$,

(iii) $\lim_{x \to c} [f(x)g(x)] = LM$, *and*

(iv) $\lim_{x \to c} f(x)/g(x) = L/M$, *provided $M \neq 0$.*

Proof. These follow from Theorem 4.2.3 and the Algebraic Limit Theorem for sequences. The details are requested in Exercise 4.2.1. $\qquad\square$

Corollary 4.2.5 (Divergence Criterion for Functional Limits). *Let f be a function defined on A, and let c be a limit point of A. If there exist two sequences (x_n) and (y_n) in A with $x_n \neq c$ and $y_n \neq c$ and*

$$\lim x_n = \lim y_n = c \quad but \quad \lim f(x_n) \neq \lim f(y_n),$$

then we can conclude that the functional limit $\lim_{x \to c} f(x)$ does not exist.

Example 4.2.6. Assuming the familiar properties of the sine function, let's show that $\lim_{x \to 0} \sin(1/x)$ does not exist (Fig. 4.5).

If $x_n = 1/2n\pi$ and $y_n = 1/(2n\pi + \pi/2)$, then $\lim(x_n) = \lim(y_n) = 0$. However, $\sin(1/x_n) = 0$ for all $n \in \mathbf{N}$ while $\sin(1/y_n) = 1$. Thus,

$$\lim \sin(1/x_n) \neq \lim \sin(1/y_n),$$

so by Corollary 4.2.5, $\lim_{x \to 0} \sin(1/x)$ does not exist.

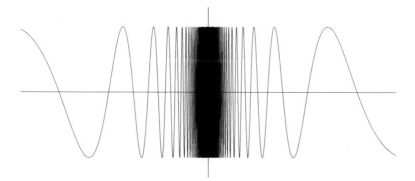

Figure 4.5: The function $\sin(1/x)$ near zero.

Exercises

Exercise 4.2.1.   (a) Supply the details for how Corollary 4.2.4 part (ii) follows from the Sequential Criterion for Functional Limits in Theorem 4.2.3 and the Algebraic Limit Theorem for sequences proved in Chapter 2.

(b) Now, write another proof of Corollary 4.2.4 part (ii) directly from Definition 4.2.1 without using the sequential criterion in Theorem 4.2.3.

(c) Repeat (a) and (b) for Corollary 4.2.4 part (iii).

Exercise 4.2.2. For each stated limit, find the largest possible δ-neighborhood that is a proper response to the given ϵ challenge.

(a) $\lim_{x \to 3}(5x - 6) = 9$, where $\epsilon = 1$.

(b) $\lim_{x \to 4} \sqrt{x} = 2$, where $\epsilon = 1$.

(c) $\lim_{x \to \pi}[[x]] = 3$, where $\epsilon = 1$. (The function $[[x]]$ returns the greatest integer less than or equal to x.)

(d) $\lim_{x \to \pi}[[x]] = 3$, where $\epsilon = .01$.

Exercise 4.2.3. Review the definition of Thomae's function $t(x)$ from Section 4.1.

(a) Construct three different sequences (x_n), (y_n), and (z_n), each of which converges to 1 without using the number 1 as a term in the sequence.

(b) Now, compute $\lim t(x_n)$, $\lim t(y_n)$, and $\lim t(z_n)$.

(c) Make an educated conjecture for $\lim_{x \to 1} t(x)$, and use Definition 4.2.1B to verify the claim. (Given $\epsilon > 0$, consider the set of points $\{x \in \mathbf{R} : t(x) \geq \epsilon\}$. Argue that all the points in this set are isolated.)

Exercise 4.2.4. Consider the reasonable but erroneous claim that

$$\lim_{x \to 10} 1/[[x]] = 1/10.$$

(a) Find the largest δ that represents a proper response to the challenge of $\epsilon = 1/2$.

(b) Find the largest δ that represents a proper response to $\epsilon = 1/50$.

(c) Find the largest ϵ challenge for which there is no suitable δ response possible.

Exercise 4.2.5. Use Definition 4.2.1 to supply a proper proof for the following limit statements.

(a) $\lim_{x \to 2}(3x + 4) = 10$.

(b) $\lim_{x \to 0} x^3 = 0$.

(c) $\lim_{x \to 2}(x^2 + x - 1) = 5$.

(d) $\lim_{x \to 3} 1/x = 1/3$.

Exercise 4.2.6. Decide if the following claims are true or false, and give short justifications for each conclusion.

(a) If a particular δ has been constructed as a suitable response to a particular ϵ challenge, then any smaller positive δ will also suffice.

(b) If $\lim_{x \to a} f(x) = L$ and a happens to be in the domain of f, then $L = f(a)$.

(c) If $\lim_{x \to a} f(x) = L$, then $\lim_{x \to a} 3[f(x) - 2]^2 = 3(L - 2)^2$.

(d) If $\lim_{x \to a} f(x) = 0$, then $\lim_{x \to a} f(x)g(x) = 0$ for any function g (with domain equal to the domain of f.)

Exercise 4.2.7. Let $g : A \to \mathbf{R}$ and assume that f is a bounded function on A in the sense that there exists $M > 0$ satisfying $|f(x)| \leq M$ for all $x \in A$.

Show that if $\lim_{x \to c} g(x) = 0$, then $\lim_{x \to c} g(x)f(x) = 0$ as well.

Exercise 4.2.8. Compute each limit or state that it does not exist. Use the tools developed in this section to justify each conclusion.

(a) $\lim_{x \to 2} \frac{|x-2|}{x-2}$

(b) $\lim_{x \to 7/4} \frac{|x-2|}{x-2}$

(c) $\lim_{x \to 0}(-1)^{[[1/x]]}$

(d) $\lim_{x \to 0} \sqrt[3]{x}(-1)^{[[1/x]]}$

Exercise 4.2.9 (Infinite Limits). The statement $\lim_{x \to 0} 1/x^2 = \infty$ certainly makes intuitive sense. To construct a rigorous definition in the challenge–response style of Definition 4.2.1 for an infinite limit statement of this form, we replace the (arbitrarily small) $\epsilon > 0$ challenge with an (arbitrarily large) $M > 0$ challenge:

Definition: $\lim_{x \to c} f(x) = \infty$ means that for all $M > 0$ we can find a $\delta > 0$ such that whenever $0 < |x - c| < \delta$, it follows that $f(x) > M$.

(a) Show $\lim_{x \to 0} 1/x^2 = \infty$ in the sense described in the previous definition.

(b) Now, construct a definition for the statement $\lim_{x \to \infty} f(x) = L$. Show $\lim_{x \to \infty} 1/x = 0$.

(c) What would a rigorous definition for $\lim_{x \to \infty} f(x) = \infty$ look like? Give an example of such a limit.

Exercise 4.2.10 (Right and Left Limits). Introductory calculus courses typically refer to the *right-hand limit* of a function as the limit obtained by "letting x approach a from the right-hand side."

(a) Give a proper definition in the style of Definition 4.2.1 for the right-hand and left-hand limit statements:

$$\lim_{x \to a^+} f(x) = L \quad \text{and} \quad \lim_{x \to a^-} f(x) = M.$$

(b) Prove that $\lim_{x \to a} f(x) = L$ if and only if both the right and left-hand limits equal L.

Exercise 4.2.11 (Squeeze Theorem). Let f, g, and h satisfy $f(x) \le g(x) \le h(x)$ for all x in some common domain A. If $\lim_{x \to c} f(x) = L$ and $\lim_{x \to c} h(x) = L$ at some limit point c of A, show $\lim_{x \to c} g(x) = L$ as well.

4.3   Continuous Functions

We now come to a significant milestone in our progress toward a rigorous theory of real-valued functions—a proper definition of the seminal concept of continuity that avoids any intuitive appeals to "unbroken curves" or functions without "jumps" or "holes."

Definition 4.3.1 (Continuity). A function $f : A \to \mathbf{R}$ is *continuous at a point* $c \in A$ if, for all $\epsilon > 0$, there exists a $\delta > 0$ such that whenever $|x - c| < \delta$ (and $x \in A$) it follows that $|f(x) - f(c)| < \epsilon$.

If f is continuous at every point in the domain A, then we say that f is *continuous on A.*

The definition of continuity looks much like the definition for functional limits, with a few subtle differences. The most important is that we require the point c to be in the domain of f. The value $f(c)$ then becomes the value of $\lim_{x \to c} f(x)$. With this observation in mind, it is tempting to shorten Definition 4.3.1 to say that f is continuous at $c \in A$ if

$$\lim_{x \to c} f(x) = f(c).$$

This is fine as long as c is a limit point of A. If c is an isolated point of A, then $\lim_{x \to c} f(x)$ isn't defined but Definition 4.3.1 can still be applied. An unremarkable but noteworthy consequence of this definition is that functions are continuous at isolated points of their domains (Exercise 4.3.5).

We saw in the previous section that, in addition to the standard ϵ–δ definition, functional limits have a useful formulation in terms of sequences. The same is true of continuity. The next theorem summarizes these various equivalent ways to characterize the continuity of a function at a given point.

Theorem 4.3.2 (Characterizations of Continuity). *Let $f : A \to \mathbf{R}$, and let $c \in A$. The function f is continuous at c if and only if any one of the following three conditions is met:*

(i) *For all $\epsilon > 0$, there exists a $\delta > 0$ such that $|x - c| < \delta$ (and $x \in A$) implies $|f(x) - f(c)| < \epsilon$;*

(ii) *For all $V_\epsilon(f(c))$, there exists a $V_\delta(c)$ with the property that $x \in V_\delta(c)$ (and $x \in A$) implies $f(x) \in V_\epsilon(f(c))$;*

(iii) *For all $(x_n) \to c$ (with $x_n \in A$), it follows that $f(x_n) \to f(c)$.*

If c is a limit point of A, then the above conditions are equivalent to

(iv) $\displaystyle\lim_{x \to c} f(x) = f(c)$.

Proof. Statement (i) is just Definition 4.3.1, and statement (ii) is the standard rewording of (i) using topological neighborhoods in place of the absolute value notation. Statement (iii) is equivalent to (i) via an argument nearly identical to that of Theorem 4.2.3, with some slight modifications for when $x_n = c$. Finally, statement (iv) is seen to be equivalent to (i) by considering Definition 4.2.1 and observing that the case $x = c$ (which is excluded in the definition of functional limits) leads to the requirement $f(c) \in V_\epsilon(f(c))$, which is trivially true. $\qquad\square$

The length of this list is somewhat deceiving. Statements (i), (ii), and (iv) are closely related and essentially remind us that functional limits have an ϵ–δ formulation as well as a topological description. Statement (iii), however, is qualitatively different from the others. As a general rule, the sequential characterization of continuity is typically the most useful for demonstrating that a function is *not* continuous at some point.

Corollary 4.3.3 (Criterion for Discontinuity). *Let $f : A \to \mathbf{R}$, and let $c \in A$ be a limit point of A. If there exists a sequence $(x_n) \subseteq A$ where $(x_n) \to c$ but such that $f(x_n)$ does not converge to $f(c)$, we may conclude that f is not continuous at c.*

The sequential characterization of continuity is also important for the other reasons that it was important for functional limits. In particular, it allows us to bring our catalog of results about the behavior of sequences to bear on the study of continuous functions. The next theorem should be compared to Corollary 4.2.3 as well as to Theorem 2.3.3.

Theorem 4.3.4 (Algebraic Continuity Theorem). *Assume $f : A \to \mathbf{R}$ and $g : A \to \mathbf{R}$ are continuous at a point $c \in A$. Then,*

(i) *$kf(x)$ is continuous at c for all $k \in \mathbf{R}$;*

(ii) *$f(x) + g(x)$ is continuous at c;*

(iii) *$f(x)g(x)$ is continuous at c; and*

(iv) *$f(x)/g(x)$ is continuous at c, provided the quotient is defined.*

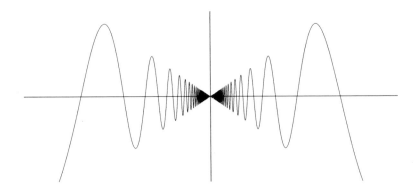

Figure 4.6: The function $x\sin(1/x)$ near zero.

Proof. All of these statements can be quickly derived from Corollary 4.2.4 and Theorem 4.3.2. $\square$

These results provide us with the tools we need to firm up our arguments in the opening section of this chapter about the behavior of Dirichlet's function and Thomae's function. The details are requested in Exercise 4.3.7. Here are some more examples of arguments for and against continuity of some familiar functions.

Example 4.3.5. All polynomials are continuous on $\mathbf{R}$. In fact, rational functions (i.e., quotients of polynomials) are continuous wherever they are defined.

To see why this is so, consider the identity function $g(x) = x$. Because $|g(x) - g(c)| = |x - c|$, we can respond to a given $\epsilon > 0$ by choosing $\delta = \epsilon$, and it follows that g is continuous on all of $\mathbf{R}$. It is even simpler to show that a constant function $f(x) = k$, is continuous. (Letting $\delta = 1$ regardless of the value of ϵ does the trick.) Because an arbitrary polynomial

$$p(x) = a_0 + a_1 x + a_2 x^2 + \cdots + a_n x^n$$

consists of sums and products of $g(x)$ with different constant functions, we may conclude from Theorem 4.3.4 that $p(x)$ is continuous.

Likewise, Theorem 4.3.4 implies that quotients of polynomials are continuous as long as the denominator is not zero.

Example 4.3.6. In Example 4.2.6, we saw that the oscillations of $\sin(1/x)$ are so rapid near the origin that $\lim_{x \to 0} \sin(1/x)$ does not exist. Now, consider the function

$$g(x) = \begin{cases} x\sin(1/x) & \text{if } x \neq 0 \\ 0 & \text{if } x = 0. \end{cases}$$

To investigate the continuity of g at $c = 0$ (Fig. 4.6), we can estimate

$$|g(x) - g(0)| = |x\sin(1/x) - 0| \leq |x|.$$

Given $\epsilon > 0$, set $\delta = \epsilon$, so that whenever $|x - 0| = |x| < \delta$ it follows that $|g(x) - g(0)| < \epsilon$. Thus, g is continuous at the origin.

Example 4.3.7. Throughout the exercises we have been using the greatest integer function $h(x) = [[x]]$ which for each $x \in \mathbf{R}$ returns the largest integer $n \in \mathbf{Z}$ satisfying $n \leq x$. This familiar step function certainly has discontinuous "jumps" at each integer value of its domain, but it is a useful exercise to try and articulate this observation in the language of analysis.

Given $m \in \mathbf{Z}$, define the sequence (x_n) by $x_n = m - 1/n$. It follows that $(x_n) \to m$, but

$$h(x_n) \to (m - 1),$$

which does not equal $m = h(m)$. By Corollary 4.3.3, we see that h fails to be continuous at each $m \in \mathbf{Z}$.

Now let's see why h is continuous at a point $c \notin \mathbf{Z}$. Given $\epsilon > 0$, we must find a δ-neighborhood $V_\delta(c)$ such that $x \in V_\delta(c)$ implies $h(x) \in V_\epsilon(h(c))$. We know that $c \in \mathbf{R}$ falls between consecutive integers $n < c < n + 1$ for some $n \in \mathbf{Z}$. If we take $\delta = \min\{c - n, (n + 1) - c\}$, then it follows from the definition of h that $h(x) = h(c)$ for all $x \in V_\delta(c)$. Thus, we certainly have

$$h(x) \in V_\epsilon(h(c))$$

whenever $x \in V_\delta(c)$.

This latter proof is quite different from the typical situation in that the value of δ does not actually depend on the choice of ϵ. Usually, a smaller ϵ requires a smaller δ in response, but here the same value of δ works no matter how small ϵ is chosen.

Example 4.3.8. Consider $f(x) = \sqrt{x}$ defined on $A = \{x \in \mathbf{R} : x \geq 0\}$. Exercise 2.3.1 outlines a sequential proof that f is continuous on A. Here, we give an ϵ–δ proof of the same fact.

Let $\epsilon > 0$. We need to argue that $|f(x) - f(c)|$ can be made less than ϵ for all values of x in some δ neighborhood around c. If $c = 0$, this reduces to the statement $\sqrt{x} < \epsilon$, which happens as long as $x < \epsilon^2$. Thus, if we choose $\delta = \epsilon^2$, we see that $|x - 0| < \delta$ implies $|f(x) - 0| < \epsilon$.

For a point $c \in A$ different from zero, we need to estimate $|\sqrt{x} - \sqrt{c}|$. This time, write

$$|\sqrt{x} - \sqrt{c}| = |\sqrt{x} - \sqrt{c}| \left(\frac{\sqrt{x} + \sqrt{c}}{\sqrt{x} + \sqrt{c}} \right) = \frac{|x - c|}{\sqrt{x} + \sqrt{c}} \leq \frac{|x - c|}{\sqrt{c}}.$$

In order to make this quantity less than ϵ, it suffices to pick $\delta = \epsilon\sqrt{c}$. Then, $|x - c| < \delta$ implies

$$|\sqrt{x} - \sqrt{c}| < \frac{\epsilon\sqrt{c}}{\sqrt{c}} = \epsilon,$$

as desired.

Although we have now shown that both polynomials and the square root function are continuous, the Algebraic Continuity Theorem does not provide the justification needed to conclude that a function such as $h(x) = \sqrt{3x^2 + 5}$ is continuous. For this, we must prove that *compositions* of continuous functions are continuous.

Theorem 4.3.9 (Composition of Continuous Functions). *Given $f : A \to \mathbf{R}$ and $g : B \to \mathbf{R}$, assume that the range $f(A) = \{f(x) : x \in A\}$ is contained in the domain B so that the composition $g \circ f(x) = g(f(x))$ is defined on A.*

If f is continuous at $c \in A$, and if g is continuous at $f(c) \in B$, then $g \circ f$ is continuous at c.

Proof. Exercise 4.3.3.   □

Exercises

Exercise 4.3.1. Let $g(x) = \sqrt[3]{x}$.

(a) Prove that g is continuous at $c = 0$.

(b) Prove that g is continuous at a point $c \neq 0$. (The identity $a^3 - b^3 = (a - b)(a^2 + ab + b^2)$ will be helpful.)

Exercise 4.3.2. To gain a deeper understanding of the relationship between ϵ and δ in the definition of continuity, let's explore some modest variations of Definition 4.3.1. In all of these, let f be a function defined on all of $\mathbf{R}$.

(a) Let's say f is *onetinuous* at c if for all $\epsilon > 0$ we can choose $\delta = 1$ and it follows that $|f(x) - f(c)| < \epsilon$ whenever $|x - c| < \delta$. Find an example of a function that is onetinuous on all of $\mathbf{R}$.

(b) Let's say f is *equaltinuous* at c if for all $\epsilon > 0$ we can choose $\delta = \epsilon$ and it follows that $|f(x) - f(c)| < \epsilon$ whenever $|x - c| < \delta$. Find an example of a function that is equaltinuous on $\mathbf{R}$ that is nowhere onetinuous, or explain why there is no such function.

(c) Let's say f is *lesstinuous* at c if for all $\epsilon > 0$ we can choose $0 < \delta < \epsilon$ and it follows that $|f(x) - f(c)| < \epsilon$ whenever $|x - c| < \delta$. Find an example of a function that is lesstinuous on $\mathbf{R}$ that is nowhere equaltinuous, or explain why there is no such function.

(d) Is every lesstinuous function continuous? Is every continuous function lesstinuous? Explain.

Exercise 4.3.3.   (a) Supply a proof for Theorem 4.3.9 using the ϵ–δ characterization of continuity.

(b) Give another proof of this theorem using the sequential characterization of continuity (from Theorem 4.3.2 (iii)).

Exercise 4.3.4. Assume f and g are defined on all of $\mathbf{R}$ and that $\lim_{x \to p} f(x) = q$ and $\lim_{x \to q} g(x) = r$.

(a) Give an example to show that it may not be true that

$$\lim_{x \to p} g(f(x)) = r.$$

(b) Show that the result in (a) does follow if we assume f and g are continuous.

(c) Does the result in (a) hold if we only assume f is continuous? How about if we only assume that g is continuous?

Exercise 4.3.5. Show using Definition 4.3.1 that if c is an isolated point of $A \subseteq \mathbf{R}$, then $f : A \to \mathbf{R}$ is continuous at c.

Exercise 4.3.6. Provide an example of each or explain why the request is impossible.

(a) Two functions f and g, neither of which is continuous at 0 but such that $f(x)g(x)$ and $f(x) + g(x)$ are continuous at 0.

(b) A function $f(x)$ continuous at 0 and $g(x)$ not continuous at 0 such that $f(x) + g(x)$ is continuous at 0.

(c) A function $f(x)$ continuous at 0 and $g(x)$ not continuous at 0 such that $f(x)g(x)$ is continuous at 0.

(d) A function $f(x)$ not continuous at 0 such that $f(x) + \frac{1}{f(x)}$ is continuous at 0.

(e) A function $f(x)$ not continuous at 0 such that $[f(x)]^3$ is continuous at 0.

Exercise 4.3.7.   (a) Referring to the proper theorems, give a formal argument that Dirichlet's function from Section 4.1 is nowhere-continuous on $\mathbf{R}$.

(b) Review the definition of Thomae's function in Section 4.1 and demonstrate that it fails to be continuous at every rational point.

(c) Use the characterization of continuity in Theorem 4.3.2 (iii) to show that Thomae's function is continuous at every irrational point in $\mathbf{R}$. (Given $\epsilon > 0$, consider the set of points $\{x \in \mathbf{R} : t(x) \geq \epsilon\}$.)

Exercise 4.3.8. Decide if the following claims are true or false, providing either a short proof or counterexample to justify each conclusion. Assume throughout that g is defined and continuous on all of $\mathbf{R}$.

(a) If $g(x) \geq 0$ for all $x < 1$, then $g(1) \geq 0$ as well.

(b) If $g(r) = 0$ for all $r \in \mathbf{Q}$, then $g(x) = 0$ for all $x \in \mathbf{R}$.

(c) If $g(x_0) > 0$ for a single point $x_0 \in \mathbf{R}$, then $g(x)$ is in fact strictly positive for uncountably many points.

Exercise 4.3.9. Assume $h : \mathbf{R} \to \mathbf{R}$ is continuous on $\mathbf{R}$ and let $K = \{x : h(x) = 0\}$. Show that K is a closed set.

Exercise 4.3.10. Observe that if a and b are real numbers, then

$$\max\{a, b\} = \frac{1}{2}[(a + b) + |a - b|].$$

(a) Show that if $f_1, f_2, \ldots, f_n$ are continuous functions, then

$$g(x) = \max\{f_1(x), f_2(x), \ldots, f_n(x)\}$$

is a continuous function.

(b) Let's explore whether the result in (a) extends to the infinite case. For each $n \in \mathbf{N}$, define f_n on $\mathbf{R}$ by

$$f_n(x) = \begin{cases} 1 & \text{if } |x| \geq 1/n \\ n|x| & \text{if } |x| < 1/n. \end{cases}$$

Now explicitly compute $h(x) = \sup\{f_1(x), f_2(x), f_3(x), \ldots\}$.

Exercise 4.3.11 (Contraction Mapping Theorem). Let f be a function defined on all of $\mathbf{R}$, and assume there is a constant c such that $0 < c < 1$ and

$$|f(x) - f(y)| \leq c|x - y|$$

for all $x, y \in \mathbf{R}$.

(a) Show that f is continuous on $\mathbf{R}$.

(b) Pick some point $y_1 \in \mathbf{R}$ and construct the sequence

$$(y_1, f(y_1), f(f(y_1)), \ldots).$$

In general, if $y_{n+1} = f(y_n)$, show that the resulting sequence (y_n) is a Cauchy sequence. Hence we may let $y = \lim y_n$.

(c) Prove that y is a fixed point of f (i.e., $f(y) = y$) and that it is unique in this regard.

(d) Finally, prove that if x is *any* arbitrary point in $\mathbf{R}$, then the sequence $(x, f(x), f(f(x)), \ldots)$ converges to y defined in (b).

Exercise 4.3.12. Let $F \subseteq \mathbf{R}$ be a nonempty closed set and define $g(x) = \inf\{|x - a| : a \in F\}$. Show that g is continuous on all of $\mathbf{R}$ and $g(x) \neq 0$ for all $x \notin F$.

Exercise 4.3.13. Let f be a function defined on all of $\mathbf{R}$ that satisfies the additive condition $f(x + y) = f(x) + f(y)$ for all $x, y \in \mathbf{R}$.

(a) Show that $f(0) = 0$ and that $f(-x) = -f(x)$ for all $x \in \mathbf{R}$.

(b) Let $k = f(1)$. Show that $f(n) = kn$ for all $n \in \mathbf{N}$, and then prove that $f(z) = kz$ for all $z \in \mathbf{Z}$. Now, prove that $f(r) = kr$ for any rational number r.

(c) Show that if f is continuous at $x = 0$, then f is continuous at every point in $\mathbf{R}$ and conclude that $f(x) = kx$ for all $x \in \mathbf{R}$. Thus, any additive function that is continuous at $x = 0$ must necessarily be a linear function through the origin.

Exercise 4.3.14. (a) Let F be a closed set. Construct a function $f : \mathbf{R} \to \mathbf{R}$ such that the set of points where f fails to be continuous is precisely F. (The concept of the interior of a set, discussed in Exercise 3.2.14, may be useful.)

(b) Now consider an open set O. Construct a function $g : \mathbf{R} \to \mathbf{R}$ whose set of discontinuous points is precisely O. (For this problem, the function in Exercise 4.3.12 may be useful.)

4.4 Continuous Functions on Compact Sets

Given a function $f : A \to \mathbf{R}$ and a subset $B \subseteq A$, the notation $f(B)$ refers to the range of f over the set B; that is,

$$f(B) = \{f(x) : x \in B\}.$$

The adjectives open, closed, bounded, compact, perfect, and connected are all used to describe subsets of the real line. An interesting question is to sort out which, if any, of these properties are preserved when a particular set B is mapped to $f(B)$ via a continuous function. For instance, if B is open and f is continuous, is $f(B)$ necessarily open? The answer to this question is no. If $f(x) = x^2$ and B is the open interval $(-1, 1)$, then $f(B)$ is the interval $[0, 1)$, which is not open.

The corresponding conjecture for closed sets also turns out to be false, although constructing a counterexample requires a little more thought. Consider the function

$$g(x) = \frac{1}{1 + x^2}$$

and the closed set $B = [0, \infty) = \{x : x \geq 0\}$. Because $g(B) = (0, 1]$ is not closed, we must conclude that continuous functions do not, in general, map closed sets to closed sets. Notice, however, that our particular counterexample required using an *unbounded* closed set B. This is not incidental. Sets that are closed and bounded—that is, compact sets—always get mapped to closed and bounded subsets by continuous functions.

Theorem 4.4.1 (Preservation of Compact Sets). *Let $f : A \to \mathbf{R}$ be continuous on A. If $K \subseteq A$ is compact, then $f(K)$ is compact as well.*

Proof. Let (y_n) be an arbitrary sequence contained in the range set $f(K)$. To prove this result, we must find a subsequence (y_{n_k}), which converges to a limit also in $f(K)$. The strategy is to take advantage of the assumption that the domain set K is compact by translating the sequence (y_n)—which is in the range of f—back to a sequence in the domain K.

To assert that $(y_n) \subseteq f(K)$ means that, for each $n \in \mathbf{N}$, we can find (at least one) $x_n \in K$ with $f(x_n) = y_n$. This yields a sequence $(x_n) \subseteq K$. Because K is compact, there exists a convergent subsequence (x_{n_k}) whose limit $x = \lim x_{n_k}$ is also in K. Finally, we make use of the fact that f is assumed to be continuous on A and so is continuous at x in particular. Given that $(x_{n_k}) \to x$, we conclude that $(y_{n_k}) \to f(x)$. Because $x \in K$, we have that $f(x) \in f(K)$, and hence $f(K)$ is compact. $\qquad \square$

An extremely important corollary is obtained by combining this result with the observation that compact sets are bounded and contain their supremums and infimums.

Theorem 4.4.2 (Extreme Value Theorem). *If $f : K \to \mathbf{R}$ is continuous on a compact set $K \subseteq \mathbf{R}$, then f attains a maximum and minimum value. In other words, there exist $x_0, x_1 \in K$ such that $f(x_0) \leq f(x) \leq f(x_1)$ for all $x \in K$.*

Proof. Because $f(K)$ is compact, we can set $\alpha = \sup f(K)$ and know $\alpha \in f(K)$ (Exercise 3.3.1). It follows that there exist $x_1 \in K$ with $\alpha = f(x_1)$. The argument for the minimum value is similar. $\qquad \square$

Uniform Continuity

Although we have proved that polynomials are always continuous on $\mathbf{R}$, there is an important lesson to be learned by constructing direct proofs that the functions $f(x) = 3x + 1$ and $g(x) = x^2$ (previously studied in Example 4.2.2) are everywhere continuous.

Example 4.4.3. (i) To show directly that $f(x) = 3x + 1$ is continuous at an arbitrary point $c \in \mathbf{R}$, we must argue that $|f(x) - f(c)|$ can be made arbitrarily small for values of x near c. Now,

$$|f(x) - f(c)| = |(3x + 1) - (3c + 1)| = 3|x - c|,$$

so, given $\epsilon > 0$, we choose $\delta = \epsilon/3$. Then, $|x - c| < \delta$ implies

$$|f(x) - f(c)| = 3|x - c| < 3 \left(\frac{\epsilon}{3} \right) = \epsilon.$$

Of particular importance for this discussion is the fact that the choice of δ is the same regardless of which point $c \in \mathbf{R}$ we are considering.

(ii) Let's contrast this with what happens when we prove $g(x) = x^2$ is continuous on $\mathbf{R}$. Given $c \in \mathbf{R}$, we have

$$|g(x) - g(c)| = |x^2 - c^2| = |x - c||x + c|.$$

As discussed in Example 4.2.2, we need an upper bound on $|x + c|$, which is obtained by insisting that our choice of δ not exceed 1. This guarantees that all values of x under consideration will necessarily fall in the interval $(c - 1, c + 1)$. It follows that

$$|x + c| \leq |x| + |c| \leq (|c| + 1) + |c| = 2|c| + 1.$$

Now, let $\epsilon > 0$. If we choose $\delta = \min\{1, \epsilon/(2|c| + 1)\}$, then $|x - c| < \delta$ implies

$$|f(x) - f(c)| = |x - c||x + c| < \left(\frac{\epsilon}{2|c| + 1}\right)(2|c| + 1) = \epsilon.$$

Now, there is nothing deficient about this argument, but it is important to notice that, in the second proof, the algorithm for choosing the response δ depends on the value of c. The statement

$$\delta = \frac{\epsilon}{2|c| + 1}$$

means that larger values of c are going to require smaller values of δ, a fact that should be evident from a consideration of the graph of $g(x) = x^2$ (Fig. 4.7). Given, say, $\epsilon = 1$, a response of $\delta = 1/3$ is sufficient for $c = 1$ because $2/3 < x < 4/3$ certainly implies $0 < x^2 < 2$. However, if $c = 10$, then the steepness of the graph of $g(x)$ means that a much smaller δ is required—$\delta = 1/21$ by our rule—to force $99 < x^2 < 101$.

The next definition is meant to distinguish between these two examples.

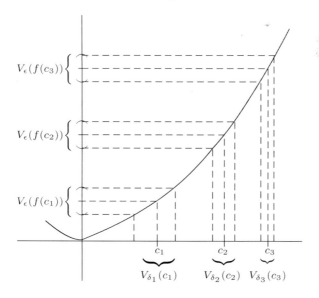

Figure 4.7: $g(x) = x^2$; **A larger c requires a smaller δ.**

Definition 4.4.4 (Uniform Continuity). A function $f : A \to \mathbf{R}$ is *uniformly continuous on A* if for every $\epsilon > 0$ there exists a $\delta > 0$ such that for all $x, y \in A$, $|x - y| < \delta$ implies $|f(x) - f(y)| < \epsilon$.

Recall that to say that "f is continuous on A" means that f is continuous at each individual point $c \in A$. In other words, given $\epsilon > 0$ and $c \in A$, we can find a $\delta > 0$ *perhaps depending on c* such that if $|x - c| < \delta$, then $|f(x) - f(c)| < \epsilon$. Uniform continuity is a strictly stronger property. The key distinction between asserting that f is "uniformly continuous on A" versus simply "continuous on A" is that, given an $\epsilon > 0$, a single $\delta > 0$ can be chosen that works simultaneously for all points c in A. To say that a function is *not* uniformly continuous on a set A, then, does not necessarily mean it is not continuous at some point. Rather, it means that there is some $\epsilon_0 > 0$ for which no single $\delta > 0$ is a suitable response for all $c \in A$.

Theorem 4.4.5 (Sequential Criterion for Absence of Uniform Continuity). *A function $f : A \to \mathbf{R}$ fails to be uniformly continuous on A if and only if there exists a particular $\epsilon_0 > 0$ and two sequences (x_n) and (y_n) in A satisfying*

$$|x_n - y_n| \to 0 \quad but \quad |f(x_n) - f(y_n)| \geq \epsilon_0.$$

Proof. The negation of Definition 4.4.4 states that f is not uniformly continuous on A if and only if there exists $\epsilon_0 > 0$ such that for all $\delta > 0$ we can find two points x and y satisfying $|x - y| < \delta$ but with $|f(x) - f(y)| \geq \epsilon_0$. Thus, if we set $\delta_1 = 1$, then there exist two points x_1 and y_1 where $|x_1 - y_1| < 1$ but $|f(x_1) - f(y_1)| \geq \epsilon_0$.

In a similar way, if we set $\delta_n = 1/n$ where $n \in \mathbf{N}$, it follows that there exist points x_n and y_n with $|x_n - y_n| < 1/n$ but where $|f(x_n) - f(y_n)| \geq \epsilon_0$. The resulting sequences (x_n) and (y_n) satisfy the requirements described in the theorem.

Conversely, if ϵ_0, (x_n) and (y_n) exist as described, it is straightforward to see that no $\delta > 0$ is a suitable response for ϵ_0. $\square$

Example 4.4.6. The function $h(x) = \sin(1/x)$ (Fig. 4.5) is continuous at every point in the open interval $(0, 1)$ but is not uniformly continuous on this interval. The problem arises near zero, where the increasingly rapid oscillations take domain values that are quite close together to range values a distance 2 apart. To illustrate Theorem 4.4.5, take $\epsilon_0 = 2$ and set

$$x_n = \frac{1}{\pi/2 + 2n\pi} \quad \text{and} \quad y_n = \frac{1}{3\pi/2 + 2n\pi}.$$

Because each of these sequences tends to zero, we have $|x_n - y_n| \to 0$, and a short calculation reveals $|h(x_n) - h(y_n)| = 2$ for all $n \in \mathbf{N}$.

Whereas continuity is defined at a single point, uniform continuity is always discussed in reference to a particular domain. In Example 4.4.3, we were not able to prove that $g(x) = x^2$ is uniformly continuous on $\mathbf{R}$ because larger values of x require smaller and smaller values of δ. (As another illustration of Theorem 4.4.5, take $x_n = n$ and $y_n = n + 1/n$.) It is true, however, that $g(x)$ *is* uniformly continuous on the bounded set $[-10, 10]$. Returning to the argument set forth in Example 4.4.3 (ii), notice that if we restrict our attention to the domain $[-10, 10]$, then $|x + y| \leq 20$ for all x and y. Given $\epsilon > 0$, we can now choose $\delta = \epsilon/20$, and verify that if $x, y \in [-10, 10]$ satisfy $|x - y| < \delta$, then

$$|f(x) - f(y)| = |x^2 - y^2| = |x - y||x + y| < \left(\frac{\epsilon}{20}\right) 20 = \epsilon.$$

In fact, it is not difficult to see how to modify this argument to show that $g(x)$ is uniformly continuous on any bounded set A in $\mathbf{R}$.

Now, Example 4.4.6 is included to keep us from jumping to the erroneous conclusion that functions that are continuous on bounded domains are necessarily uniformly continuous. A general result does follow, however, if we assume that the domain is compact.

Theorem 4.4.7 (Uniform Continuity on Compact Sets). *A function that is continuous on a compact set K is uniformly continuous on K.*

Proof. Assume $f : K \to \mathbf{R}$ is continuous at every point of a compact set $K \subseteq \mathbf{R}$. To prove that f is uniformly continuous on K we argue by contradiction.

By the criterion in Theorem 4.4.5, if f is not uniformly continuous on K, then there exist two sequences (x_n) and (y_n) in K such that

$$\lim |x_n - y_n| = 0 \quad \text{while} \quad |f(x_n) - f(y_n)| \geq \epsilon_0$$

for some particular $\epsilon_0 > 0$. Because K is compact, the sequence (x_n) has a convergent subsequence (x_{n_k}) with $x = \lim x_{n_k}$ also in K.

We could use the compactness of K again to produce a convergent subsequence of (y_n), but notice what happens when we consider the particular subsequence (y_{n_k}) consisting of those terms in (y_n) that correspond to the terms in the convergent subsequence (x_{n_k}). By the Algebraic Limit Theorem,

$$\lim(y_{n_k}) = \lim((y_{n_k} - x_{n_k}) + x_{n_k}) = 0 + x.$$

The conclusion is that both (x_{n_k}) and (y_{n_k}) converge to $x \in K$. Because f is assumed to be continuous at x, we have $\lim f(x_{n_k}) = f(x)$ and $\lim f(y_{n_k}) = f(x)$, which implies

$$\lim(f(x_{n_k}) - f(y_{n_k})) = 0.$$

A contradiction arises when we recall that (x_n) and (y_n) were chosen to satisfy

$$|f(x_n) - f(y_n)| \geq \epsilon_0$$

for all $n \in \mathbf{N}$. We conclude, then, that f is indeed uniformly continuous on K. $\qquad\square$

Exercises

Exercise 4.4.1. (a) Show that $f(x) = x^3$ is continuous on all of $\mathbf{R}$.

(b) Argue, using Theorem 4.4.5, that f is not uniformly continuous on $\mathbf{R}$.

(c) Show that f is uniformly continuous on any bounded subset of $\mathbf{R}$.

Exercise 4.4.2. (a) Is $f(x) = 1/x$ uniformly continuous on $(0, 1)$?

(b) Is $g(x) = \sqrt{x^2 + 1}$ uniformly continuous on $(0, 1)$?

(c) Is $h(x) = x \sin(1/x)$ uniformly continuous on $(0, 1)$?

Exercise 4.4.3. Show that $f(x) = 1/x^2$ is uniformly continuous on the set $[1, \infty)$ but not on the set $(0, 1]$.

Exercise 4.4.4. Decide whether each of the following statements is true or false, justifying each conclusion.

(a) If f is continuous on $[a, b]$ with $f(x) > 0$ for all $a \leq x \leq b$, then $1/f$ is bounded on $[a, b]$ (meaning $1/f$ has bounded range).

(b) If f is uniformly continuous on a bounded set A, then $f(A)$ is bounded.

(c) If f is defined on $\mathbf{R}$ and $f(K)$ is compact whenever K is compact, then f is continuous on $\mathbf{R}$.

Exercise 4.4.5. Assume that g is defined on an open interval (a, c) and it is known to be uniformly continuous on $(a, b]$ and $[b, c)$, where $a < b < c$. Prove that g is uniformly continuous on (a, c).

Exercise 4.4.6. Give an example of each of the following, or state that such a request is impossible. For any that are impossible, supply a short explanation for why this is the case.

(a) A continuous function $f : (0, 1) \to \mathbf{R}$ and a Cauchy sequence (x_n) such that $f(x_n)$ is not a Cauchy sequence;

(b) A uniformly continuous function $f : (0, 1) \to \mathbf{R}$ and a Cauchy sequence (x_n) such that $f(x_n)$ is not a Cauchy sequence;

(c) A continuous function $f : [0, \infty) \to \mathbf{R}$ and a Cauchy sequence (x_n) such that $f(x_n)$ is not a Cauchy sequence;

Exercise 4.4.7. Prove that $f(x) = \sqrt{x}$ is uniformly continuous on $[0, \infty)$.

Exercise 4.4.8. Give an example of each of the following, or provide a short argument for why the request is impossible.

(a) A continuous function defined on $[0, 1]$ with range $(0, 1)$.

(b) A continuous function defined on $(0, 1)$ with range $[0, 1]$.

(c) A continuous function defined on $(0, 1]$ with range $(0, 1)$.

Exercise 4.4.9 (Lipschitz Functions). A function $f : A \to \mathbf{R}$ is called *Lipschitz* if there exists a bound $M > 0$ such that

$$\left| \frac{f(x) - f(y)}{x - y} \right| \leq M$$

for all $x \neq y \in A$. Geometrically speaking, a function f is Lipschitz if there is a uniform bound on the magnitude of the slopes of lines drawn through any two points on the graph of f.

(a) Show that if $f : A \to \mathbf{R}$ is Lipschitz, then it is uniformly continuous on A.

(b) Is the converse statement true? Are all uniformly continuous functions necessarily Lipschitz?

Exercise 4.4.10. Assume that f and g are uniformly continuous functions defined on a common domain A. Which of the following combinations are necessarily uniformly continuous on A:

$$f(x) + g(x), \quad f(x)g(x), \quad \frac{f(x)}{g(x)}, \quad f(g(x)) \,?$$

(Assume that the quotient and the composition are properly defined and thus at least continuous.)

Exercise 4.4.11 (Topological Characterization of Continuity). Let g be defined on all of $\mathbf{R}$. If B is a subset of $\mathbf{R}$, define the set $g^{-1}(B)$ by

$$g^{-1}(B) = \{x \in \mathbf{R} : g(x) \in B\}.$$

Show that g is continuous if and only if $g^{-1}(O)$ is open whenever $O \subseteq \mathbf{R}$ is an open set.

Exercise 4.4.12. Review Exercise 4.4.11, and then determine which of the following statements is true about a continuous function defined on $\mathbf{R}$:

(a) $f^{-1}(B)$ is finite whenever B is finite.

(b) $f^{-1}(K)$ is compact whenever K is compact.

(c) $f^{-1}(A)$ is bounded whenever A is bounded.

(d) $f^{-1}(F)$ is closed whenever F is closed.

Exercise 4.4.13 (Continuous Extension Theorem). (a) Show that a uniformly continuous function preserves Cauchy sequences; that is, if $f : A \to \mathbf{R}$ is uniformly continuous and $(x_n) \subseteq A$ is a Cauchy sequence, then show $f(x_n)$ is a Cauchy sequence.

(b) Let g be a continuous function on the open interval (a, b). Prove that g is uniformly continuous on (a, b) if and only if it is possible to define values $g(a)$ and $g(b)$ at the endpoints so that the extended function g is continuous on $[a, b]$. (In the forward direction, first produce candidates for $g(a)$ and $g(b)$, and then show the extended g is continuous.)

Exercise 4.4.14. Construct an alternate proof of Theorem 4.4.7 using the open cover characterization of compactness from the Heine–Borel Theorem (Theorem 3.3.8 (iii)).

4.5   The Intermediate Value Theorem

The Intermediate Value Theorem (IVT) is the name given to the very intuitive observation that a continuous function f on a closed interval $[a, b]$ attains every value that falls between the range values $f(a)$ and $f(b)$ (Fig. 4.8).

Here is this observation in the language of analysis.

Theorem 4.5.1 (Intermediate Value Theorem). *Let $f : [a, b] \to \mathbf{R}$ be continuous. If L is a real number satisfying $f(a) < L < f(b)$ or $f(a) > L > f(b)$, then there exists a point $c \in (a, b)$ where $f(c) = L$.*

This theorem was freely used by mathematicians of the 18th century (including Euler and Gauss) without any consideration of its validity. In fact, the first analytical proof was not offered until 1817 by Bolzano in a paper that also contains the first appearance of a somewhat modern definition of continuity. This emphasizes the significance of this result. As discussed in Section 4.1, Bolzano and his contemporaries had arrived at a point in the evolution of mathematics where it was becoming increasingly important to firm up the foundations of the subject. Doing so, however, was not simply a matter of going back and supplying the missing proofs. The real battle lay in first obtaining a thorough and mutually agreed-upon understanding of the relevant concepts. The importance of the Intermediate Value Theorem for us is similar in that our understanding of continuity and the nature of the real line is now mature enough for a proof *to be possible.* Indeed, there are several satisfying arguments for this simple result, each one isolating, in a slightly different way, the interplay between continuity and completeness.

Preservation of Connected Sets

The most potentially useful way to understand the Intermediate Value Theorem (IVT) is as a special case of the fact that continuous functions map connected sets to connected sets. In Theorem 4.4.1, we saw that if f is a continuous function on a compact set K, then the range set $f(K)$ is also compact. The analogous observation holds for connected sets.

Theorem 4.5.2 (Preservation of Connected Sets). *Let $f : G \to \mathbf{R}$ be continuous. If $E \subseteq G$ is connected, then $f(E)$ is connected as well.*

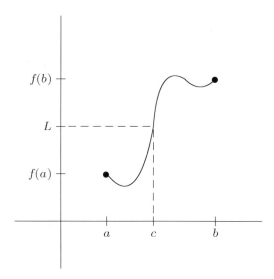

Figure 4.8: Intermediate Value Theorem.

Proof. Intending to use the characterization of connected sets in Theorem 3.4.6, let $f(E) = A \cup B$ where A and B are disjoint and nonempty. Our goal is to produce a sequence contained in one of these sets that converges to a limit in the other.

Let

$$C = \{x \in E : f(x) \in A\} \quad \text{and} \quad D = \{x \in E : f(x) \in B\}.$$

The sets C and D are called the *preimages* of A and B, respectively. Using the properties of A and B, it is straightforward to check that C and D are nonempty and disjoint and satisfy $E = C \cup D$. Now, we are assuming E is a connected set, so by Theorem 3.4.6, there exists a sequence (x_n) contained in one of C or D with $x = \lim x_n$ contained in the other. Finally, because f is continuous at x, we get $f(x) = \lim f(x_n)$. Thus, it follows that $f(x_n)$ is a convergent sequence contained in either A or B while the limit $f(x)$ is an element of the other. With another nod to Theorem 3.4.6, the proof is complete.   □

In $\mathbf{R}$, a set is connected if and only if it is a (possibly unbounded) interval. This fact, together with Theorem 4.5.2, leads to a short proof of the Intermediate Value Theorem (Exercise 4.5.1). We should point out that the proof of Theorem 4.5.2 does not make use of the equivalence between connected sets and intervals in $\mathbf{R}$ but relies only on the general definitions. The previous comment that this is the most *useful* way to approach IVT stems from the fact that, although it is not discussed here, the definitions of continuity and connectedness can be easily adapted to higher-dimensional settings. Theorem 4.5.2, then, remains a valid conclusion in higher dimensions, whereas the Intermediate Value Theorem is essentially a one-dimensional result.

Completeness

A typical way the Intermediate Value Theorem is applied is to prove the existence of roots. Given $f(x) = x^2 - 2$, for instance, we see that $f(1) = -1$ and $f(2) = 2$. Therefore, there exists a point $c \in (1, 2)$ where $f(c) = 0$.

In this case, we can easily compute $c = \sqrt{2}$, meaning that we really did not need IVT to show that f has a root. We spent a good deal of time in Chapter 1 proving that $\sqrt{2}$ exists, which was only possible once we insisted on the Axiom of Completeness as part of our assumptions about the real numbers. The fact that the Intermediate Value Theorem has just asserted that $\sqrt{2}$ exists suggests that another way to understand this result is in terms of the relationship between the continuity of f and the completeness of $\mathbf{R}$.

The Axiom of Completeness (AoC) from the first chapter states that "Nonempty sets that are bounded above have least upper bounds." Later, we saw that the Nested Interval Property (NIP) is an equivalent way to assert that the real numbers have no "gaps." Either of these characterizations of completeness can be used as the cornerstone for an alternate proof of Theorem 4.5.1.

Proof. **I.** (*First approach using AoC.*) To simplify matters a bit, let's consider the special case where f is a continuous function satisfying $f(a) < 0 < f(b)$ and show that $f(c) = 0$ for some $c \in (a, b)$. First let

$$K = \{x \in [a, b] : f(x) \leq 0\}.$$

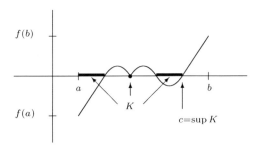

Notice that K is bounded above by b, and $a \in K$ so K is not empty. Thus we may appeal to the Axiom of Completeness to assert that $c = \sup K$ exists.

There are three cases to consider:

$$f(c) > 0, \quad f(c) < 0, \quad \text{and} \quad f(c) = 0.$$

The fact that c is the least upper bound of K can be used to rule out the first two cases, resulting in the desired conclusion that $f(c) = 0$. The details are requested in Exercise 4.5.5(a).

II. (*Second approach using NIP.*) Again, consider the special case where $L = 0$ and $f(a) < 0 < f(b)$. Let $I_0 = [a, b]$, and consider the midpoint

$$z = (a + b)/2.$$

If $f(z) \geq 0$, then set $a_1 = a$ and $b_1 = z$. If $f(z) < 0$, then set $a_1 = z$ and $b_1 = b$. In either case, the interval $I_1 = [a_1, b_1]$ has the property that f is negative at the left endpoint and nonnegative at the right.

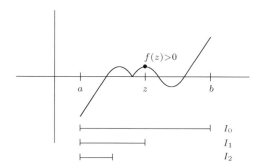

This procedure can be inductively repeated, setting the stage for an application of the Nested Interval Property. The remainder of the argument is left as Exercise 4.5.5(b). □

The Intermediate Value Property

Does the Intermediate Value Theorem have a converse?

Definition 4.5.3. A function f has the *intermediate value property* on an interval $[a, b]$ if for all $x < y$ in $[a, b]$ and all L between $f(x)$ and $f(y)$, it is always possible to find a point $c \in (x, y)$ where $f(c) = L$.

Another way to summarize the Intermediate Value Theorem is to say that every continuous function on $[a, b]$ has the intermediate value property. There is an understandable temptation to suspect that any function that has the intermediate value property must necessarily be continuous, but that is not the case. We have seen that

$$g(x) = \begin{cases} \sin(1/x) & \text{if } x \neq 0 \\ 0 & \text{if } x = 0 \end{cases}$$

is not continuous at zero (Example 4.2.6), but it does have the intermediate value property on $[0, 1]$.

The intermediate value property *does* imply continuity if we insist that our function is monotone (Exercise 4.5.3).

Exercises

Exercise 4.5.1. Show how the Intermediate Value Theorem follows as a corollary to Theorem 4.5.2.

Exercise 4.5.2. Provide an example of each of the following, or explain why the request is impossible

(a) A continuous function defined on an open interval with range equal to a closed interval.

(b) A continuous function defined on a closed interval with range equal to an open interval.

(c) A continuous function defined on an open interval with range equal to an unbounded closed set different from $\mathbf{R}$.

(d) A continuous function defined on all of $\mathbf{R}$ with range equal to $\mathbf{Q}$.

Exercise 4.5.3. A function f is *increasing* on A if $f(x) \leq f(y)$ for all $x < y$ in A. Show that if f is increasing on $[a, b]$ and satisfies the intermediate value property (Definition 4.5.3), then f is continuous on $[a, b]$.

Exercise 4.5.4. Let g be continuous on an interval A and let F be the set of points where g fails to be one-to-one; that is,

$$F = \{x \in A : f(x) = f(y) \text{ for some } y \neq x \text{ and } y \in A\}.$$

Show F is either empty or uncountable.

Exercise 4.5.5.   (a) Finish the proof of the Intermediate Value Theorem using the Axiom of Completeness started previously.

(b) Finish the proof of the Intermediate Value Theorem using the Nested Interval Property started previously.

Exercise 4.5.6. Let $f : [0, 1] \to \mathbf{R}$ be continuous with $f(0) = f(1)$.

(a) Show that there must exist $x, y \in [0, 1]$ satisfying $|x - y| = 1/2$ and $f(x) = f(y)$.

(b) Show that for each $n \in \mathbf{N}$ there exist $x_n, y_n \in [0, 1]$ with $|x_n - y_n| = 1/n$ and $f(x_n) = f(y_n)$.

(c) If $h \in (0, 1/2)$ is not of the form $1/n$, there does not necessarily exist $|x - y| = h$ satisfying $f(x) = f(y)$. Provide an example that illustrates this using $h = 2/5$.

Exercise 4.5.7. Let f be a continuous function on the closed interval $[0, 1]$ with range also contained in $[0, 1]$. Prove that f must have a fixed point; that is, show $f(x) = x$ for at least one value of $x \in [0, 1]$.

Exercise 4.5.8 (Inverse functions). If a function $f : A \to \mathbf{R}$ is one-to-one, then we can define the inverse function f^{-1} on the range of f in the natural way: $f^{-1}(y) = x$ where $y = f(x)$.

Show that if f is continuous on an interval $[a, b]$ and one-to-one, then f^{-1} is also continuous.

4.6 Sets of Discontinuity

Given a function $f : \mathbf{R} \to \mathbf{R}$, define $D_f \subseteq \mathbf{R}$ to be the set of points where the function f fails to be continuous. In Section 4.1, we saw that Dirichlet's function $g(x)$ had $D_g = \mathbf{R}$. The modification $h(x)$ of Dirichlet's function had $D_h = \mathbf{R}\backslash\{0\}$, zero being the only point of continuity. Finally, for Thomae's function $t(x)$, we saw that $D_t = \mathbf{Q}$.

Exercise 4.6.1. Using modifications of these functions, construct a function $f : \mathbf{R} \to \mathbf{R}$ so that

(a) $D_f = \mathbf{Z}^c$.

(b) $D_f = \{x : 0 < x \leq 1\}$.

Exercise 4.6.2. Given a countable set $A = \{a_1, a_2, a_3, \ldots\}$, define $f(a_n) = 1/n$ and $f(x) = 0$ for all $x \notin A$. Find D_f.

We concluded the introduction with a question about whether D_f could take the form of *any* arbitrary subset of the real line. As it turns out, this is not the case. The set of discontinuities of a real-valued function on $\mathbf{R}$ has a specific topological structure that is not possessed by every subset of $\mathbf{R}$. Specifically, D_f, no matter how f is chosen, can always be written as the countable union of closed sets. In the case where f is *monotone*, these closed sets can be taken to be single points.

Monotone Functions

Classifying D_f for an arbitrary f is somewhat involved, so it is interesting that describing D_f is fairly straightforward for the class of monotone functions.

Definition 4.6.1. A function $f : A \to \mathbf{R}$ is *increasing on* A if $f(x) \leq f(y)$ whenever $x < y$ and *decreasing* if $f(x) \geq f(y)$ whenever $x < y$ in A. A *monotone* function is one that is either increasing or decreasing.

Continuity of f at a point c means that $\lim_{x \to c} f(x) = f(c)$. One particular way for a discontinuity to occur is if the limit from the right at c is different from the limit from the left at c. As always with new terminology, we need to be precise about what we mean by "from the left" and "from the right."

Definition 4.6.2. Given a limit point c of a set A and a function $f : A \to \mathbf{R}$, we write

$$\lim_{x \to c^+} f(x) = L$$

if for all $\epsilon > 0$ there exists a $\delta > 0$ such that $|f(x) - L| < \epsilon$ whenever $0 < x - c < \delta$.

Equivalently, in terms of sequences, $\lim_{x \to c^+} f(x) = L$ if $\lim f(x_n) = L$ for all sequences (x_n) satisfying $x_n > c$ and $\lim(x_n) = c$.

Exercise 4.6.3. State a similar definition for the left-hand limit

$$\lim_{x \to c^-} f(x) = L.$$

Theorem 4.6.3. *Given* $f : A \to \mathbf{R}$ *and a limit point* c *of* A, $\lim_{x \to c} f(x) = L$ *if and only if*

$$\lim_{x \to c^-} f(x) = L \quad and \quad \lim_{x \to c^+} f(x) = L.$$

Exercise 4.6.4. Supply a proof for this proposition.

Generally speaking, discontinuities can be divided into three categories:

(i) If $\lim_{x \to c} f(x)$ exists but has a value different from $f(c)$, the discontinuity at c is called *removable*.

(ii) If $\lim_{x \to c^+} f(x) \neq \lim_{x \to c^-} f(x)$, then f has a *jump* discontinuity at c.

(iii) If $\lim_{x \to c} f(x)$ does not exist for some other reason, then the discontinuity at c is called an *essential* discontinuity.

We are now equipped to characterize the set D_f for an arbitrary monotone function f.

Exercise 4.6.5. Prove that the only type of discontinuity a monotone function can have is a jump discontinuity.

Exercise 4.6.6. Construct a bijection between the set of jump discontinuities of a monotone function f and a subset of $\mathbf{Q}$. Conclude that D_f for a monotone function f must either be finite or countable, but not uncountable.

D_f for an Arbitrary Function

Recall that the intersection of an infinite collection of closed sets is closed, but for unions we must restrict ourselves to *finite* collections of closed sets in order to ensure the union is closed. For open sets the situation is reversed. The arbitrary union of open sets is open, but only finite intersections of open sets are necessarily open.

Definition 4.6.4. A set that can be written as the countable union of closed sets is in the class F_σ. (This definition also appeared in Section 3.5.)

In Section 4.1 we constructed functions where the set of discontinuity was $\mathbf{R}$ (Dirichlet's function), $\mathbf{R}\backslash\{0\}$ (modified Dirichlet function), and $\mathbf{Q}$ (Thomae's function).

Exercise 4.6.7.   (a) Show that in each of the above cases we get an F_σ set as the set where the function is discontinuous.

(b) Show that the two sets of discontinuity in Exercise 4.6.1 are F_σ sets.

The upcoming argument depends on a concept called α-continuity.

Definition 4.6.5. Let f be defined on $\mathbf{R}$, and let $\alpha > 0$. The function f is *α-continuous at $x \in \mathbf{R}$* if there exists a $\delta > 0$ such that for all $y, z \in (x - \delta, x + \delta)$ it follows that $|f(y) - f(z)| < \alpha$.

The most important thing to note about this definition is that there is no "for all" in front of the $\alpha > 0$. As we will investigate, adding this quantifier would make this definition equivalent to our definition of continuity. In a sense, α-continuity is a measure of the variation of the function in the neighborhood of a particular point. A function is α-continuous at a point c if there is some interval centered at c in which the variation of the function never exceeds the value $\alpha > 0$.

Given a function f on $\mathbf{R}$, define D_f^α to be the set of points where the function f fails to be α-continuous. In other words,

$$D_f^\alpha = \{x \in \mathbf{R} : f \text{ is not } \alpha\text{-continuous at } x\}.$$

Exercise 4.6.8. Prove that, for a fixed $\alpha > 0$, the set D_f^α is closed.

The stage is set. It is time to characterize the set of discontinuity for an arbitrary function f on $\mathbf{R}$.

Theorem 4.6.6. *Let $f : \mathbf{R} \to \mathbf{R}$ be an arbitrary function. Then, D_f is an F_σ set.*

Proof. Recall that

$$D_f = \{x \in \mathbf{R} : f \text{ is not continuous at } x\}.$$

Exercise 4.6.9. If $\alpha < \alpha'$, show that $D_f^{\alpha'} \subseteq D_f^\alpha$.

Exercise 4.6.10. Let $\alpha > 0$ be given. Show that if f is continuous at x, then it is α-continuous at x as well. Explain how it follows that $D_f^\alpha \subseteq D_f$.

Exercise 4.6.11. Show that if f is not continuous at x, then f is not α-continuous for some $\alpha > 0$. Now explain why this guarantees that

$$D_f = \bigcup_{n=1}^\infty D_f^{\alpha_n},$$

where $\alpha_n = 1/n$.

Because each $D_f^{\alpha_n}$ is closed, the proof is complete. $\qquad\square$

4.7   Epilogue

Theorem 4.6.6 is only interesting if we can demonstrate that not every subset of **R** is in an F_σ set. This takes some effort and was included as an exercise in Section 3.5 on the Baire Category Theorem. Baire's Theorem states that if **R** is written as the countable union of closed sets, then at least one of these sets must contain a nonempty open interval. Now **Q** is the countable union of singleton points, and we can view each point as a closed set that obviously contains no intervals. If the set of irrationals **I** were a countable union of closed sets, it would have to be that none of these closed sets contained any open intervals or else they would then contain some rational numbers. But this leads to a contradiction to Baire's Theorem. Thus, **I** is not the countable union of closed sets, and consequently it is not an F_σ set. We may therefore conclude that there is no function f that is continuous at every rational point and discontinuous at every irrational point. This should be compared with Thomae's function discussed earlier.

The converse question is interesting as well. Given an arbitrary F_σ set, W.H. Young showed in 1903 that it is always possible to construct a function that has discontinuities precisely on this set. Exercise 4.3.14 gives some clues for how to do this in the simpler case of an arbitrary closed set, and Exercise 4.6.2 handles the case of an arbitrary countable set. Combining the techniques in these two exercises with the Dirichlet-type definitions we have seen leads to a proof of Young's result. (Try it!) A function demonstrating the converse for the monotone case described in Exercise 4.6.6 is also not too difficult to describe. Let

$$D = \{x_1, x_2, x_3, x_4, \ldots\}$$

be an arbitrary countable set of real numbers. In order to construct a monotone function that has discontinuities precisely on D, we first consider a particular $x_n \in D$ and define the step function

$$u_n(x) = \begin{cases} 1/2^n & \text{for } x > x_n \\ 0 & \text{for } x \le x_n. \end{cases}$$

Observing that each $u_n(x)$ is monotone and everywhere continuous except for a single discontinuity at x_n, we now set

$$f(x) = \sum_{n=1}^{\infty} u_n(x).$$

The convergence of the series $\sum 1/2^n$ guarantees that our function f is defined on all of **R**, and intuition certainly suggests that f is monotone with jump discontinuities precisely on D. Providing a rigorous proof for this conclusion is one of the many pleasures that awaits in Chapter 6, where we take up the study of infinite series of functions.

Chapter 5

The Derivative

5.1   Discussion: Are Derivatives Continuous?

The geometric motivation for the derivative is most likely familiar territory. Given a function $g(x)$, the derivative $g'(x)$ is understood to be the slope of the graph of g at each point x in the domain. A graphical picture (Fig. 5.1) reveals the impetus behind the mathematical definition

$$g'(c) = \lim_{x \to c} \frac{g(x) - g(c)}{x - c}.$$

The difference quotient $(g(x) - g(c))/(x - c)$ represents the slope of the line through the two points $(x, g(x))$ and $(c, g(c))$. By taking the limit as x approaches c, we arrive at a well-defined mathematical meaning for the slope of the tangent line at $x = c$.

The myriad applications of the derivative function are the topic of much of the calculus sequence, as well as several other upper-level courses in mathematics. None of these applied questions are pursued here in any length, but it should be pointed out that the rigorous underpinnings for differentiation worked out in this chapter are an essential foundation for any applied study. Eventually, as the derivative is subjected to more and more complex manipulations, it becomes crucial to know precisely how differentiation is defined and how it interacts with other mathematical operations.

Although physical applications are not explicitly discussed, we will encounter several questions of a more abstract quality as we develop the theory. Many of these are concerned with the relationship between differentiation and continuity. Are continuous functions always differentiable? If not, how nondifferentiable can a continuous function be? Are differentiable functions continuous? Given that

© Springer Science+Business Media New York 2015
S. Abbott, *Understanding Analysis*, Undergraduate Texts
in Mathematics, DOI 10.1007/978-1-4939-2712-8_5

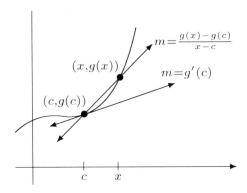

Figure 5.1: DEFINITION OF $g'(c)$.

a function f has a derivative at every point in its domain, what can we say about the function f'? Is f' continuous? How accurately can we describe the set of all possible derivatives, or are there no restrictions? Put another way, if we are given an arbitrary function g, is it always possible to find a differentiable function f such that $f' = g$, or are there some properties that g must possess for this to occur? In our study of continuity, we saw that restricting our attention to monotone functions had a significant impact on the answers to questions about sets of discontinuity. What effect, if any, does this same restriction have on our questions about potential sets of nondifferentiable points? Some of these issues are harder to resolve than others, and some remain unanswered in any satisfactory way.

A particularly useful class of examples for this discussion are functions of the form

$$g_n(x) = \begin{cases} x^n \sin(1/x) & \text{if } x \neq 0 \\ 0 & \text{if } x = 0. \end{cases}$$

When $n = 0$, we have seen (Example 4.2.6) that the oscillations of $\sin(1/x)$ prevent $g_0(x)$ from being continuous at $x = 0$. When $n = 1$, these oscillations are squeezed between $|x|$ and $-|x|$, the result being that g_1 is continuous at $x = 0$ (Example 4.3.6). Is $g_1'(0)$ defined? Using the preceding definition, we get

$$g_1'(0) = \lim_{x \to 0} \frac{g_1(x)}{x} = \lim_{x \to 0} \sin(1/x),$$

which, as we now know, does not exist. Thus, g_1 is not differentiable at $x = 0$. On the other hand, the same calculation shows that g_2 *is* differentiable at zero. In fact, we have

$$g_2'(0) = \lim_{x \to 0} x \sin(1/x) = 0.$$

At points different from zero, we can use the familiar rules of differentiation (soon to be justified) to conclude that g_2 is differentiable everywhere in $\mathbf{R}$ with

$$g_2'(x) = \begin{cases} -\cos(1/x) + 2x \sin(1/x) & \text{if } x \neq 0 \\ 0 & \text{if } x = 0. \end{cases}$$

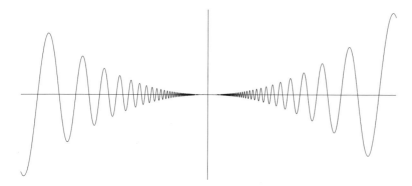

Figure 5.2: THE FUNCTION $g_2(x) = x^2 \sin(1/x)$ NEAR ZERO.

But now consider

$$\lim_{x \to 0} g_2'(x).$$

Because the $\cos(1/x)$ term is not preceded by a factor of x, we must conclude that this limit does not exist and that, consequently, the derivative function is not continuous. To summarize, the function $g_2(x)$ is continuous and differentiable everywhere on $\mathbf{R}$ (Fig. 5.2), the derivative function g_2' is thus defined everywhere on $\mathbf{R}$, but g_2' has a discontinuity at zero. The conclusion is that derivatives need not, in general, be continuous!

The discontinuity in g_2' is *essential*, meaning $\lim_{x \to 0} g'(x)$ does not exist as a one-sided limit. But, what about a function with a simple jump discontinuity? For example, does there exist a function h such that

$$h'(x) = \begin{cases} -1 & \text{if } x \leq 0 \\ 1 & \text{if } x > 0. \end{cases}$$

A first impression may bring to mind the absolute value function, which has slopes of -1 at points to the left of zero and slopes of 1 to the right. However, the absolute value function is *not* differentiable at zero. We are seeking a function that is differentiable everywhere, including the point zero, where we are insisting that the slope of the graph be -1. The degree of difficulty of this request should start to become apparent. Without sacrificing differentiability at any point, we are demanding that the slopes jump from -1 to 1 and not attain any value in between.

Although we have seen that continuity is not a required property of derivatives, the intermediate value property will prove a more stubborn quality to ignore.

5.2 Derivatives and the Intermediate Value Property

Although the definition would technically make sense for more complicated domains, all of the interesting results about the relationship between a function and its derivative require that the domain of the given function be an interval. Thinking geometrically of the derivative as a rate of change, it should not be too surprising that we would want to confine the independent variable to move about a connected domain.

The theory of functional limits from Section 4.2 is all that is needed to supply a rigorous definition for the derivative.

Definition 5.2.1 (Differentiability). Let $g : A \to \mathbf{R}$ be a function defined on an interval A. Given $c \in A$, the *derivative of g at c* is defined by

$$g'(c) = \lim_{x \to c} \frac{g(x) - g(c)}{x - c},$$

provided this limit exists. In this case we say g *is differentiable at c*. If g' exists for all points $c \in A$, we say that g *is differentiable on A*.

Example 5.2.2.   (i) Consider $f(x) = x^n$, where $n \in \mathbf{N}$, and let c be any arbitrary point in $\mathbf{R}$. Using the algebraic identity

$$x^n - c^n = (x - c)(x^{n-1} + cx^{n-2} + c^2 x^{n-3} + \cdots + c^{n-1}),$$

we can calculate the familiar formula

$$
\begin{aligned}
f'(c) = \lim_{x \to c} \frac{x^n - c^n}{x - c} &= \lim_{x \to c} (x^{n-1} + cx^{n-2} + c^2 x^{n-3} + \cdots + c^{n-1}) \\
&= c^{n-1} + c^{n-1} + \cdots + c^{n-1} = nc^{n-1}.
\end{aligned}
$$

(ii) If $g(x) = |x|$, then attempting to compute the derivative at $c = 0$ produces the limit

$$g'(0) = \lim_{x \to 0} \frac{|x|}{x},$$

which is $+1$ or -1 depending on whether x approaches zero from the right or left. Consequently, this limit does not exist, and we conclude that g is not differentiable at zero.

Example 5.2.2 (ii) is a reminder that continuity of g does not imply that g is necessarily differentiable. On the other hand, if g is differentiable at a point, then it is true that g must be continuous at this point.

Theorem 5.2.3. *If $g : A \to \mathbf{R}$ is differentiable at a point $c \in A$, then g is continuous at c as well.*

Proof. We are assuming that

$$g'(c) = \lim_{x \to c} \frac{g(x) - g(c)}{x - c}$$

exists, and we want to prove that $\lim_{x \to c} g(x) = g(c)$. But notice that the Algebraic Limit Theorem for functional limits allows us to write

$$\lim_{x \to c} (g(x) - g(c)) = \lim_{x \to c} \left(\frac{g(x) - g(c)}{x - c} \right) (x - c) = g'(c) \cdot 0 = 0.$$

It follows that $\lim_{x \to c} g(x) = g(c)$. $\qquad\qquad\qquad\qquad\qquad\qquad\qquad\square$

Combinations of Differentiable Functions

The Algebraic Limit Theorem (Theorem 2.3.3) led easily to the conclusion that algebraic combinations of continuous functions are continuous. With only slightly more work, we arrive at a similar conclusion for sums, products, and quotients of differentiable functions.

Theorem 5.2.4 (Algebraic Differentiability Theorem). *Let f and g be functions defined on an interval A, and assume both are differentiable at some point $c \in A$. Then,*

(i) $(f + g)'(c) = f'(c) + g'(c)$,

(ii) $(kf)'(c) = kf'(c)$, *for all* $k \in \mathbf{R}$,

(iii) $(fg)'(c) = f'(c)g(c) + f(c)g'(c)$, *and*

(iv) $(f/g)'(c) = \frac{g(c)f'(c) - f(c)g'(c)}{[g(c)]^2}$, *provided that* $g(c) \neq 0$.

Proof. Statements (i) and (ii) are left as exercises. To prove (iii), we rewrite the difference quotient as

$$\frac{(fg)(x) - (fg)(c)}{x - c} = \frac{f(x)g(x) - f(x)g(c) + f(x)g(c) - f(c)g(c)}{x - c}$$

$$= f(x) \left[\frac{g(x) - g(c)}{x - c} \right] + g(c) \left[\frac{f(x) - f(c)}{x - c} \right].$$

Because f is differentiable at c, it is continuous there and thus $\lim_{x \to c} f(x) = f(c)$. This fact, together with the functional-limit version of the Algebraic Limit Theorem (Theorem 4.2.4), justifies the conclusion

$$\lim_{x \to c} \frac{(fg)(x) - (fg)(c)}{x - c} = f(c)g'(c) + f'(c)g(c).$$

A similar proof of (iv) is possible, or we can use an argument based on the next result. Each of these options is discussed in Exercise 5.2.3. $\qquad\qquad\square$

The composition of two differentiable functions also fortunately results in another differentiable function. This fact is referred to as the *Chain Rule*. To discover the proper formula for the derivative of the composition $g \circ f$, we can write

$$
\begin{aligned}
(g \circ f)'(c) = \lim_{x \to c} \frac{g(f(x)) - g(f(c))}{x - c} &= \lim_{x \to c} \frac{g(f(x)) - g(f(c))}{f(x) - f(c)} \cdot \frac{f(x) - f(c)}{x - c} \\
&= g'(f(c)) \cdot f'(c).
\end{aligned}
$$

With a little polish, this string of equations could qualify as a proof except for the pesky fact that the $f(x) - f(c)$ expression causes problems in the denominator if $f(x) = f(c)$ for x values in arbitrarily small neighborhoods of c. (The function $g_2(x)$ discussed in Section 5.1 exhibits this behavior near $c = 0$.) The upcoming proof of the Chain Rule manages to finesse this problem but in content is essentially the argument just given. Another approach is sketched in Exercise 5.2.4.

Theorem 5.2.5 (Chain Rule). *Let $f : A \to \mathbf{R}$ and $g : B \to \mathbf{R}$ satisfy $f(A) \subseteq B$ so that the composition $g \circ f$ is defined. If f is differentiable at $c \in A$ and if g is differentiable at $f(c) \in B$, then $g \circ f$ is differentiable at c with $(g \circ f)'(c) = g'(f(c)) \cdot f'(c)$.*

Proof. Because g is differentiable at $f(c)$, we know that

$$
g'(f(c)) = \lim_{y \to f(c)} \frac{g(y) - g(f(c))}{y - f(c)}.
$$

Another way to assert this same fact is to let $d(y)$ be the difference quotient

$$
(1) \qquad\qquad d(y) = \frac{g(y) - g(f(c))}{y - f(c)},
$$

and observe that $\lim_{y \to f(c)} d(y) = g'(f(c))$. At the moment, $d(y)$ is not defined when $y = f(c)$, but it should seem natural to declare that $d(f(c)) = g'(f(c))$, so that d is continuous at $f(c)$.

Now, we come to the finesse. Equation (1) can be rewritten as

$$
(2) \qquad\qquad g(y) - g(f(c)) = d(y)(y - f(c)).
$$

Observe that this equation holds *for all $y \in B$ including $y = f(c)$.* Thus, we are free to substitute $y = f(t)$ for any arbitrary $t \in A$. If $t \neq c$, we can divide equation (2) by $(t - c)$ to get

$$
\frac{g(f(t)) - g(f(c))}{t - c} = d(f(t)) \frac{(f(t) - f(c))}{t - c}
$$

for all $t \neq c$. Finally, taking the limit as $t \to c$ and applying the Algebraic Limit Theorem together with Theorem 4.3.9 yields the desired formula. $\qquad\square$

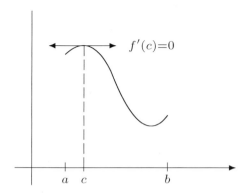

Figure 5.3: THE INTERIOR EXTREMUM THEOREM.

Darboux's Theorem

One conclusion from this chapter's introduction is that although continuity is necessary for the derivative to exist, it is not the case that the derivative function itself will always be continuous. Our specific example was $g_2(x) = x^2 \sin(1/x)$, where we set $g_2(0) = 0$. By tinkering with the exponent of the leading x^2 factor, it is possible to construct examples of differentiable functions with derivatives that are unbounded, or twice-differentiable functions that have discontinuous second derivatives (Exercise 5.2.7). The underlying principle in all of these examples is that by controlling the size of the oscillations of the original function, we can make the corresponding oscillations of the slopes volatile enough to prevent the existence of the relevant limits.

It is significant that for this class of examples, the discontinuities that arise are never simple jump discontinuities. (A precise definition of "jump discontinuity" is presented in Section 4.6.) We are now ready to confirm our earlier suspicions that although derivatives do not in general have to be continuous, they do possess the intermediate value property. (See Definition 4.5.3.) This surprising observation is a fairly straightforward corollary to the more obvious observation that differentiable functions attain maximums and minimums only at points where the derivative is equal to zero (Fig. 5.3).

Theorem 5.2.6 (Interior Extremum Theorem). *Let f be differentiable on an open interval (a, b). If f attains a maximum value at some point $c \in (a, b)$ (i.e., $f(c) \geq f(x)$ for all $x \in (a, b)$), then $f'(c) = 0$. The same is true if $f(c)$ is a minimum value.*

Proof. Because c is in the open interval (a, b), we can construct two sequences (x_n) and (y_n), which converge to c and satisfy $x_n < c < y_n$ for all $n \in \mathbf{N}$. The fact that $f(c)$ is a maximum implies that $f(y_n) - f(c) \leq 0$ for all n, and thus

$$f'(c) = \lim_{n \to \infty} \frac{f(y_n) - f(c)}{y_n - c} \leq 0$$

by the Order Limit Theorem (Theorem 2.3.4). In a similar way,

$$\frac{f(x_n) - f(c)}{x_n - c} \geq 0$$

for each x_n because both numerator and denominator are negative. This implies that

$$f'(c) = \lim_{n \to \infty} \frac{f(x_n) - f(c)}{x_n - c} \geq 0,$$

and therefore $f'(c) = 0$, as desired. $\square$

The Interior Extremum Theorem is the fundamental fact behind the use of the derivative as a tool for solving applied optimization problems. This idea, discovered and exploited by Pierre de Fermat, is as old as the derivative itself. In a sense, finding maximums and minimums is arguably why Fermat invented his method of finding slopes of tangent lines. It was 200 years later that the French mathematician Gaston Darboux (1842–1917) pointed out that Fermat's method of finding maximums and minimums carries with it the implication that if a derivative function attains two distinct values $f'(a)$ and $f'(b)$, then it must also attain every value in between.

Theorem 5.2.7 (Darboux's Theorem). *If f is differentiable on an interval $[a, b]$, and if α satisfies $f'(a) < \alpha < f'(b)$ (or $f'(a) > \alpha > f'(b)$), then there exists a point $c \in (a, b)$ where $f'(c) = \alpha$.*

Proof. We first simplify matters by defining a new function $g(x) = f(x) - \alpha x$ on $[a, b]$. Notice that g is differentiable on $[a, b]$ with $g'(x) = f'(x) - \alpha$. In terms of g, our hypothesis states that $g'(a) < 0 < g'(b)$, and we hope to show that $g'(c) = 0$ for some $c \in (a, b)$.

The remainder of the argument is outlined in Exercise 5.2.11. $\square$

Exercises

Exercise 5.2.1. Supply proofs for parts (i) and (ii) of Theorem 5.2.4.

Exercise 5.2.2. Exactly one of the following requests is impossible. Decide which it is, and provide examples for the other three. In each case, let's assume the functions are defined on all of **R**.

(a) Functions f and g not differentiable at zero but where fg is differentiable at zero.

(b) A function f not differentiable at zero and a function g differentiable at zero where fg is differentiable at zero.

(c) A function f not differentiable at zero and a function g differentiable at zero where $f + g$ is differentiable at zero.

(d) A function f differentiable at zero but not differentiable at any other point.

Exercise 5.2.3. (a) Use Definition 5.2.1 to produce the proper formula for the derivative of $h(x) = 1/x$.

(b) Combine the result in part (a) with the Chain Rule (Theorem 5.2.5) to supply a proof for part (iv) of Theorem 5.2.4.

(c) Supply a direct proof of Theorem 5.2.4 (iv) by algebraically manipulating the difference quotient for (f/g) in a style similar to the proof of Theorem 5.2.4 (iii).

Exercise 5.2.4. Follow these steps to provide a slightly modified proof of the Chain Rule.

(a) Show that a function $h : A \to \mathbf{R}$ is differentiable at $a \in A$ if and only if there exists a function $l : A \to \mathbf{R}$ which is continuous at a and satisfies

$$h(x) - h(a) = l(x)(x - a) \qquad \text{for all } x \in A.$$

(b) Use this criterion for differentiability (in both directions) to prove Theorem 5.2.5.

Exercise 5.2.5. Let $f_a(x) = \begin{cases} x^a & \text{if } x > 0 \\ 0 & \text{if } x \le 0. \end{cases}$

(a) For which values of a is f continuous at zero?

(b) For which values of a is f differentiable at zero? In this case, is the derivative function continuous?

(c) For which values of a is f twice-differentiable?

Exercise 5.2.6. Let g be defined on an interval A, and let $c \in A$.

(a) Explain why $g'(c)$ in Definition 5.2.1 could have been given by

$$g'(c) = \lim_{h \to 0} \frac{g(c + h) - g(c)}{h}.$$

(b) Assume A is open. If g is differentiable at $c \in A$, show

$$g'(c) = \lim_{h \to 0} \frac{g(c + h) - g(c - h)}{2h}.$$

Exercise 5.2.7. Let

$$g_a(x) = \begin{cases} x^a \sin(1/x) & \text{if } x \ne 0 \\ 0 & \text{if } x = 0. \end{cases}$$

Find a particular (potentially noninteger) value for a so that

(a) g_a is differentiable on $\mathbf{R}$ but such that g_a' is unbounded on $[0, 1]$.

(b) g_a is differentiable on $\mathbf{R}$ with g'_a continuous but not differentiable at zero.

(c) g_a is differentiable on $\mathbf{R}$ and g'_a is differentiable on $\mathbf{R}$, but such that g''_a is not continuous at zero.

Exercise 5.2.8. Review the definition of uniform continuity (Definition 4.4.4). Given a differentiable function $f : A \to \mathbf{R}$, let's say that f is *uniformly differentiable* on A if, given $\epsilon > 0$ there exists a $\delta > 0$ such that

$$\left| \frac{f(x) - f(y)}{x - y} - f'(y) \right| < \epsilon \quad \text{whenever } 0 < |x - y| < \delta.$$

(a) Is $f(x) = x^2$ uniformly differentiable on $\mathbf{R}$? How about $g(x) = x^3$?

(b) Show that if a function is uniformly differentiable on an interval A, then the derivative must be continuous on A.

(c) Is there a theorem analogous to Theorem 4.4.7 for differentiation? Are functions that are differentiable on a closed interval $[a, b]$ necessarily uniformly differentiable?

Exercise 5.2.9. Decide whether each conjecture is true or false. Provide an argument for those that are true and a counterexample for each one that is false.

(a) If f' exists on an interval and is not constant, then f' must take on some irrational values.

(b) If f' exists on an open interval and there is some point c where $f'(c) > 0$, then there exists a δ-neighborhood $V_\delta(c)$ around c in which $f'(x) > 0$ for all $x \in V_\delta(c)$.

(c) If f is differentiable on an interval containing zero and if $\lim_{x \to 0} f'(x) = L$, then it must be that $L = f'(0)$.

Exercise 5.2.10. Recall that a function $f : (a, b) \to \mathbf{R}$ is *increasing* on (a, b) if $f(x) \leq f(y)$ whenever $x < y$ in (a, b). A familiar mantra from calculus is that a differentiable function is increasing if its derivative is positive, but this statement requires some sharpening in order to be completely accurate.

Show that the function

$$g(x) = \begin{cases} x/2 + x^2 \sin(1/x) & \text{if } x \neq 0 \\ 0 & \text{if } x = 0 \end{cases}$$

is differentiable on $\mathbf{R}$ and satisfies $g'(0) > 0$. Now, prove that g is *not* increasing over any open interval containing 0.

In the next section we will see that f is indeed increasing on (a, b) if and only if $f'(x) \geq 0$ for all $x \in (a, b)$.

Exercise 5.2.11. Assume that g is differentiable on $[a, b]$ and satisfies $g'(a) < 0 < g'(b)$.

(a) Show that there exists a point $x \in (a, b)$ where $g(a) > g(x)$, and a point $y \in (a, b)$ where $g(y) < g(b)$.

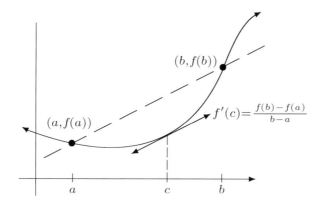

Figure 5.4: THE MEAN VALUE THEOREM.

(b) Now complete the proof of Darboux's Theorem started earlier.

Exercise 5.2.12 (Inverse functions). If $f : [a, b] \to \mathbf{R}$ is one-to-one, then there exists an inverse function f^{-1} defined on the range of f given by $f^{-1}(y) = x$ where $y = f(x)$. In Exercise 4.5.8 we saw that if f is continuous on $[a, b]$, then f^{-1} is continuous on its domain. Let's add the assumption that f is differentiable on $[a, b]$ with $f'(x) \neq 0$ for all $x \in [a, b]$. Show f^{-1} is differentiable with

$$\left(f^{-1}\right)'(y) = \frac{1}{f'(x)} \qquad \text{where } y = f(x).$$

5.3 The Mean Value Theorems

The Mean Value Theorem (Fig. 5.4) makes the geometrically plausible assertion that a differentiable function f on an interval $[a, b]$ will, at some point, attain a slope equal to the slope of the line through the endpoints $(a, f(a))$ and $(b, f(b))$. More tersely put,

$$f'(c) = \frac{f(b) - f(a)}{b - a}$$

for at least one point $c \in (a, b)$.

On the surface, there does not appear to be anything especially remarkable about this observation. Its validity appears undeniable—much like the Intermediate Value Theorem for continuous functions—and its proof is rather short. The ease of the proof, however, is misleading, as it is built on top of some hard-fought accomplishments from the study of limits and continuity. In this regard, the Mean Value Theorem is a kind of reward for a job well done. As we will see, it is a prize of exceptional value. Although the result itself is geometrically unsurprising, the Mean Value Theorem is the cornerstone of the proof for almost every major theorem pertaining to differentiation. We will use it to prove L'Hospital's rules regarding limits of quotients of differentiable functions.

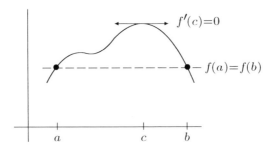

Figure 5.5: ROLLE'S THEOREM.

A rigorous analysis of how infinite series of functions behave when differentiated requires the Mean Value Theorem (Theorem 6.4.3), and it is the crucial step in the proof of the Fundamental Theorem of Calculus (Theorem 7.5.1). It is also the fundamental concept underlying Lagrange's Remainder Theorem (Theorem 6.6.3) which approximates the error between a Taylor polynomial and the function that generates it.

The Mean Value Theorem can be stated in various degrees of generality, each one important enough to be given its own special designation. Recall that the Extreme Value Theorem (Theorem 4.4.2) states that continuous functions on compact sets always attain maximum and minimum values. Combining this observation with the Interior Extremum Theorem for differentiable functions (Theorem 5.2.6) yields a special case of the Mean Value Theorem first noted by the mathematician Michel Rolle (1652–1719) (Fig. 5.5).

Theorem 5.3.1 (Rolle's Theorem). *Let $f : [a, b] \to \mathbf{R}$ be continuous on $[a, b]$ and differentiable on (a, b). If $f(a) = f(b)$, then there exists a point $c \in (a, b)$ where $f'(c) = 0$.*

Proof. Because f is continuous on a compact set, f attains a maximum and a minimum. If both the maximum and minimum occur at the endpoints, then f is necessarily a constant function and $f'(x) = 0$ on all of (a, b). In this case, we can choose c to be any point we like. On the other hand, if either the maximum or minimum occurs at some point c in the interior (a, b), then it follows from the Interior Extremum Theorem (Theorem 5.2.6) that $f'(c) = 0$. ☐

Theorem 5.3.2 (Mean Value Theorem). *If $f : [a, b] \to \mathbf{R}$ is continuous on $[a, b]$ and differentiable on (a, b), then there exists a point $c \in (a, b)$ where*

$$f'(c) = \frac{f(b) - f(a)}{b - a}.$$

Proof. Notice that the Mean Value Theorem reduces to Rolle's Theorem in the case where $f(a) = f(b)$. The strategy of the proof is to reduce the more general statement to this special case.

The equation of the line through $(a, f(a))$ and $(b, f(b))$ is

$$y = \left(\frac{f(b) - f(a)}{b - a} \right)(x - a) + f(a).$$

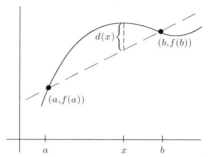

We want to consider the difference between this line and the function $f(x)$. To this end, let

$$d(x) = f(x) - \left[\left(\frac{f(b) - f(a)}{b - a} \right)(x - a) + f(a) \right],$$

and observe that d is continuous on $[a, b]$, differentiable on (a, b), and satisfies $d(a) = 0 = d(b)$. Thus, by Rolle's Theorem, there exists a point $c \in (a, b)$ where $d'(c) = 0$. Because

$$d'(x) = f'(x) - \frac{f(b) - f(a)}{b - a},$$

we get

$$0 = f'(c) - \frac{f(b) - f(a)}{b - a},$$

which completes the proof. $\qquad \square$

The point has been made that the Mean Value Theorem manages to find its way into nearly every proof of any statement related to the geometrical nature of the derivative. As a simple example, if f is a constant function $f(x) = k$ on some interval A, then a straightforward calculation of f' using Definition 5.2.1 shows that $f'(x) = 0$ for all $x \in A$. But how do we prove the converse statement? If we know that a differentiable function g satisfies $g'(x) = 0$ everywhere on A, our intuition suggests that we should be able to prove $g(x)$ is constant. It is the Mean Value Theorem that provides us with a way to articulate rigorously what seems geometrically valid.

Corollary 5.3.3. *If $g : A \to \mathbf{R}$ is differentiable on an interval A and satisfies $g'(x) = 0$ for all $x \in A$, then $g(x) = k$ for some constant $k \in \mathbf{R}$.*

Proof. Take $x, y \in A$ and assume $x < y$. Applying the Mean Value Theorem to g on the interval $[x, y]$, we see that

$$g'(c) = \frac{g(y) - g(x)}{y - x}$$

for some $c \in A$. Now, $g'(c) = 0$, so we conclude that $g(y) = g(x)$. Set k equal to this common value. Because x and y are arbitrary, it follows that $g(x) = k$ for all $x \in A$.   □

Corollary 5.3.4. *If f and g are differentiable functions on an interval A and satisfy $f'(x) = g'(x)$ for all $x \in A$, then $f(x) = g(x) + k$ for some constant $k \in \mathbf{R}$.*

Proof. Let $h(x) = f(x) - g(x)$ and apply Corollary 5.3.3 to the differentiable function h.   □

The Mean Value Theorem has a more general form due to Cauchy. It is this generalized version of the theorem that is needed to analyze L'Hospital's rules and Lagrange's Remainder Theorem.

Theorem 5.3.5 (Generalized Mean Value Theorem). *If f and g are continuous on the closed interval $[a, b]$ and differentiable on the open interval (a, b), then there exists a point $c \in (a, b)$ where*

$$[f(b) - f(a)]g'(c) = [g(b) - g(a)]f'(c).$$

If g' is never zero on (a, b), then the conclusion can be stated as

$$\frac{f'(c)}{g'(c)} = \frac{f(b) - f(a)}{g(b) - g(a)}.$$

Proof. This result follows by applying the Mean Value Theorem to the function $h(x) = [f(b) - f(a)]g(x) - [g(b) - g(a)]f(x)$. The details are requested in Exercise 5.3.5.   □

L'Hospital's Rules

The Algebraic Limit Theorem asserts that when taking a limit of a quotient of functions we can write

$$\lim_{x \to c} \frac{f(x)}{g(x)} = \frac{\lim_{x \to c} f(x)}{\lim_{x \to c} g(x)},$$

provided that each individual limit exists and $\lim_{x \to c} g(x)$ is not zero. If the denominator does converge to zero and the numerator has a nonzero limit, then it is not difficult to argue that the quotient $f(x)/g(x)$ grows in absolute value without bound as x approaches c. L'Hospital's Rules are named for the Marquis de L'Hospital (1661–1704), who learned the results from his tutor, Johann Bernoulli (1667–1748), and published them in 1696 in what is regarded as the first calculus text. Stated in different levels of generality, they are an effective tool for handling the indeterminant cases when either numerator and denominator both tend to zero or both tend simultaneously to infinity.

Theorem 5.3.6 (L'Hospital's Rule: 0/0 case). *Let f and g be continuous on an interval containing a, and assume f and g are differentiable on this interval with the possible exception of the point a. If $f(a) = g(a) = 0$ and $g'(x) \neq 0$ for all $x \neq a$, then*

$$\lim_{x \to a} \frac{f'(x)}{g'(x)} = L \quad implies \quad \lim_{x \to a} \frac{f(x)}{g(x)} = L.$$

Proof. This argument follows from a straightforward application of the Generalized Mean Value Theorem. It is requested as Exercise 5.3.11. $\qquad\square$

L'Hospital's Rule remains true if we replace the assumption that $f(a) = g(a) = 0$ with the hypothesis that $\lim_{x \to a} g(x) = \infty$. To this point we have not been explicit about what it means to say that a limit equals ∞. The logical structure of such a definition is precisely the same as it is for finite functional limits. The difference is that rather than trying to force the function to take on values in some small ϵ-neighborhood around a proposed limit, we must show that $g(x)$ eventually *exceeds* any proposed upper bound. The arbitrarily small $\epsilon > 0$ is replaced by an arbitrarily large $M > 0$.

Definition 5.3.7. Given $g : A \to \mathbf{R}$ and a limit point c of A, we say that $\lim_{x \to c} g(x) = \infty$ if, for every $M > 0$, there exists a $\delta > 0$ such that whenever $0 < |x - c| < \delta$ it follows that $g(x) \geq M$.
    We can define $\lim_{x \to c} g(x) = -\infty$ in a similar way.

The following version of L'Hospital's Rule is typically referred to as the ∞/∞ case even though the hypothesis only requires that the function in the denominator tend to infinity. To simplify the notation of the proof, we state the result using a one-sided limit.

Theorem 5.3.8 (L'Hospital's Rule: ∞/∞ case). *Assume f and g are differentiable on (a, b) and that $g'(x) \neq 0$ for all $x \in (a, b)$. If $\lim_{x \to a} g(x) = \infty$ (or $-\infty$), then*

$$\lim_{x \to a} \frac{f'(x)}{g'(x)} = L \quad implies \quad \lim_{x \to a} \frac{f(x)}{g(x)} = L.$$

Proof. Let $\epsilon > 0$. Because $\lim_{x \to a} \frac{f'(x)}{g'(x)} = L$, there exists a $\delta_1 > 0$ such that

$$\left| \frac{f'(x)}{g'(x)} - L \right| < \frac{\epsilon}{2}$$

for all $a < x < a + \delta_1$. For convenience of notation, let $t = a + \delta_1$ and note that t is fixed for the remainder of the argument.
    Our functions are not defined at a, but for any $x \in (a, t)$ we can apply the Generalized Mean Value Theorem on the interval $[x, t]$ to get

$$\frac{f(x) - f(t)}{g(x) - g(t)} = \frac{f'(c)}{g'(c)}$$

for some $c \in (x, t)$. Our choice of t then implies

(1)
$$L - \frac{\epsilon}{2} < \frac{f(x) - f(t)}{g(x) - g(t)} < L + \frac{\epsilon}{2}$$

for all x in (a, t).

In an effort to isolate the fraction $\frac{f(x)}{g(x)}$, the strategy is to multiply inequality (1) by $(g(x) - g(t))/g(x)$. We need to be sure, however, that this quantity is positive, which amounts to insisting that $1 \geq g(t)/g(x)$. Because t is fixed and $\lim_{x \to a} g(x) = \infty$, we can choose $\delta_2 > 0$ so that $g(x) \geq g(t)$ for all $a < x < a + \delta_2$. Carrying out the desired multiplication results in

$$\left(L - \frac{\epsilon}{2}\right)\left(1 - \frac{g(t)}{g(x)}\right) < \frac{f(x) - f(t)}{g(x)} < \left(L + \frac{\epsilon}{2}\right)\left(1 - \frac{g(t)}{g(x)}\right),$$

which after some algebraic manipulations yields

$$L - \frac{\epsilon}{2} + \frac{-Lg(t) + \frac{\epsilon}{2}g(t) + f(t)}{g(x)} < \frac{f(x)}{g(x)} < L + \frac{\epsilon}{2} + \frac{-Lg(t) - \frac{\epsilon}{2}g(t) + f(t)}{g(x)}.$$

Again, let's remind ourselves that t is fixed and that $\lim_{x \to a} g(x) = \infty$. Thus, we can choose a δ_3 such that $a < x < a + \delta_3$ implies that $g(x)$ is large enough to ensure that both

$$\frac{-Lg(t) + \frac{\epsilon}{2}g(t) + f(t)}{g(x)} \quad \text{and} \quad \frac{-Lg(t) - \frac{\epsilon}{2}g(t) + f(t)}{g(x)}$$

are less than $\epsilon/2$ in absolute value. Putting this all together and choosing $\delta = \min\{\delta_1, \delta_2, \delta_3\}$ guarantees that

$$\left| \frac{f(x)}{g(x)} - L \right| < \epsilon$$

for all $a < x < a + \delta$. $\square$

Exercises

Exercise 5.3.1. Recall from Exercise 4.4.9 that a function $f : A \to \mathbf{R}$ is Lipschitz on A if there exists an $M > 0$ such that

$$\left| \frac{f(x) - f(y)}{x - y} \right| \leq M$$

for all $x \neq y$ in A.

(a) Show that if f is differentiable on a closed interval $[a, b]$ and if f' is continuous on $[a, b]$, then f is Lipschitz on $[a, b]$.

(b) Review the definition of a contractive function in Exercise 4.3.11. If we add the assumption that $|f'(x)| < 1$ on $[a, b]$, does it follow that f is contractive on this set?

Exercise 5.3.2. Let f be differentiable on an interval A. If $f'(x) \neq 0$ on A, show that f is one-to-one on A. Provide an example to show that the converse statement need not be true.

Exercise 5.3.3. Let h be a differentiable function defined on the interval $[0, 3]$, and assume that $h(0) = 1$, $h(1) = 2$, and $h(3) = 2$.

(a) Argue that there exists a point $d \in [0, 3]$ where $h(d) = d$.

(b) Argue that at some point c we have $h'(c) = 1/3$.

(c) Argue that $h'(x) = 1/4$ at some point in the domain.

Exercise 5.3.4. Let f be differentiable on an interval A containing zero, and assume (x_n) is a sequence in A with $(x_n) \to 0$ and $x_n \neq 0$.

(a) If $f(x_n) = 0$ for all $n \in N$, show $f(0) = 0$ and $f'(0) = 0$.

(b) Add the assumption that f is twice-differentiable at zero and show that $f''(0) = 0$ as well.

Exercise 5.3.5. (a) Supply the details for the proof of Cauchy's Generalized Mean Value Theorem (Theorem 5.3.5).

(b) Give a graphical interpretation of the Generalized Mean Value Theorem analogous to the one given for the Mean Value Theorem at the beginning of Section 5.3. (Consider f and g as parametric equations for a curve.)

Exercise 5.3.6. (a) Let $g : [0, a] \to \mathbf{R}$ be differentiable, $g(0) = 0$, and $|g'(x)| \leq M$ for all $x \in [0, a]$. Show $|g(x)| \leq Mx$ for all $x \in [0, a]$.

(b) Let $h : [0, a] \to \mathbf{R}$ be twice differentiable, $h'(0) = h(0) = 0$ and $|h''(x)| \leq M$ for all $x \in [0, a]$. Show $|h(x)| \leq Mx^2/2$ for all $x \in [0, a]$.

(c) Conjecture and prove an analogous result for a function that is differentiable three times on $[0, a]$.

Exercise 5.3.7. A *fixed point* of a function f is a value x where $f(x) = x$. Show that if f is differentiable on an interval with $f'(x) \neq 1$, then f can have at most one fixed point.

Exercise 5.3.8. Assume f is continuous on an interval containing zero and differentiable for all $x \neq 0$. If $\lim_{x \to 0} f'(x) = L$, show $f'(0)$ exists and equals L.

Exercise 5.3.9. Assume f and g are as described in Theorem 5.3.6, but now add the assumption that f and g are differentiable at a, and f' and g' are continuous at a with $g'(a) \neq 0$. Find a short proof for the 0/0 case of L'Hospital's Rule under this stronger hypothesis.

Exercise 5.3.10. Let $f(x) = x \sin(1/x^4)e^{-1/x^2}$ and $g(x) = e^{-1/x^2}$. Using the familiar properties of these functions, compute the limit as x approaches zero of $f(x)$, $g(x)$, $f(x)/g(x)$, and $f'(x)/g'(x)$. Explain why the results are surprising but not in conflict with the content of Theorem 5.3.6.[1]

Exercise 5.3.11.   (a) Use the Generalized Mean Value Theorem to furnish a proof of the 0/0 case of L'Hospital's Rule (Theorem 5.3.6).

(b) If we keep the first part of the hypothesis of Theorem 5.3.6 the same but we assume that

$$\lim_{x \to a} \frac{f'(x)}{g'(x)} = \infty,$$

does it necessarily follow that

$$\lim_{x \to a} \frac{f(x)}{g(x)} = \infty?$$

Exercise 5.3.12. If f is twice differentiable on an open interval containing a and f'' is continuous at a, show

$$\lim_{h \to 0} \frac{f(a+h) - 2f(a) + f(a-h)}{h^2} = f''(a).$$

(Compare this to Exercise 5.2.6(b).)

5.4   A Continuous Nowhere-Differentiable Function

Exploring the relationship between continuity and differentiability has led to both fruitful results and pathological counterexamples. The bulk of discussion to this point has focused on the continuity of derivatives, but historically a significant amount of debate revolved around the question of whether continuous functions were necessarily differentiable. Early in the chapter, we saw that continuity was a requirement for differentiability, but, as the absolute value function demonstrates, the converse of this proposition is not true. A function can be continuous but not differentiable at some point. But just how nondifferentiable can a continuous function be? Given a *finite* set of points, it is not difficult to imagine how to construct a graph with corners at each of these points, so that the corresponding function fails to be differentiable on this finite set. The trick gets more difficult, however, when the set becomes infinite. For instance, is it possible to construct a function that is continuous on all of **R** but fails to be differentiable at every rational point? Not only is this possible, but the situation is even more disconcerting. In 1872, Karl Weierstrass presented an example of a continuous function that was not differentiable at *any* point. (It seems to be

---

[1] A large class of "counterexamples" of this sort to L'Hospital's Rule are explored in [4].

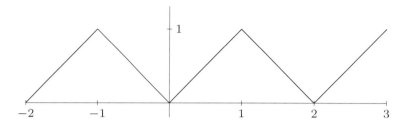

Figure 5.6: THE FUNCTION $h(x)$.

the case that Bernhard Bolzano had his own example of such a beast as early as 1830, but it was not published until much later.)

Weierstrass actually discovered a class of nowhere-differentiable functions of the form

$$f(x) = \sum_{n=0}^{\infty} a^n \cos(b^n x)$$

where the values of a and b are carefully chosen. Such functions are specific examples of Fourier series discussed in Section 8.5. The details of Weierstrass' argument are simplified if we replace the cosine function with a piecewise linear function that has oscillations qualitatively like $\cos(x)$.

Define

$$h(x) = |x|$$

on the interval $[-1, 1]$ and extend the definition of h to all of $\mathbf{R}$ by requiring that $h(x + 2) = h(x)$. The result is a periodic "sawtooth" function (Fig. 5.6).

Exercise 5.4.1. Sketch a graph of $(1/2)h(2x)$ on $[-2, 3]$. Give a qualitative description of the functions

$$h_n(x) = \frac{1}{2^n} h(2^n x)$$

as n gets larger.

Now, define

$$g(x) = \sum_{n=0}^{\infty} h_n(x) = \sum_{n=0}^{\infty} \frac{1}{2^n} h(2^n x).$$

The claim is that $g(x)$ is continuous on all of $\mathbf{R}$ but fails to be differentiable at any point.

Infinite Series of Functions and Continuity

The definition of $g(x)$ is a significant departure from the way we usually define functions. For each $x \in \mathbf{R}$, $g(x)$ is defined to be the value of an infinite series.

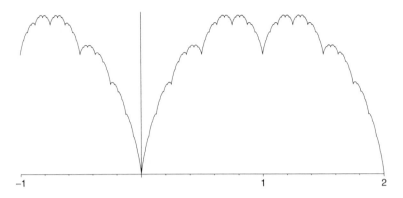

Figure 5.7:   A SKETCH OF $g(x) = \sum_{n=0}^{\infty}(1/2^n)h(2^n x)$.

Exercise 5.4.2. Fix $x \in \mathbf{R}$. Argue that the series

$$\sum_{n=0}^{\infty}\frac{1}{2^n}h(2^n x)$$

converges and thus $g(x)$ is properly defined.

Exercise 5.4.3. Taking the continuity of $h(x)$ as given, reference the proper theorems from Chapter 4 that imply that the *finite* sum

$$g_m(x) = \sum_{n=0}^{m}\frac{1}{2^n}h(2^n x)$$

is continuous on $\mathbf{R}$.

   This brings us to an archetypical question in analysis: When do conclusions that are valid in finite settings extend to infinite ones? A finite sum of continuous functions is certainly continuous, but does this necessarily hold for an infinite sum of continuous functions? In general, we will see that this is *not* always the case. For this particular sum, however, the continuity of the limit function $g(x)$ can be proved. Deciphering when results about finite sums of functions extend to infinite sums is one of the fundamental themes of Chapter 6. Although a self-contained argument for the continuity of g is not beyond our means at this point, we will nevertheless postpone the proof (see, for example, Exercise 6.4.3), leaving it as an enticement for the upcoming study of uniform convergence.

Exercise 5.4.4. As the graph in Figure 5.7 suggests, the structure of $g(x)$ is quite intricate. Answer the following questions, assuming that $g(x)$ is indeed continuous.

   (a) How do we know g attains a maximum value M on $[0, 2]$? What is this
       value?

(b) Let D be the set of points in $[0, 2]$ where g attains its maximum. That is $D = \{x \in [0, 2] : g(x) = M\}$. Find one point in D.

(c) Is D finite, countable, or uncountable?

Nondifferentiability

When the proper tools are in place, the proof that g is continuous is quite straightforward. The more difficult task is to show that g is not differentiable at any point in $\mathbf{R}$.

Let's first look at the point $x = 0$. Our function g does not appear to be differentiable here, and a rigorous proof is not too difficult. Consider the sequence $x_m = 1/2^m$, where $m = 0, 1, 2, \ldots$.

Exercise 5.4.5. Show that

$$\frac{g(x_m) - g(0)}{x_m - 0} = m + 1,$$

and use this to prove that $g'(0)$ does not exist.

Any temptation to say something like $g'(0) = \infty$ should be resisted. Setting $x_m = -(1/2^m)$ in the previous argument produces difference quotients heading toward $-\infty$. The geometric manifestation of this is the "cusp" that appears at $x = 0$ in the graph of g.

Exercise 5.4.6. (a) Modify the previous argument to show that $g'(1)$ does not exist. Show that $g'(1/2)$ does not exist.

(b) Show that $g'(x)$ does not exist for any rational number of the form $x = p/2^k$ where $p \in \mathbf{Z}$ and $k \in \mathbf{N} \cup \{0\}$.

The points described in Exercise 5.4.6 (b) are called *dyadic* points. If $x = p/2^k$ is a dyadic rational number, then the function h_n has a corner at x as long as $n \geq k$. Thus, it should not be too surprising that g fails to be differentiable at points of this form. The argument is more delicate at points between the dyadic points.

Assume x is *not* a dyadic number. For a fixed value of $m \in \mathbf{N} \cup \{0\}$, x falls between two adjacent dyadic points,

$$\frac{p_m}{2^m} < x < \frac{p_m + 1}{2^m}.$$

Set $x_m = p_m/2^m$ and $y_m = (p_m + 1)/2^m$. Repeating this for each m yields two sequences (x_m) and (y_m) satisfying

$$\lim x_m = \lim y_m = x \quad \text{and} \quad x_m < x < y_m.$$

Exercise 5.4.7. (a) First prove the following general lemma: Let f be defined on an open interval J and assume f is differentiable at $a \in J$. If (a_n) and (b_n) are sequences satisfying $a_n < a < b_n$ and $\lim a_n = \lim b_n = a$, show

$$f'(a) = \lim_{n \to \infty} \frac{f(b_n) - f(a_n)}{b_n - a_n}.$$

(b) Now use this lemma to show that $g'(x)$ does not exist.

Weierstrass's original 1872 paper contained a demonstration that the infinite sum

$$f(x) = \sum_{n=0}^{\infty} a^n \cos(b^n x)$$

defined a continuous nowhere-differentiable function provided $0 < a < 1$ and b was an odd integer satisfying $ab > 1 + 3\pi/2$. The condition on a is easy to understand. If $0 < a < 1$, then $\sum_{n=0}^{\infty} a^n$ is a convergent geometric series, and the forthcoming Weierstrass M-Test (Theorem 6.4.5) can be used to conclude that f is continuous. The restriction on b is more mysterious. In 1916, G.H. Hardy extended Weierstrass' result to include any value of b for which $ab \geq 1$. Without looking at the details of either of these arguments, we nevertheless get a sense that the lack of a derivative is intricately tied to the relationship between the compression factor (the parameter a) and the rate at which the frequency of the oscillations increases (the parameter b).

Exercise 5.4.8. Review the argument for the nondifferentiability of $g(x)$ at nondyadic points. Does the argument still work if we replace $g(x)$ with the summation $\sum_{n=0}^{\infty} (1/2^n) h(3^n x)$? Does the argument work for the function $\sum_{n=0}^{\infty} (1/3^n) h(2^n x)$?

5.5   Epilogue

Far from being an anomaly to be relegated to the margins of our understanding of continuous functions, Weierstrass's example and those like it should actually serve as a guide to our intuition. The image of continuity as a smooth curve in our mind's eye severely misrepresents the situation and is the result of a bias stemming from an overexposure to the much smaller class of differentiable functions. The lesson here is that continuity is a strictly weaker notion than differentiability. In Section 3.6, we alluded to a corollary of the Baire Category Theorem, which asserts that Weierstrass's construction is actually typical of continuous functions. We will see that most continuous functions are nowhere-differentiable, so that it is really the differentiable functions that are the exceptions rather than the rule. The details of how to phrase this observation more rigorously are spelled out in Section 8.2.

To say that the nowhere-differentiable function g constructed in the previous section has "corners" at every point of its domain misses the mark. Weierstrass's original class of nowhere-differentiable functions was constructed from infinite sums of *smooth* trigonometric functions. It is the densely nested oscillating structure that makes the definition of a tangent line impossible. So what happens when we restrict our attention to monotone functions? How nondifferentiable can an increasing function be? Given a finite set of points, it is not difficult to piece together a monotone function which has actual corners—and thus is not differentiable—at each point in the given set. A natural question is whether there exists a continuous, monotone function that is nowhere-differentiable. Weierstrass suspected that such a function existed but only managed to produce an example of a continuous, increasing function which failed to be differentiable on a countable dense set (Exercise 7.5.11). In 1903, the French mathematician Henri Lebesgue (1875–1941) demonstrated that Weierstrass's intuition had failed on this account. Lebesgue proved that a continuous, monotone function would have to be differentiable at "almost" every point in its domain. To be specific, Lebesgue showed that, for every $\epsilon > 0$, the set of points where such a function fails to be differentiable can be covered by a countable union of intervals whose lengths sum to less than ϵ. This notion of "zero length," or "measure zero" as it is called, was encountered in our discussion of the Cantor set and is explored more fully in Section 7.6, where Lebesgue's substantial contribution to the theory of integration is discussed.

With the relationship between the continuity of f and the existence of f' somewhat in hand, we once more return to the question of characterizing the set of all derivatives. Not every function is a derivative. Darboux's Theorem forces us to conclude that there are some functions—those with jump discontinuities in particular—that cannot appear as the derivative of some other function. Another way to phrase Darboux's Theorem is to say that all derivatives must satisfy the intermediate value property. Continuous functions do possess the intermediate value property, and it is natural to ask whether every continuous function is necessarily a derivative. For this smaller class of functions, the answer is yes. The Fundamental Theorem of Calculus, treated in Chapter 7, states that, given a continuous function f, the function $F(x) = \int_a^x f$ satisfies $F' = f$. This does the trick. The collection of derivatives at least contains the continuous functions. The search for a concise characterization of *all* possible derivatives, however, remains largely unsuccessful.

As a final remark, we will see that by cleverly choosing f, this technique of defining F via $F(x) = \int_a^x f$ can be used to produce examples of continuous functions which fail to be differentiable on interesting sets, *provided we can show that $\int_a^x f$ is defined*. The question of just how to define integration became a central theme in analysis in the latter half of the 19th century and has continued on to the present. Much of this story is discussed in detail in Chapter 7 and Section 8.1.

Chapter 6

Sequences and Series of Functions

6.1 Discussion: The Power of Power Series

In 1689, Jakob Bernoulli published his *Tractatus de seriebus infinitis* summarizing what was known about infinite series toward the end of the 17th century. Full of clever calculations and conclusions, this publication was also notable for one particular question that it didn't answer; namely, what is the precise value of the series

$$\sum_{n=1}^{\infty} \frac{1}{n^2} = 1 + \frac{1}{4} + \frac{1}{9} + \frac{1}{16} + \cdots.$$

Bernoulli convincingly argued that $\sum 1/n^2$ converged to something less than 2 (see Example 2.4.4) but he was unable to find an explicit expression for the limit. Generally speaking, it is much harder to sum a series than it is to determine whether or not it converges. In fact, being able to find the sum of a convergent series is the exception rather than the rule. In this case, however, the series $\sum 1/n^2$ seemed so elementary; more elementary than, say, $\sum_{n=1}^{\infty} n^2/2^n$ or $\sum_{n=1}^{\infty} 1/n(n+1)$, both of which Bernoulli was able to handle. "If anyone finds and communicates to us that which has so far eluded our efforts," Bernoulli wrote, "great will be our gratitude." [1]

Geometric series are the most prominent class of examples that can be readily summed. In Example 2.7.5 we proved that

$$(1) \qquad \frac{1}{1-x} = 1 + x + x^2 + x^3 + \cdots$$

---

[1] As quoted in [12], which contains a much more thorough account of this story.

© Springer Science+Business Media New York 2015
S. Abbott, *Understanding Analysis*, Undergraduate Texts in Mathematics, DOI 10.1007/978-1-4939-2712-8_6

for all $|x| < 1$. Thus, for example, $\sum_{n=0}^{\infty} 1/2^n = 2$ and $\sum_{n=0}^{\infty}(-1/3)^n = 3/4$. Geometric series were part of mathematical folklore long before Bernoulli: however, what was relatively novel in Bernoulli's time was the idea of operating on infinite series such as (1) with tools from the budding theory of calculus. For instance, what happens if we take the derivative on each side of equation (1)? The left side is easy enough—we just get $1/(1 - x)^2$. But what about the right side? Adopting a 17th century mindset, a natural way to proceed is to treat the infinite series as a polynomial, albeit of infinite degree. Differentiation across equation (1) in this fashion gives

$$(2) \qquad \frac{1}{(1 - x)^2} = 0 + 1 + 2x + 3x^2 + 4x^3 + \cdots .$$

Is this a valid formula, at least for values of x in $(-1, 1)$? Empirical evidence suggests it is. Setting $x = 1/2$ we get

$$4 = \sum_{n=1}^{\infty} \frac{n}{2^{n-1}} = 1 + 1 + \frac{3}{4} + \frac{4}{8} + \frac{5}{16} + \cdots ,$$

which feels plausible, and is in fact true. Although not Bernoulli's requested series, this does suggest a possible new line of attack.

Manipulations of this sort can be used to create a wide assortment of new series representations for familiar functions. Substituting $-x^2$ for x in (1) gives

$$(3) \qquad \frac{1}{1 + x^2} = 1 - x^2 + x^4 - x^6 + x^8 - \cdots ,$$

for all $x \in (-1, 1)$.

Once again closing our eyes to the potential danger of treating an infinite series as though it were a polynomial, let's see what happens when we take antiderivatives. Using the fact that

$$(\arctan(x))' = \frac{1}{1 + x^2} \qquad \text{and} \qquad \arctan(0) = 0,$$

equation (3) becomes

$$(4) \qquad \arctan(x) = x - \frac{x^3}{3} + \frac{x^5}{5} - \frac{x^7}{7} + \cdots .$$

Plugging $x = 1$ into equation (4) yields the striking relationship

$$(5) \qquad \frac{\pi}{4} = 1 - \frac{1}{3} + \frac{1}{5} - \frac{1}{7} + \frac{1}{9} - \cdots .$$

The constant π, which arises from the geometry of circles, has somehow found its way into an equation involving the reciprocals of the odd integers. Is this a valid formula? Can we really treat the infinite series in (3) like a finite polynomial? Even if the answer is yes there is still another mystery to solve in this example.

Plugging $x = 1$ into equations (1), (2), or (3) yields mathematical gibberish, so is it prudent to anticipate something meaningful arising from equation (4) at this same value? Will any of these ideas get us closer to computing $\sum_{n=1}^{\infty} 1/n^2$?

As it turned out, Bernoulli's plea for help was answered in an unexpected way by Leonard Euler. At a young age, Euler was a student of Jakob Bernoulli's brother Johann, and the stellar pupil quickly rose to become the preeminent mathematician of his age. Euler's solution is impossible to anticipate. In 1735, he announced that

$$1 + \frac{1}{4} + \frac{1}{9} + \frac{1}{16} + \cdots = \frac{\pi^2}{6},$$

a provocative formula that, even more than equation (5), hints at deep connections between geometry, number theory and analysis. Euler's argument is quite short, but it needs to be viewed in the context of the time in which it was created. The "infinite polynomials" in this discussion are examples of *power series*, and a major catalyst for the expanding power of calculus in the 17th and 18th centuries was a proliferation of techniques like the ones used to generate formulas (2), (3), and (4). The machinations of both algebra and calculus are relatively straightforward when restricted to the class of polynomials. So, if in fact power series could be treated more or less like unending polynomials, then there was a great incentive to try to find power series representations for familiar functions like e^x, $\sqrt{1+x}$, or $\sin(x)$.

The appearance of $\arctan(x)$ in (4) is an encouraging sign that this might indeed always be possible. One of Isaac Newton's more significant achievements was to produce a generalization of the binomial formula. If $n \in \mathbf{N}$, then old-fashioned finite algebra leads to the formula

$$(1+x)^n = 1 + nx + \frac{n(n-1)}{2!}x^2 + \frac{n(n-1)(n-2)}{3!}x^3 + \cdots + x^n.$$

Through a process of experimentation and intuition Newton realized that for $r \notin \mathbf{N}$, the infinite series

$$(1+x)^r = 1 + rx + \frac{r(r-1)}{2!}x^2 + \frac{r(r-1)(r-2)}{3!}x^3 + \cdots$$

was meaningful, at least for $x \in (-1, 1)$. Setting $r = -1$, for example, yields

$$\frac{1}{1+x} = 1 - x + x^2 - x^3 + x^4 - \cdots,$$

which is easily seen to be equivalent to equation (1). Setting $r = 1/2$ we get

$$\sqrt{1+x} = 1 + \frac{1}{2}x - \frac{1}{2^2 2!}x^2 + \frac{3}{2^3 3!}x^3 - \frac{3 \cdot 5}{2^4 4!}x^4 + \cdots .$$

One way to lend a little credence to this formula for $\sqrt{1+x}$ is to focus on the first few terms and square the series:

$$\left(\sqrt{1+x}\right)^2 = \left(1 + \frac{1}{2}x - \frac{1}{8}x^2 + \cdots\right)\left(1 + \frac{1}{2}x - \frac{1}{8}x^2 + \cdots\right)$$

$$= 1 + \left(\frac{1}{2} + \frac{1}{2}\right)x + \left(-\frac{1}{8} + \frac{1}{4} - \frac{1}{8}\right)x^2 + \cdots$$

$$= 1 + x + 0x^2 + 0x^3 + \cdots .$$

Amid all of the unfounded assumptions we are making about infinity, calculations like this induce a feeling of optimism about the legitimacy of our search for power series representations.

Newton's binomial series is the starting point for a modern proof of Euler's famous sum, which is sketched out in detail in Section 8.3. Euler's original 1735 argument, however, started from the power series representation for $\sin(x)$. The formula

$$\sin x = x - \frac{x^3}{3!} + \frac{x^5}{5!} - \frac{x^7}{7!} + \cdots$$

was known to Newton, Bernoulli, and Euler alike. In contrast to equation (1), we will see that this formula is valid for all $x \in \mathbf{R}$. Factoring out x and dividing yields a power series with leading coefficient equal to 1:

(6)
$$\frac{\sin x}{x} = 1 - \frac{x^2}{3!} + \frac{x^4}{5!} - \frac{x^6}{7!} + \cdots .$$

Euler's idea was to continue factoring the power series in (6), and his strategy for doing this was very much in keeping with what we have seen so far—treat the power series as though it were a polynomial and then extend the pattern to infinity.

Factoring a polynomial of, say, degree three is straightforward if we know its roots. If $p(x) = 1 + ax + bx^2 + cx^3$ has roots r_1, r_2, and r_3, then

$$p(x) = \left(1 - \frac{x}{r_1}\right)\left(1 - \frac{x}{r_2}\right)\left(1 - \frac{x}{r_3}\right).$$

To see this just directly substitute to get $p(0) = 1$ and $p(r_1) = p(r_2) = p(r_3) = 0$.

The roots of the power series in (6) are the nonzero roots of $\sin x$, or $x = \pm\pi, \pm 2\pi, \pm 3\pi$, and so on. All right then—relying on his fabled intuition, Euler surmised that

(7)
$$1 - \frac{x^2}{3!} + \frac{x^4}{5!} - \frac{x^6}{7!} + \cdots$$
$$= \left(1 - \frac{x}{\pi}\right)\left(1 + \frac{x}{\pi}\right)\left(1 - \frac{x}{2\pi}\right)\left(1 + \frac{x}{2\pi}\right)\left(1 - \frac{x}{3\pi}\right)\left(1 + \frac{x}{3\pi}\right)\cdots$$
$$= \left(1 - \frac{x^2}{\pi^2}\right)\left(1 - \frac{x^2}{4\pi^2}\right)\left(1 - \frac{x^2}{9\pi^2}\right)\cdots ,$$

where in the last step adjacent pairs of factors have been multiplied together. What happens if we continue to multiply out the factors on the right? Well, the constant term comes out to be 1 which happily matches the constant term on the left. The magic comes when we compare the x^2 term on each side of (7). Multiplying out the infinite number of factors on the right (using our imagination as necessary) and collecting like powers of x, equation (7) becomes

$$1 - \frac{x^2}{3!} + \frac{x^4}{5!} - \frac{x^6}{7!} + \cdots$$
$$= 1 + \left(-\frac{1}{\pi^2} - \frac{1}{4\pi^2} - \frac{1}{9\pi^2} - \cdots\right)x^2 + \left(\frac{1}{4\pi^4} + \frac{1}{9\pi^4} + \cdots\right)x^4 + \cdots.$$

Equating the coefficients of x^2 on each side yields

$$-\frac{1}{3!} = -\frac{1}{\pi^2} - \frac{1}{4\pi^2} - \frac{1}{9\pi^2} - \cdots,$$

which when we multiply by $-\pi^2$ becomes

$$\frac{\pi^2}{6} = 1 + \frac{1}{4} + \frac{1}{9} + \frac{1}{16} + \cdots.$$

Numerical approximations of each side of this equation confirmed for Euler that, despite the audacious leaps in his argument, he had landed on solid ground. By our standards, this derivation falls well short of being a proper proof, and we will have to tend to this in the upcoming chapters. The takeaway of this discussion is that the hard work ahead is worth the effort. Infinite series representations of functions are both useful and surprisingly elegant, and can lead to remarkable conclusions when they are properly handled.

The evidence so far suggests power series are quite robust when treated as if they were finite in nature. Term-by-term differentiation produced a valid conclusion in equation (2), and taking antiderivatives fared similarly well in (4). We will see that these manipulations are *not* always justified for infinite series of more general types of functions. What is it about power series in particular that makes them so impervious to the dangers of the infinite? Of the many unanswered questions in this discussion, this last one is probably the most central, and the most important to understanding series of functions in general.

6.2 Uniform Convergence of a Sequence of Functions

Adopting the same strategy we used in Chapter 2, we will initially concern ourselves with the behavior and properties of converging *sequences* of functions. Because convergence of infinite series is defined in terms of the associated sequence of partial sums, the results from our study of sequences will be immediately applicable to the questions we have raised about both power series and more general infinite series of functions.

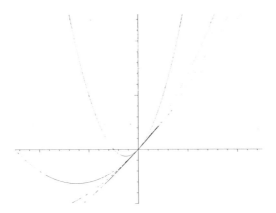

Figure 6.1: f_1, f_5, f_{10}, AND f_{20} WHERE $f_n = (x^2 + nx)/n$.

Pointwise Convergence

Definition 6.2.1. For each $n \in \mathbf{N}$, let f_n be a function defined on a set $A \subseteq \mathbf{R}$. The sequence (f_n) of functions *converges pointwise on A* to a function f if, for all $x \in A$, the sequence of real numbers $f_n(x)$ converges to $f(x)$.

In this case, we write $f_n \to f$, $\lim f_n = f$, or $\lim_{n \to \infty} f_n(x) = f(x)$. This last expression is helpful if there is any confusion as to whether x or n is the limiting variable.

Example 6.2.2.   (i) Consider

$$f_n(x) = (x^2 + nx)/n$$

on all of $\mathbf{R}$. Graphs of f_1, f_5, f_{10}, and f_{20} (Fig. 6.1) give an indication of what is happening as n gets larger. Algebraically, we can compute

$$\lim_{n \to \infty} f_n(x) = \lim_{n \to \infty} \frac{x^2 + nx}{n} = \lim_{n \to \infty} \frac{x^2}{n} + x = x.$$

Thus, (f_n) converges pointwise to $f(x) = x$ on $\mathbf{R}$.

(ii) Let $g_n(x) = x^n$ on the set $[0, 1]$, and consider what happens as n tends to infinity (Fig. 6.2). If $0 \le x < 1$, then we have seen that $x^n \to 0$. On the other hand, if $x = 1$, then $x^n \to 1$. It follows that $g_n \to g$ pointwise on $[0, 1]$, where

$$g(x) = \begin{cases} 0 & \text{for } 0 \le x < 1 \\ 1 & \text{for } x = 1. \end{cases}$$

(iii) Consider $h_n(x) = x^{1 + \frac{1}{2n-1}}$ on the set $[-1, 1]$ (Fig. 6.3). For a fixed $x \in [-1, 1]$ we have

$$\lim_{n \to \infty} h_n(x) = x \lim_{n \to \infty} x^{\frac{1}{2n-1}} = |x|.$$

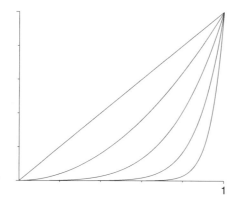

Figure 6.2: $g(x) = \lim_{n \to \infty} x^n$ IS NOT CONTINUOUS ON $[0, 1]$.

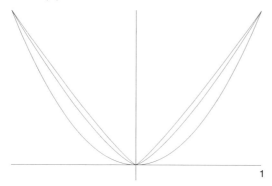

Figure 6.3: $h_n \to |x|$ ON $[-1, 1]$; LIMIT IS NOT DIFFERENTIABLE.

Examples 6.2.2 (ii) and (iii) are our first indication that there is some difficult work ahead of us. The central theme of this chapter is analyzing which properties the limit function inherits from the approximating sequence. In Example 6.2.2 (iii) we have a sequence of differentiable functions converging pointwise to a limit that is not differentiable at the origin. In Example 6.2.2 (ii), we see an even more fundamental problem of a sequence of continuous functions converging to a limit that is not continuous.

Continuity of the Limit Function

With Example 6.2.2 (ii) firmly in mind, we begin this discussion with a doomed attempt to prove that the pointwise limit of continuous functions is continuous. Upon discovering the problem in the argument, we will be in a better position to understand the need for a stronger notion of convergence for sequences of functions.

Assume (f_n) is a sequence of continuous functions on a set $A \subseteq \mathbf{R}$, and assume (f_n) converges pointwise to a limit f. To argue that f is continuous, fix a point $c \in A$, and let $\epsilon > 0$. We need to find a $\delta > 0$ such that

$$|x - c| < \delta \quad \text{implies} \quad |f(x) - f(c)| < \epsilon.$$

By the triangle inequality,

$$
\begin{aligned}
|f(x) - f(c)| &= |f(x) - f_n(x) + f_n(x) - f_n(c) + f_n(c) - f(c)| \\
&\leq |f(x) - f_n(x)| + |f_n(x) - f_n(c)| + |f_n(c) - f(c)|.
\end{aligned}
$$

Our first, optimistic impression is that each term in the sum on the right-hand side can be made small—the first and third by the fact that $f_n \to f$, and the middle term by the continuity of f_n. In order to use the continuity of f_n, we must first establish which particular f_n we are talking about. Because $c \in A$ is fixed, choose $N \in \mathbf{N}$ so that

$$|f_N(c) - f(c)| < \frac{\epsilon}{3}.$$

Now that N is chosen, the continuity of f_N implies that there exists a $\delta > 0$ such that

$$|f_N(x) - f_N(c)| < \frac{\epsilon}{3}$$

for all x satisfying $|x - c| < \delta$.

But here is the problem. We also need

$$|f_N(x) - f(x)| < \frac{\epsilon}{3} \quad \text{for all } x \text{ satisfying } |x - c| < \delta.$$

The values of x depend on δ, which depends on the choice of N. Thus, we cannot go back and simply choose a different N. More to the point, the variable x is not fixed the way c is in this discussion but represents any point in the interval $(c-\delta, c+\delta)$. Pointwise convergence implies that we can make $|f_n(x)-f(x)| < \epsilon/3$ for large enough values of n, but *the value of n depends on the point x.* It is possible that different values for x will result in the need for different—larger—choices for n. This phenomenon is apparent in Example 6.2.2 (ii). To achieve the inequality

$$|g_n(1/2) - g(1/2)| < \frac{1}{3},$$

we need $n \geq 2$, whereas

$$|g_n(9/10) - g(9/10)| < \frac{1}{3}$$

is true only after $n \geq 11$.

Uniform Convergence

To resolve this dilemma, we define a new, stronger notion of convergence of functions.

Definition 6.2.3 (Uniform Convergence). Let (f_n) be a sequence of functions defined on a set $A \subseteq \mathbf{R}$. Then, (f_n) *converges uniformly on* A to a limit function f defined on A if, for every $\epsilon > 0$, there exists an $N \in \mathbf{N}$ such that $|f_n(x) - f(x)| < \epsilon$ whenever $n \geq N$ and $x \in A$.

To emphasize the difference between uniform convergence and pointwise convergence, we restate Definition 6.2.1, being more explicit about the relationship between $\epsilon, N,$ and x. In particular, notice where the domain point x is referenced in each definition and consequently how the choice of N then does or does not depend on this value.

Definition 6.2.1B. Let (f_n) be a sequence of functions defined on a set $A \subseteq \mathbf{R}$. Then, (f_n) *converges pointwise on* A to a limit f defined on A if, for every $\epsilon > 0$ and $x \in A$, there exists an $N \in \mathbf{N}$ (perhaps dependent on x) such that $|f_n(x) - f(x)| < \epsilon$ whenever $n \geq N$.

The use of the adverb *uniformly* here should be reminiscent of its use in the phrase "uniformly continuous" from Chapter 4. In both cases, the term "uniformly" is employed to express the fact that the response (δ or N) to a prescribed ϵ can be chosen to work simultaneously for all values of x in the relevant domain.

Example 6.2.4. (i) Let

$$g_n(x) = \frac{1}{n(1 + x^2)}.$$

For any fixed $x \in \mathbf{R}$, we can see that $\lim g_n(x) = 0$ so that $g(x) = 0$ is the pointwise limit of the sequence (g_n) on $\mathbf{R}$. Is this convergence uniform? The observation that $1/(1 + x^2) \leq 1$ for all $x \in \mathbf{R}$ implies that

$$|g_n(x) - g(x)| = \left| \frac{1}{n(1 + x^2)} - 0 \right| \leq \frac{1}{n}.$$

Thus, given $\epsilon > 0$, we can choose $N > 1/\epsilon$ (which does not depend on x), and it follows that

$$n \geq N \quad \text{implies} \quad |g_n(x) - g(x)| < \epsilon$$

for all $x \in \mathbf{R}$. By Definition 6.2.3, $g_n \to 0$ uniformly on $\mathbf{R}$.

(ii) Look back at Example 6.2.2 (i), where we saw that $f_n(x) = (x^2 + nx)/n$ converges pointwise on $\mathbf{R}$ to $f(x) = x$. On $\mathbf{R}$, the convergence is not uniform. To see this write

$$|f_n(x) - f(x)| = \left| \frac{x^2 + nx}{n} - x \right| = \frac{x^2}{n},$$

and notice that in order to force $|f_n(x) - f(x)| < \epsilon$, we are going to have to choose

$$N > \frac{x^2}{\epsilon}.$$

Although this is possible to do for each $x \in \mathbf{R}$, there is no way to choose a single value of N that will work for all values of x at the same time.

On the other hand, we can show that $f_n \to f$ uniformly on the set $[-b, b]$. By restricting our attention to a bounded interval, we may now assert that

$$\frac{x^2}{n} \le \frac{b^2}{n}.$$

Given $\epsilon > 0$, then, we can choose

$$N > \frac{b^2}{\epsilon}$$

independently of $x \in [-b, b]$.

Graphically speaking, the uniform convergence of f_n to a limit f on a set A can be visualized by constructing a band of radius $\pm\epsilon$ around the limit function f. If $f_n \to f$ uniformly, then there exists a point in the sequence after which each f_n is *completely* contained in this ϵ-strip (Fig. 6.4). This image should be compared with the graphs in Figures 6.1–6.2 from Example 6.2.2 and the one in Figure 6.5.

Cauchy Criterion

Recall that the Cauchy Criterion for convergent sequences of real numbers was an equivalent characterization of convergence which, unlike the definition, did not make explicit mention of the limit. The usefulness of the Cauchy Criterion suggests the need for an analogous characterization of uniformly convergent sequences of functions. As with all statements about uniformity, pay special attention to the relationship between the response variable ($N \in \mathbf{N}$) and the domain variable ($x \in A$).

Theorem 6.2.5 (Cauchy Criterion for Uniform Convergence). *A sequence of functions (f_n) defined on a set $A \subseteq \mathbf{R}$ converges uniformly on A if and only if for every $\epsilon > 0$ there exists an $N \in \mathbf{N}$ such that $|f_n(x) - f_m(x)| < \epsilon$ whenever $m, n \ge N$ and $x \in A$.*

Proof. Exercise 6.2.5.   □

Continuity Revisited

The stronger assumption of uniform convergence is precisely what is required to remove the flaws from our attempted proof that the limit of continuous functions is continuous.

Theorem 6.2.6 (Continuous Limit Theorem). *Let (f_n) be a sequence of functions defined on $A \subseteq \mathbf{R}$ that converges uniformly on A to a function f. If each f_n is continuous at $c \in A$, then f is continuous at c.*

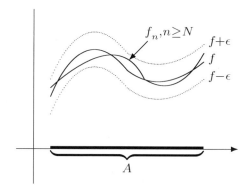

Figure 6.4: $f_n \to f$ UNIFORMLY ON A.

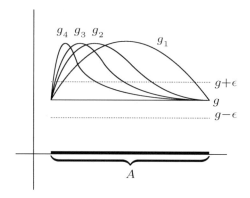

Figure 6.5: $g_n \to g$ POINTWISE, BUT NOT UNIFORMLY.

Proof. Fix $c \in A$ and let $\epsilon > 0$. Choose N so that

$$|f_N(x) - f(x)| < \frac{\epsilon}{3}$$

for all $x \in A$. Because f_N is continuous, there exists a $\delta > 0$ for which

$$|f_N(x) - f_N(c)| < \frac{\epsilon}{3}$$

is true whenever $|x - c| < \delta$. But this implies

$$
\begin{aligned}
|f(x) - f(c)| &= |f(x) - f_N(x) + f_N(x) - f_N(c) + f_N(c) - f(c)| \\
&\leq |f(x) - f_N(x)| + |f_N(x) - f_N(c)| + |f_N(c) - f(c)| \\
&< \frac{\epsilon}{3} + \frac{\epsilon}{3} + \frac{\epsilon}{3} = \epsilon.
\end{aligned}
$$

Thus, f is continuous at $c \in A$. $\qquad \square$

Exercises

Exercise 6.2.1. Let

$$f_n(x) = \frac{nx}{1 + nx^2}.$$

(a) Find the pointwise limit of (f_n) for all $x \in (0, \infty)$.

(b) Is the convergence uniform on $(0, \infty)$?

(c) Is the convergence uniform on $(0, 1)$?

(d) Is the convergence uniform on $(1, \infty)$?

Exercise 6.2.2.   (a) Define a sequence of functions on $\mathbf{R}$ by

$$f_n(x) = \begin{cases} 1 & \text{if } x = 1, \frac{1}{2}, \frac{1}{3}, \ldots, \frac{1}{n} \\ 0 & \text{otherwise} \end{cases}$$

and let f be the pointwise limit of f_n.

Is each f_n continuous at zero? Does $f_n \to f$ uniformly on $\mathbf{R}$? Is f continuous at zero?

(b) Repeat this exercise using the sequence of functions

$$g_n(x) = \begin{cases} x & \text{if } x = 1, \frac{1}{2}, \frac{1}{3}, \ldots, \frac{1}{n} \\ 0 & \text{otherwise.} \end{cases}$$

(c) Repeat the exercise once more with the sequence

$$h_n(x) = \begin{cases} 1 & \text{if } x = \frac{1}{n} \\ x & \text{if } x = 1, \frac{1}{2}, \frac{1}{3}, \ldots, \frac{1}{n-1} \\ 0 & \text{otherwise.} \end{cases}$$

In each case, explain how the results are consistent with the content of the Continuous Limit Theorem (Theorem 6.2.6).

Exercise 6.2.3. For each $n \in \mathbf{N}$ and $x \in [0, \infty)$, let

$$g_n(x) = \frac{x}{1 + x^n} \quad \text{and} \quad h_n(x) = \begin{cases} 1 & \text{if } x \geq 1/n \\ nx & \text{if } 0 \leq x < 1/n. \end{cases}$$

Answer the following questions for the sequences (g_n) and (h_n):

(a) Find the pointwise limit on $[0, \infty)$.

(b) Explain how we know that the convergence *cannot* be uniform on $[0, \infty)$.

(c) Choose a smaller set over which the convergence is uniform and supply an argument to show that this is indeed the case.

Exercise 6.2.4. Review Exercise 5.2.8 which includes the definition for a uniformly differentiable function. Use the results discussed in Section 6.2 to show that if f is uniformly differentiable, then f' is continuous.

Exercise 6.2.5. Using the Cauchy Criterion for convergent sequences of real numbers (Theorem 2.6.4), supply a proof for Theorem 6.2.5. (First, define a candidate for $f(x)$, and then argue that $f_n \to f$ uniformly.)

Exercise 6.2.6. Assume $f_n \to f$ on a set A. Theorem 6.2.6 is an example of a typical type of question which asks whether a trait possessed by each f_n is inherited by the limit function. Provide an example to show that *all* of the following propositions are false if the convergence is only assumed to be pointwise on A. Then go back and decide which are true under the stronger hypothesis of uniform convergence.

(a) If each f_n is uniformly continuous, then f is uniformly continuous.

(b) If each f_n is bounded, then f is bounded.

(c) If each f_n has a finite number of discontinuities, then f has a finite number of discontinuities.

(d) If each f_n has fewer than M discontinuities (where $M \in \mathbf{N}$ is fixed), then f has fewer than M discontinuities.

(e) If each f_n has at most a countable number of discontinuities, then f has at most a countable number of discontinuities.

Exercise 6.2.7. Let f be uniformly continuous on all of $\mathbf{R}$, and define a sequence of functions by $f_n(x) = f(x + \frac{1}{n})$. Show that $f_n \to f$ uniformly. Give an example to show that this proposition fails if f is only assumed to be continuous and not uniformly continuous on $\mathbf{R}$.

Exercise 6.2.8. Let (g_n) be a sequence of continuous functions that converges uniformly to g on a compact set K. If $g(x) \neq 0$ on K, show $(1/g_n)$ converges uniformly on K to $1/g$.

Exercise 6.2.9. Assume (f_n) and (g_n) are uniformly convergent sequences of functions.

(a) Show that $(f_n + g_n)$ is a uniformly convergent sequence of functions.

(b) Give an example to show that the product $(f_n g_n)$ may not converge uniformly.

(c) Prove that if there exists an $M > 0$ such that $|f_n| \leq M$ and $|g_n| \leq M$ for all $n \in \mathbf{N}$, then $(f_n g_n)$ does converge uniformly.

Exercise 6.2.10. This exercise and the next explore partial converses of the Continuous Limit Theorem (Theorem 6.2.6). Assume $f_n \to f$ pointwise on $[a, b]$ and the limit function f is continuous on $[a, b]$. If each f_n is increasing (but not necessarily continuous), show $f_n \to f$ uniformly.

Exercise 6.2.11 (Dini's Theorem). Assume $f_n \to f$ pointwise on a compact set K and assume that for each $x \in K$ the sequence $f_n(x)$ is increasing. Follow these steps to show that if f_n and f are continuous on K, then the convergence is uniform.

(a) Set $g_n = f - f_n$ and translate the preceding hypothesis into statements about the sequence (g_n).

(b) Let $\epsilon > 0$ be arbitrary, and define $K_n = \{x \in K : g_n(x) \geq \epsilon\}$. Argue that $K_1 \supseteq K_2 \supseteq K_3 \supseteq \cdots$, and use this observation to finish the argument.

Exercise 6.2.12 (Cantor Function). Review the construction of the Cantor set $C \subseteq [0, 1]$ from Section 3.1. This exercise makes use of results and notation from this discussion.

(a) Define $f_0(x) = x$ for all $x \in [0, 1]$. Now, let

$$f_1(x) = \begin{cases} (3/2)x & \text{for } 0 \leq x \leq 1/3 \\ 1/2 & \text{for } 1/3 < x < 2/3 \\ (3/2)x - 1/2 & \text{for } 2/3 \leq x \leq 1. \end{cases}$$

Sketch f_0 and f_1 over $[0, 1]$ and observe that f_1 is continuous, increasing, and constant on the middle third $(1/3, 2/3) = [0, 1] \backslash C_1$.

(b) Construct f_2 by imitating this process of flattening out the middle third of each nonconstant segment of f_1. Specifically, let

$$f_2(x) = \begin{cases} (1/2)f_1(3x) & \text{for } 0 \leq x \leq 1/3 \\ f_1(x) & \text{for } 1/3 < x < 2/3 \\ (1/2)f_1(3x - 2) + 1/2 & \text{for } 2/3 \leq x \leq 1. \end{cases}$$

If we continue this process, show that the resulting sequence (f_n) converges uniformly on $[0, 1]$.

(c) Let $f = \lim f_n$. Prove that f is a continuous, increasing function on $[0, 1]$ with $f(0) = 0$ and $f(1) = 1$ that satisfies $f'(x) = 0$ for all x in the open set $[0, 1] \backslash C$. Recall that the "length" of the Cantor set C is 0. Somehow, f manages to increase from 0 to 1 while remaining constant on a set of "length 1."

Exercise 6.2.13. Recall that the Bolzano–Weierstrass Theorem (Theorem 2.5.5) states that every bounded sequence of real numbers has a convergent subsequence. An analogous statement for bounded sequences of functions is not true in general, but under stronger hypotheses several different conclusions are possible. One avenue is to assume the common domain for all of the functions in the sequence is countable. (Another is explored in the next two exercises.)

Let $A = \{x_1, x_2, x_3, \ldots\}$ be a countable set. For each $n \in \mathbf{N}$, let f_n be defined on A and assume there exists an $M > 0$ such that $|f_n(x)| \leq M$ for all $n \in \mathbf{N}$ and $x \in A$. Follow these steps to show that there exists a subsequence of (f_n) that converges pointwise on A.

(a) Why does the sequence of real numbers $f_n(x_1)$ necessarily contain a convergent subsequence (f_{n_k})? To indicate that the subsequence of functions (f_{n_k}) is generated by considering the values of the functions at x_1, we will use the notation $f_{n_k} = f_{1,k}$.

(b) Now, explain why the sequence $f_{1,k}(x_2)$ contains a convergent subsequence.

(c) Carefully construct a nested family of subsequences $(f_{m,k})$, and show how this can be used to produce a single subsequence of (f_n) that converges at every point of A.

Exercise 6.2.14. A sequence of functions (f_n) defined on a set $E \subseteq \mathbf{R}$ is called *equicontinuous* if for every $\epsilon > 0$ there exists a $\delta > 0$ such that $|f_n(x) - f_n(y)| < \epsilon$ for all $n \in \mathbf{N}$ and $|x - y| < \delta$ in E.

(a) What is the difference between saying that a sequence of functions (f_n) is equicontinuous and just asserting that each f_n in the sequence is individually uniformly continuous?

(b) Give a qualitative explanation for why the sequence $g_n(x) = x^n$ is not equicontinuous on $[0, 1]$. Is each g_n uniformly continuous on $[0, 1]$?

Exercise 6.2.15 (Arzela–Ascoli Theorem). For each $n \in \mathbf{N}$, let f_n be a function defined on $[0, 1]$. If (f_n) is bounded on $[0, 1]$—that is, there exists an $M > 0$ such that $|f_n(x)| \leq M$ for all $n \in \mathbf{N}$ and $x \in [0, 1]$—and if the collection of functions (f_n) is equicontinuous (Exercise 6.2.14), follow these steps to show that (f_n) contains a uniformly convergent subsequence.

(a) Use Exercise 6.2.13 to produce a subsequence (f_{n_k}) that converges at every rational point in $[0, 1]$. To simplify the notation, set $g_k = f_{n_k}$. It remains to show that (g_k) converges uniformly on all of $[0, 1]$.

(b) Let $\epsilon > 0$. By equicontinuity, there exists a $\delta > 0$ such that

$$|g_k(x) - g_k(y)| < \frac{\epsilon}{3}$$

for all $|x - y| < \delta$ and $k \in \mathbf{N}$. Using this δ, let $r_1, r_2, \ldots, r_m$ be a *finite* collection of rational points with the property that the union of the neighborhoods $V_\delta(r_i)$ contains $[0,1]$.

Explain why there must exist an $N \in \mathbf{N}$ such that

$$|g_s(r_i) - g_t(r_i)| < \frac{\epsilon}{3}$$

for all $s, t \geq N$ and r_i in the finite subset of $[0, 1]$ just described. Why does having the set $\{r_1, r_2, \ldots, r_m\}$ be finite matter?

(c) Finish the argument by showing that, for an arbitrary $x \in [0, 1]$,

$$|g_s(x) - g_t(x)| < \epsilon$$

for all $s, t \geq N$.

6.3   Uniform Convergence and Differentiation

Example 6.2.2 (iii) imposes some significant restrictions on what we might hope to be true regarding differentiation and uniform convergence. If $h_n \to h$ uniformly and each h_n is differentiable, we should not anticipate that $h_n' \to h'$ because in this example $h'(x)$ does not even exist at $x = 0$. There are also examples (see Exercise 6.3.4) where $f_n \to f$ uniformly with (f_n) and f all differentiable, but the sequence (f_n') diverges at every point of the domain.

The key assumption necessary to be able to prove any facts about the derivative of the limit function is that the *sequence of derivatives* be uniformly convergent. This may sound as though we are assuming what it is we would like to prove, and there is some validity to this complaint. The more hypotheses a proposition has, the more difficult it is to apply. The content of the next theorem is that if we are given a pointwise convergent sequence of differentiable functions, and if we know that the sequence of derivatives converges uniformly *to something*, then the limit of the derivatives is indeed the derivative of the limit.

Theorem 6.3.1 (Differentiable Limit Theorem). *Let $f_n \to f$ pointwise on the closed interval $[a, b]$, and assume that each f_n is differentiable. If (f_n') converges uniformly on $[a, b]$ to a function g, then the function f is differentiable and $f' = g$.*

Proof. Fix $c \in [a, b]$ and let $\epsilon > 0$. We want to argue that $f'(c)$ exists and equals $g(c)$. Because f' is defined by the limit

$$f'(c) = \lim_{x \to c} \frac{f(x) - f(c)}{x - c},$$

our task is to produce a $\delta > 0$ so that

$$\left| \frac{f(x) - f(c)}{x - c} - g(c) \right| < \epsilon$$

whenever $0 < |x - c| < \delta$.

To motivate the strategy of the proof, observe that for all $x \neq c$ and all $n \in \mathbf{N}$, the triangle inequality implies

$$
\begin{aligned}
\left| \frac{f(x) - f(c)}{x - c} - g(c) \right| \;\leq\; & \left| \frac{f(x) - f(c)}{x - c} - \frac{f_n(x) - f_n(c)}{x - c} \right| \\
& + \left| \frac{f_n(x) - f_n(c)}{x - c} - f_n'(c) \right| + |f_n'(c) - g(c)|.
\end{aligned}
$$

Our intent is to first find an f_n that forces the first and third terms on the right-hand side to be less than $\epsilon/3$. Once we establish which f_n we want, we can then use the differentiability of f_n to produce a δ that makes the middle term less than $\epsilon/3$ for all x satisfying $0 < |x - c| < \delta$.

Let's start by choosing an N_1 such that

(1)
$$|f'_m(c) - g(c)| < \frac{\epsilon}{3}$$

for all $m \geq N_1$. We now invoke the uniform convergence of (f'_n) to assert (via Theorem 6.2.5) that there exists an N_2 such that $m, n \geq N_2$ implies

$$|f'_m(x) - f'_n(x)| < \frac{\epsilon}{3} \quad \text{for all } x \in [a, b].$$

Set $N = \max\{N_1, N_2\}$.

The function f_N is differentiable at c, and so there exists a $\delta > 0$ for which

(2)
$$\left| \frac{f_N(x) - f_N(c)}{x - c} - f'_N(c) \right| < \frac{\epsilon}{3}$$

whenever $0 < |x - c| < \delta$. This is our sought after δ, but it takes some effort to show that it has the desired property.

Fix an x satisfying $0 < |x - c| < \delta$, let $m \geq N$, and apply the Mean Value Theorem to $f_m - f_N$ on the interval $[c, x]$, (If $x < c$ the argument is the same.) By MVT, there exists an $\alpha \in (c, x)$ such that

$$f'_m(\alpha) - f'_N(\alpha) = \frac{(f_m(x) - f_N(x)) - (f_m(c) - f_N(c))}{x - c}.$$

Recall that our choice of N implies

$$|f'_m(\alpha) - f'_N(\alpha)| < \frac{\epsilon}{3},$$

and so it follows that

$$\left| \frac{f_m(x) - f_m(c)}{x - c} - \frac{f_N(x) - f_N(c)}{x - c} \right| < \frac{\epsilon}{3}.$$

Because $f_m \to f$ we can take the limit as $m \to \infty$, and the Order Limit Theorem (Theorem 2.3.4) asserts that

(3)
$$\left| \frac{f(x) - f(c)}{x - c} - \frac{f_N(x) - f_N(c)}{x - c} \right| \leq \frac{\epsilon}{3}.$$

Finally, the inequalities in (1), (2), and (3), together imply that for x satisfying $0 < |x - c| < \delta$,

$$\left| \frac{f(x) - f(c)}{x - c} - g(c) \right| \leq \left| \frac{f(x) - f(c)}{x - c} - \frac{f_N(x) - f_N(c)}{x - c} \right|$$
$$+ \left| \frac{f_N(x) - f_N(c)}{x - c} - f'_N(c) \right| + |f'_N(c) - g(c)|$$
$$< \frac{\epsilon}{3} + \frac{\epsilon}{3} + \frac{\epsilon}{3} = \epsilon. \qquad \square$$

The hypothesis in the Differentiable Limit Theorem is unnecessarily strong. We actually do not need to assume that $f_n(x) \to f(x)$ at each point in the domain because the assumption that the sequence of derivatives (f_n') converges uniformly is nearly strong enough to *prove* that (f_n) converges, uniformly in fact. Two functions with the same derivative may differ by a constant, so we must assume that there is at least one point x_0 where $f_n(x_0) \to f(x_0)$.

Theorem 6.3.2. *Let (f_n) be a sequence of differentiable functions defined on the closed interval $[a, b]$, and assume (f_n') converges uniformly on $[a, b]$. If there exists a point $x_0 \in [a, b]$ where $f_n(x_0)$ is convergent, then (f_n) converges uniformly on $[a, b]$.*

Proof. Exercise 6.3.7.   □

Combining the last two results produces a stronger version of Theorem 6.3.1.

Theorem 6.3.3. *Let (f_n) be a sequence of differentiable functions defined on the closed interval $[a, b]$, and assume (f_n') converges uniformly to a function g on $[a, b]$. If there exists a point $x_0 \in [a, b]$ for which $f_n(x_0)$ is convergent, then (f_n) converges uniformly. Moreover, the limit function $f = \lim f_n$ is differentiable and satisfies $f' = g$.*

Exercises

Exercise 6.3.1. Consider the sequence of functions defined by

$$g_n(x) = \frac{x^n}{n}.$$

(a) Show (g_n) converges uniformly on $[0, 1]$ and find $g = \lim g_n$. Show that g is differentiable and compute $g'(x)$ for all $x \in [0, 1]$.

(b) Now, show that (g_n') converges on $[0, 1]$. Is the convergence uniform? Set $h = \lim g_n'$ and compare h and g'. Are they the same?

Exercise 6.3.2. Consider the sequence of functions

$$h_n(x) = \sqrt{x^2 + \frac{1}{n}}.$$

(a) Compute the pointwise limit of (h_n) and then prove that the convergence is uniform on $\mathbf{R}$.

(b) Note that each h_n is differentiable. Show $g(x) = \lim h_n'(x)$ exists for all x, and explain how we can be certain that the convergence is *not* uniform on any neighborhood of zero.

Exercise 6.3.3. Consider the sequence of functions

$$f_n(x) = \frac{x}{1 + nx^2}.$$

(a) Find the points on $\mathbf{R}$ where each $f_n(x)$ attains its maximum and minimum value. Use this to prove (f_n) converges uniformly on $\mathbf{R}$. What is the limit function?

(b) Let $f = \lim f_n$. Compute $f'_n(x)$ and find all the values of x for which $f'(x) = \lim f'_n(x)$.

Exercise 6.3.4. Let
$$h_n(x) = \frac{\sin(nx)}{\sqrt{n}}.$$

Show that $h_n \to 0$ uniformly on $\mathbf{R}$ but that the sequence of derivatives (h'_n) diverges for every $x \in \mathbf{R}$.

Exercise 6.3.5. Let
$$g_n(x) = \frac{nx + x^2}{2n},$$

and set $g(x) = \lim g_n(x)$. Show that g is differentiable in two ways:

(a) Compute $g(x)$ by algebraically taking the limit as $n \to \infty$ and then find $g'(x)$.

(b) Compute $g'_n(x)$ for each $n \in \mathbf{N}$ and show that the sequence of derivatives (g'_n) converges uniformly on every interval $[-M, M]$. Use Theorem 6.3.3 to conclude $g'(x) = \lim g'_n(x)$.

(c) Repeat parts (a) and (b) for the sequence $f_n(x) = (nx^2 + 1)/(2n + x)$.

Exercise 6.3.6. Provide an example or explain why the request is impossible. Let's take the domain of the functions to be all of $\mathbf{R}$.

(a) A sequence (f_n) of nowhere differentiable functions with $f_n \to f$ uniformly and f everywhere differentiable.

(b) A sequence (f_n) of differentiable functions such that (f'_n) converges uniformly but the original sequence (f_n) does not converge for any $x \in \mathbf{R}$.

(c) A sequence (f_n) of differentiable functions such that both (f_n) and (f'_n) converge uniformly but $f = \lim f_n$ is not differentiable at some point.

Exercise 6.3.7. Use the Mean Value Theorem to supply a proof for Theorem 6.3.2. To get started, observe that the triangle inequality implies that, for any $x \in [a, b]$ and $m, n \in \mathbf{N}$,

$$|f_n(x) - f_m(x)| \le |(f_n(x) - f_m(x)) - (f_n(x_0) - f_m(x_0))| + |f_n(x_0) - f_m(x_0)|.$$

6.4   Series of Functions

Definition 6.4.1. For each $n \in \mathbf{N}$, let f_n and f be functions defined on a set $A \subseteq \mathbf{R}$. The infinite series

$$\sum_{n=1}^{\infty} f_n(x) = f_1(x) + f_2(x) + f_3(x) + \cdots$$

converges pointwise on A to $f(x)$ if the sequence $s_k(x)$ of partial sums defined by

$$s_k(x) = f_1(x) + f_2(x) + \cdots + f_k(x)$$

converges pointwise to $f(x)$. The series *converges uniformly on A* to f if the sequence $s_k(x)$ converges uniformly on A to $f(x)$.

   In either case, we write $f = \sum_{n=1}^{\infty} f_n$ or $f(x) = \sum_{n=1}^{\infty} f_n(x)$, always being explicit about the type of convergence involved.

   If we have a series $\sum_{n=1}^{\infty} f_n$ where the functions f_n are continuous, then the Algebraic Continuity Theorem (Theorem 4.3.4) guarantees that the partial sums—because they are finite sums—will be continuous as well. A corresponding observation is true if we are dealing with differentiable functions. As a consequence, we can immediately translate the results for sequences in the previous sections into statements about the behavior of infinite series of functions.

Theorem 6.4.2 (Term-by-term Continuity Theorem). *Let f_n be continuous functions defined on a set $A \subseteq \mathbf{R}$, and assume $\sum_{n=1}^{\infty} f_n$ converges uniformly on A to a function f. Then, f is continuous on A.*

Proof. Apply the Continuous Limit Theorem (Theorem 6.2.6) to the partial sums $s_k = f_1 + f_2 + \cdots + f_k$.   □

Theorem 6.4.3 (Term-by-term Differentiability Theorem). *Let f_n be differentiable functions defined on an interval A, and assume $\sum_{n=1}^{\infty} f_n'(x)$ converges uniformly to a limit $g(x)$ on A. If there exists a point $x_0 \in [a, b]$ where $\sum_{n=1}^{\infty} f_n(x_0)$ converges, then the series $\sum_{n=1}^{\infty} f_n(x)$ converges uniformly to a differentiable function $f(x)$ satisfying $f'(x) = g(x)$ on A. In other words,*

$$f(x) = \sum_{n=1}^{\infty} f_n(x) \quad and \quad f'(x) = \sum_{n=1}^{\infty} f_n'(x).$$

Proof. Apply the stronger form of the Differentiable Limit Theorem (Theorem 6.3.3) to the partial sums $s_k = f_1 + f_2 + \cdots + f_k$. Observe that Theorem 5.2.4 implies that $s_k' = f_1' + f_2' + \cdots + f_k'$.   □

   In the vocabulary of infinite series, the Cauchy Criterion takes the following form.

Theorem 6.4.4 (Cauchy Criterion for Uniform Convergence of Series). *A series $\sum_{n=1}^{\infty} f_n$ converges uniformly on $A \subseteq \mathbf{R}$ if and only if for every $\epsilon > 0$ there exists an $N \in \mathbf{N}$ such that*

$$|f_{m+1}(x) + f_{m+2}(x) + f_{m+3}(x) + \cdots + f_n(x)| < \epsilon$$

whenever $n > m \geq N$ and $x \in A$.

The benefits of uniform convergence over pointwise convergence suggest the need for some ways of determining when a series converges uniformly. The following corollary to the Cauchy Criterion is the most common such tool. In particular, it will be quite useful in our upcoming investigations of power series.

Corollary 6.4.5 (Weierstrass M-Test). *For each $n \in \mathbf{N}$, let f_n be a function defined on a set $A \subseteq \mathbf{R}$, and let $M_n > 0$ be a real number satisfying*

$$|f_n(x)| \leq M_n$$

for all $x \in A$. If $\sum_{n=1}^{\infty} M_n$ converges, then $\sum_{n=1}^{\infty} f_n$ converges uniformly on A.

Proof. Exercise 6.4.1. □

Exercises

Exercise 6.4.1. Supply the details for the proof of the Weierstrass M-Test (Corollary 6.4.5).

Exercise 6.4.2. Decide whether each proposition is true or false, providing a short justification or counterexample as appropriate.

(a) If $\sum_{n=1}^{\infty} g_n$ converges uniformly, then (g_n) converges uniformly to zero.

(b) If $0 \leq f_n(x) \leq g_n(x)$ and $\sum_{n=1}^{\infty} g_n$ converges uniformly, then $\sum_{n=1}^{\infty} f_n$ converges uniformly.

(c) If $\sum_{n=1}^{\infty} f_n$ converges uniformly on A, then there exist constants M_n such that $|f_n(x)| \leq M_n$ for all $x \in A$ and $\sum_{n=1}^{\infty} M_n$ converges.

Exercise 6.4.3. (a) Show that

$$g(x) = \sum_{n=0}^{\infty} \frac{\cos(2^n x)}{2^n}$$

is continuous on all of $\mathbf{R}$.

(b) The function g was cited in Section 5.4 as an example of a continuous nowhere differentiable function. What happens if we try to use Theorem 6.4.3 to explore whether g is differentiable?

Exercise 6.4.4. Define

$$g(x) = \sum_{n=0}^{\infty} \frac{x^{2n}}{(1+x^{2n})}.$$

Find the values of x where the series converges and show that we get a continuous function on this set.

Exercise 6.4.5.   (a) Prove that

$$h(x) = \sum_{n=1}^{\infty} \frac{x^n}{n^2} = x + \frac{x^2}{4} + \frac{x^3}{9} + \frac{x^4}{16} + \cdots$$

is continuous on $[-1, 1]$.

(b) The series

$$f(x) = \sum_{n=1}^{\infty} \frac{x^n}{n} = x + \frac{x^2}{2} + \frac{x^3}{3} + \frac{x^4}{4} + \cdots$$

converges for every x in the half-open interval $[-1, 1)$ but does not converge when $x = 1$. For a fixed $x_0 \in (-1, 1)$, explain how we can still use the Weierstrass M-Test to prove that f is continuous at x_0.

Exercise 6.4.6. Let

$$f(x) = \frac{1}{x} - \frac{1}{x+1} + \frac{1}{x+2} - \frac{1}{x+3} + \frac{1}{x+4} - \cdots.$$

Show f is defined for all $x > 0$. Is f continuous on $(0, \infty)$? How about differentiable?

Exercise 6.4.7. Let

$$f(x) = \sum_{k=1}^{\infty} \frac{\sin(kx)}{k^3}.$$

(a) Show that $f(x)$ is differentiable and that the derivative $f'(x)$ is continuous.

(b) Can we determine if f is twice-differentiable?

Exercise 6.4.8. Consider the function

$$f(x) = \sum_{k=1}^{\infty} \frac{\sin(x/k)}{k}.$$

Where is f defined? Continuous? Differentiable? Twice-differentiable?

Exercise 6.4.9. Let

$$h(x) = \sum_{n=1}^{\infty} \frac{1}{x^2 + n^2}.$$

(a) Show that h is a continuous function defined on all of $\mathbf{R}$.

(b) Is h differentiable? If so, is the derivative function h' continuous?

Exercise 6.4.10. Let $\{r_1, r_2, r_3, \ldots\}$ be an enumeration of the set of rational numbers. For each $r_n \in \mathbf{Q}$, define

$$u_n(x) = \begin{cases} 1/2^n & \text{for } x > r_n \\ 0 & \text{for } x \leq r_n. \end{cases}$$

Now, let $h(x) = \sum_{n=1}^{\infty} u_n(x)$. Prove that h is a monotone function defined on all of $\mathbf{R}$ that is continuous at every irrational point.

6.5   Power Series

It is time to put some mathematical teeth into our understanding of functions expressed in the form of a power series; that is, functions of the form

$$f(x) = \sum_{n=0}^{\infty} a_n x^n = a_0 + a_1 x + a_2 x^2 + a_3 x^3 + \cdots .$$

The first order of business is to determine the points $x \in \mathbf{R}$ for which the resulting series on the right-hand side converges. This set certainly contains $x = 0$, and, as the next result demonstrates, it takes a very predictable form.

Theorem 6.5.1. *If a power series $\sum_{n=0}^{\infty} a_n x^n$ converges at some point $x_0 \in \mathbf{R}$, then it converges absolutely for any x satisfying $|x| < |x_0|$.*

Proof. If $\sum_{n=0}^{\infty} a_n x_0^n$ converges, then the sequence of terms $(a_n x_0^n)$ is bounded. (In fact, it converges to 0.) Let $M > 0$ satisfy $|a_n x_0^n| \leq M$ for all $n \in \mathbf{N}$. If $x \in \mathbf{R}$ satisfies $|x| < |x_0|$, then

$$|a_n x^n| = |a_n x_0^n| \left| \frac{x}{x_0} \right|^n \leq M \left| \frac{x}{x_0} \right|^n .$$

But notice that

$$\sum_{n=0}^{\infty} M \left| \frac{x}{x_0} \right|^n$$

is a geometric series with ratio $|x/x_0| < 1$ and so converges. By the Comparison Test, $\sum_{n=0}^{\infty} a_n x^n$ converges absolutely. $\qquad \square$

The main implication of Theorem 6.5.1 is that the set of points for which a given power series converges must necessarily be $\{0\}$, $\mathbf{R}$, or a bounded interval centered around $x = 0$. Because of the strict inequality in Theorem 6.5.1, there is some ambiguity about the endpoints of the interval, and it is possible that the set of convergent points may be of the form $(-R, R)$, $[-R, R)$, $(-R, R]$, or $[-R, R]$.

The value of R is referred to as the *radius of convergence* of a power series, and it is customary to assign R the value 0 or ∞ to represent the set $\{0\}$ or $\mathbf{R}$, respectively. Some of the standard devices for computing the radius of convergence for a power series are explored in the exercises. Of more interest to us here is the investigation of the properties of functions defined in this way. Are they continuous? Are they differentiable? If so, can we differentiate the series term-by-term? What happens at the endpoints?

Establishing Uniform Convergence

The positive answers to the preceding questions, and the usefulness of power series in general, are largely due to the fact that they converge uniformly on compact sets contained in their domain of convergent points. As we are about to see, a complete proof of this fact requires a fairly delicate argument attributed to the Norwegian mathematician Niels Henrik Abel. A significant amount of progress, however, can be made with the Weierstrass M-Test (Corollary 6.4.5).

Theorem 6.5.2. *If a power series $\sum_{n=0}^{\infty} a_n x^n$ converges absolutely at a point x_0, then it converges uniformly on the closed interval $[-c, c]$, where $c = |x_0|$.*

Proof. This proof requires a straightforward application of the Weierstrass M-Test. The details are requested in Exercise 6.5.3. $\square$

For many applications, Theorem 6.5.2 is good enough. For instance, because any $x \in (-R, R)$ is contained in the interior of a closed interval $[-c, c] \subseteq (-R, R)$, it now follows that a power series that converges on an open interval is necessarily continuous on this interval.

But what happens if we know that a series converges at an endpoint of its interval of convergence? Does the good behavior of the series on $(-R, R)$ necessarily extend to the endpoint $x = R$? If the convergence of the series at $x = R$ is absolute convergence, then we can again rely on Theorem 6.5.2 to conclude that the series converges uniformly on the set $[-R, R]$. The remaining interesting open question is what happens if a series converges *conditionally* at a point $x = R$. We may still use Theorem 6.5.1 to conclude that we have pointwise convergence on the interval $(-R, R]$, but more work is needed to establish uniform convergence on compact sets containing $x = R$.

Abel's Theorem

We should remark that if the power series $g(x) = \sum_{n=0}^{\infty} a_n x^n$ converges conditionally at $x = R$, then it is possible for it to diverge when $x = -R$. The series

$$\sum_{n=1}^{\infty} \frac{(-1)^n x^n}{n}$$

with $R = 1$ is an example. To keep our attention fixed on the convergent endpoint, we will prove uniform convergence on the set $[0, R]$.

The first step in the argument is an estimate that should be compared to Abel's Test for convergence of series, developed back in Chapter 2 (Exercise 2.7.13).

Lemma 6.5.3 (Abel's Lemma). *Let b_n satisfy $b_1 \geq b_2 \geq b_3 \geq \cdots \geq 0$, and let $\sum_{n=1}^{\infty} a_n$ be a series for which the partial sums are bounded. In other words, assume there exists $A > 0$ such that*

$$|a_1 + a_2 + \cdots + a_n| \leq A$$

for all $n \in N$. Then, for all $n \in \mathbf{N}$,

$$|a_1 b_1 + a_2 b_2 + a_3 b_3 + \cdots + a_n b_n| \leq A b_1.$$

Proof. Let $s_n = a_1 + a_2 + \cdots + a_n$. Using the summation-by-parts formula derived in Exercise 2.7.12, we can write

$$
\begin{aligned}
\left| \sum_{k=1}^{n} a_k b_k \right| &= \left| s_n b_{n+1} + \sum_{k=1}^{n} s_k (b_k - b_{k+1}) \right| \\
&\leq A b_{n+1} + \sum_{k=1}^{n} A(b_k - b_{k+1}) \\
&= A b_{n+1} + (A b_1 - A b_{n+1}) = A b_1. \qquad \square
\end{aligned}
$$

It is worth observing that if A were an upper bound on the partial sums of $\sum |a_n|$ (note the absolute value bars), then the proof of Lemma 6.5.3 would be a simple exercise in the triangle inequality. The point of the matter is that because we are only assuming conditional convergence, the triangle inequality is not going to be of any use in proving Abel's Theorem, but we are now in possession of an inequality that we can use in its place.

Theorem 6.5.4 (Abel's Theorem). *Let $g(x) = \sum_{n=0}^{\infty} a_n x^n$ be a power series that converges at the point $x = R > 0$. Then the series converges uniformly on the interval $[0, R]$. A similar result holds if the series converges at $x = -R$.*

Proof. To set the stage for an application of Lemma 6.5.3, we first write

$$g(x) = \sum_{n=0}^{\infty} a_n x^n = \sum_{n=0}^{\infty} (a_n R^n) \left(\frac{x}{R} \right)^n.$$

Let $\epsilon > 0$. By the Cauchy Criterion for Uniform Convergence of Series (Theorem 6.4.4), we will be done if we can produce an N such that $n > m \geq N$ implies

$$
\text{(1)} \quad \left| (a_{m+1} R^{m+1}) \left(\frac{x}{R} \right)^{m+1} + (a_{m+2} R^{m+2}) \left(\frac{x}{R} \right)^{m+2} + \cdots \right.
$$
$$
\left. + (a_n R^n) \left(\frac{x}{R} \right)^n \right| < \epsilon.
$$

Because we are assuming that $\sum_{n=0}^{\infty} a_n R^n$ converges, the Cauchy Criterion for convergent series of real numbers guarantees that there exists an N such that

$$|a_{m+1}R^{m+1} + a_{m+2}R^{m+2} + \cdots + a_n R^n| < \frac{\epsilon}{2}$$

whenever $n > m \geq N$. But now, for any fixed $m \in \mathbf{N}$, we can apply Abel's Lemma (Lemma 6.5.3) to the sequences obtained by omitting the first m terms. Using $\epsilon/2$ as a bound on the partial sums of $\sum_{j=1}^{\infty} a_{m+j}R^{m+j}$ and observing that $(x/R)^{m+j}$ is monotone decreasing, an application of Abel's Lemma to equation (1) yields

$$\left| (a_{m+1}R^{m+1}) \left(\frac{x}{R}\right)^{m+1} + (a_{m+2}R^{m+2}) \left(\frac{x}{R}\right)^{m+2} + \cdots \right.$$
$$\left. + (a_n R^n) \left(\frac{x}{R}\right)^{n} \right| \leq \frac{\epsilon}{2} \left(\frac{x}{R}\right)^{m+1} < \epsilon.$$

$\square$

The Success of Power Series

An economical way to summarize the conclusions of Theorem 6.5.2 and Abel's Theorem is with the following statement.

Theorem 6.5.5. *If a power series converges pointwise on the set $A \subseteq \mathbf{R}$, then it converges uniformly on any compact set $K \subseteq A$.*

Proof. A compact set contains both a maximum x_1 and a minimum x_0, which by hypothesis must be in A. Abel's Theorem implies the series converges uniformly on the interval $[x_0, x_1]$ and thus also on K. $\square$

This fact leads to the desirable conclusion that a power series is continuous at every point at which it converges. To make an argument for differentiability, we would like to appeal to Theorem 6.4.3; however, this result has a slightly more involved set of hypotheses. In order to conclude that a power series $\sum_{n=0}^{\infty} a_n x^n$ is differentiable, and that term-by-term differentiation is allowed, we need to know beforehand that the differentiated series $\sum_{n=1}^{\infty} n a_n x^{n-1}$ converges uniformly.

Theorem 6.5.6. *If $\sum_{n=0}^{\infty} a_n x^n$ converges for all $x \in (-R, R)$, then the differentiated series $\sum_{n=1}^{\infty} n a_n x^{n-1}$ converges at each $x \in (-R, R)$ as well. Consequently, the convergence is uniform on compact sets contained in $(-R, R)$.*

Proof. Exercise 6.5.5. $\square$

We should point out that it is possible for a series to converge at an endpoint $x = R$ but for the differentiated series to diverge at this point. The series $\sum_{n=1}^{\infty} x^n/n$ has this property when $x = -1$. On the other hand, if the differentiated series does converge at the point $x = R$, then Abel's Theorem applies and the convergence of the differentiated series is uniform on compact sets that contain R.

With all the pieces in place, we summarize the impressive conclusions of this section.

Theorem 6.5.7. *Assume*

$$f(x) = \sum_{n=0}^{\infty} a_n x^n$$

converges on an interval $A \subseteq \mathbf{R}$. The function f is continuous on A and differentiable on any open interval $(-R, R) \subseteq A$. The derivative is given by

$$f'(x) = \sum_{n=1}^{\infty} n a_n x^{n-1}.$$

Moreover, f is infinitely differentiable on $(-R, R)$, and the successive derivatives can be obtained via term-by-term differentiation of the appropriate series.

Proof. The details for why f is continuous have been discussed. Theorem 6.5.6 justifies the application of the Term-by-term Differentiability Theorem (Theorem 6.4.3), which verifies the formula for f'.

A differentiated power series is a power series in its own right, and Theorem 6.5.6 implies that, although the series may no longer converge at a particular endpoint, the radius of convergence does not change. By induction, then, power series are differentiable an infinite number of times. $\square$

Exercises

Exercise 6.5.1. Consider the function g defined by the power series

$$g(x) = x - \frac{x^2}{2} + \frac{x^3}{3} - \frac{x^4}{4} + \frac{x^5}{5} - \cdots .$$

(a) Is g defined on $(-1, 1)$? Is it continuous on this set? Is g defined on $(-1, 1]$? Is it continuous on this set? What happens on $[-1, 1]$? Can the power series for $g(x)$ possibly converge for any other points $|x| > 1$? Explain.

(b) For what values of x is $g'(x)$ defined? Find a formula for g'.

Exercise 6.5.2. Find suitable coefficients (a_n) so that the resulting power series $\sum a_n x^n$ has the given properties, or explain why such a request is impossible.

(a) Converges for every value of $x \in \mathbf{R}$.

(b) Diverges for every value of $x \in \mathbf{R}$.

(c) Converges absolutely for all $x \in [-1, 1]$ and diverges off of this set.

(d) Converges conditionally at $x = -1$ and converges absolutely at $x = 1$.

(e) Converges conditionally at both $x = -1$ and $x = 1$.

Exercise 6.5.3. Use the Weierstrass M-Test to prove Theorem 6.5.2.

Exercise 6.5.4 (Term-by-term Antidifferentiation). Assume $f(x) = \sum_{n=0}^{\infty} a_n x^n$ converges on $(-R, R)$.

(a) Show

$$F(x) = \sum_{n=0}^{\infty} \frac{a_n}{n+1} x^{n+1}$$

is defined on $(-R, R)$ and satisfies $F'(x) = f(x)$.

(b) Antiderivatives are not unique. If g is an arbitrary function satisfying $g'(x) = f(x)$ on $(-R, R)$, find a power series representation for g.

Exercise 6.5.5.   (a) If s satisfies $0 < s < 1$, show ns^{n-1} is bounded for all $n \geq 1$.

(b) Given an arbitrary $x \in (-R, R)$, pick t to satisfy $|x| < t < R$. Use this start to construct a proof for Theorem 6.5.6.

Exercise 6.5.6. Previous work on geometric series (Example 2.7.5) justifies the formula

$$\frac{1}{1-x} = 1 + x + x^2 + x^3 + x^4 + \cdots, \qquad \text{for all } |x| < 1.$$

Use the results about power series proved in this section to find values for $\sum_{n=1}^{\infty} n/2^n$ and $\sum_{n=1}^{\infty} n^2/2^n$. The discussion in Section 6.1 may be helpful.

Exercise 6.5.7. Let $\sum a_n x^n$ be a power series with $a_n \neq 0$, and assume

$$L = \lim_{n \to \infty} \left| \frac{a_{n+1}}{a_n} \right|$$

exists.

(a) Show that if $L \neq 0$, then the series converges for all x in $(-1/L, 1/L)$. (The advice in Exercise 2.7.9 may be helpful.)

(b) Show that if $L = 0$, then the series converges for all $x \in \mathbf{R}$.

(c) Show that (a) and (b) continue to hold if L is replaced by the limit

$$L' = \lim_{n \to \infty} s_n \quad \text{where} \quad s_n = \sup \left\{ \left| \frac{a_{k+1}}{a_k} \right| : k \geq n \right\}.$$

(General properties of the *limit superior* are discussed in Exercise 2.4.7.)

Exercise 6.5.8. (a) Show that power series representations are unique. If we have

$$\sum_{n=0}^{\infty} a_n x^n = \sum_{n=0}^{\infty} b_n x^n$$

for all x in an interval $(-R, R)$, prove that $a_n = b_n$ for all $n = 0, 1, 2, \ldots$.

(b) Let $f(x) = \sum_{n=0}^{\infty} a_n x^n$ converge on $(-R, R)$, and assume $f'(x) = f(x)$ for all $x \in (-R, R)$ and $f(0) = 1$. Deduce the values of a_n.

Exercise 6.5.9. Review the definitions and results from Section 2.8 concerning products of series and Cauchy products in particular. At the end of Section 2.9, we mentioned the following result: If both $\sum a_n$ and $\sum b_n$ converge conditionally to A and B respectively, then it is possible for the Cauchy product,

$$\sum d_n \quad \text{where} \quad d_n = a_0 b_n + a_1 b_{n-1} + \cdots + a_n b_0,$$

to diverge. However, if $\sum d_n$ does converge, then it must converge to AB. To prove this, set

$$f(x) = \sum a_n x^n, \quad g(x) = \sum b_n x^n, \quad \text{and} \quad h(x) = \sum d_n x^n.$$

Use Abel's Theorem and the result in Exercise 2.8.7 to establish this result.

Exercise 6.5.10. Let $g(x) = \sum_{n=0}^{\infty} b_n x^n$ converge on $(-R, R)$, and assume $(x_n) \to 0$ with $x_n \neq 0$. If $g(x_n) = 0$ for all $n \in \mathbf{N}$, show that $g(x)$ must be identically zero on all of $(-R, R)$.

Exercise 6.5.11. A series $\sum_{n=0}^{\infty} a_n$ is said to be *Abel-summable to L* if the power series

$$f(x) = \sum_{n=0}^{\infty} a_n x^n$$

converges for all $x \in [0, 1)$ and $L = \lim_{x \to 1^-} f(x)$.

(a) Show that any series that converges to a limit L is also Abel-summable to L.

(b) Show that $\sum_{n=0}^{\infty} (-1)^n$ is Abel-summable and find the sum.

6.6 Taylor Series

Our study of power series has led to some enthusiastic conclusions about the nature of functions of the form

$$f(x) = a_0 + a_1 x + a_2 x^2 + a_3 x^3 + a_4 x^4 + \cdots.$$

Despite their infinite character, power series can be manipulated more or less as though they are polynomials. On its interval of convergence, a power series is continuous and infinitely differentiable, and successive derivatives or antiderivatives can be computed by performing the desired operation on each individual term in the series—just as it is done for polynomials.

In Section 6.1 we informally encountered the powerful idea that familiar functions such as $\arctan(x)$ and $\sqrt{1+x}$ can be represented as power series. This is a game changing revelation. If a function can be represented as a power series, and a power series can be treated like a polynomial, then vast new possibilities are suddenly available for the kinds of calculations that can be undertaken. Given this state of affairs, it is natural to wonder whether *all* of the well-behaved—i.e., infinitely differentiable—functions of calculus might have representations as power series.

In the examples and exercises in this section, we will assume the familiar properties of the trigonometric, inverse trigonometric, exponential, and logarithmic functions. Rigorously defining these functions is an important exercise in analysis. In fact, one of the most common methods for providing proper definitions is through power series, a point of view that is explored in Section 8.4. The point of this discussion, however, is to come at this question from the other direction. Assuming we are in possession of an infinitely differentiable function such as $\sin(x)$, can we find suitable coefficients a_n so that

$$\sin(x) = a_0 + a_1 x + a_2 x^2 + a_3 x^3 + a_4 x^4 + \cdots$$

for at least some nonzero values of x?

Manipulating Series

In Section 6.1 we generated several new series representations starting from the formula

$$(1) \qquad \frac{1}{1-x} = 1 + x + x^2 + x^3 + x^4 + \cdots, \quad \text{for all } |x| < 1$$

proved in Example 2.7.5. At the time, we were not concerned with supplying rigorous proofs, but we have since done the bulk of the work necessary to confidently assert that the manipulations in Section 6.1 are perfectly valid.

Example 6.6.1. Theorem 6.5.7 applied to equation (1) gives

$$\frac{1}{(1-x)^2} = 1 + 2x + 3x^2 + 4x^3 + 5x^4 + \cdots, \quad \text{for all } |x| < 1.$$

What about the series we generated for $\arctan(x)$? The substitution of $-x^2$ for x in (1) doesn't cause any problem:

$$\frac{1}{1+x^2} = 1 - x^2 + x^4 - x^6 + x^8 - \cdots, \quad \text{for all } |x| < 1.$$

The content of Exercise 6.5.4 is that we can take the term-by-term antiderivative of this series and arrive at an antiderivative for $1/(1 + x^2)$. Noting that $\arctan(0) = 0$, it follows that

$$(2) \qquad \arctan(x) = x - \frac{1}{3}x^3 + \frac{1}{5}x^5 - \frac{1}{7}x^7 + \cdots,$$

for all $x \in (-1, 1)$. In fact, this formula is also valid for $x = \pm 1$. (Exercise 6.6.1.) Similar methods can be used to find series representations for functions such as $\log(1 + x)$ and $x/(1 + x^2)^2$.

Taylor's Formula for the Coefficients

Manipulating old series to produce new ones was a well-honed craft in the 17th and 18th centuries, but there also emerged a formula for producing the coefficients from "scratch"—a recipe for generating a power series representation using only the function in question and its derivatives. The technique is named after the mathematician Brook Taylor (1685–1731) who published it in 1715, although it was certainly known previous to this date.

Given an infinitely differentiable function f defined on some interval centered at zero, the idea is to assume that f has a power series expansion and deduce what the coefficients must be.

Theorem 6.6.2 (Taylor's Formula). *Let*

$$(3) \qquad f(x) = a_0 + a_1 x + a_2 x^2 + a_3 x^3 + a_4 x^4 + a_5 x^5 + \cdots$$

be defined on some nontrivial interval centered at zero. Then,

$$a_n = \frac{f^{(n)}(0)}{n!}.$$

Proof. Exercise 6.6.3 □

Let's use Taylor's formula to produce the so-called *Taylor series* for $\sin(x)$. For the constant term we get $a_0 = \sin(0) = 0$. Then, $a_1 = \cos(0) = 1$, $a_2 = -\sin(0)/2! = 0$, and $a_3 = -\cos(0)/3! = -1/3!$. Continuing on, we are led to the series

$$x - \frac{x^3}{3!} + \frac{x^5}{5!} - \frac{x^7}{7!} + \cdots.$$

So can we say that this series *equals* $\sin(x)$? Well, we need to be very clear about what we have proved to this point. To derive Taylor's formula, *we assumed that f actually had a power series representation*. The conclusion is that if f can be expressed in the form

$$f(x) = \sum_{n=0}^{\infty} a_n x^n,$$

then it must be that

$$a_n = \frac{f^{(n)}(0)}{n!}.$$

But what about the converse question? Assume f is infinitely differentiable in a neighborhood of zero. If we let

$$a_n = \frac{f^{(n)}(0)}{n!},$$

does the resulting series

$$\sum_{n=0}^{\infty} a_n x^n$$

converge to $f(x)$ on some nontrivial set of points? Does it converge at all? If it does converge, we know that the limit function is an infinitely differentiable function whose derivatives at zero are exactly the same as the derivatives of f. Is it possible for this limit to be different from f? In other words, might the Taylor series of a function converge to the wrong thing?

Let

$$S_N(x) = a_0 + a_1 x + a_2 x^2 + \cdots + a_N x^N.$$

The polynomial $S_N(x)$ is a partial sum of the Taylor series expansion for the function $f(x)$. Thus, we are interested in whether or not

$$\lim_{N \to \infty} S_N(x) = f(x)$$

for some values of x besides zero.

Lagrange's Remainder Theorem

A powerful tool for analyzing this question was provided by Joseph Louis Lagrange (1736–1813). The idea is to consider the difference

$$E_N(x) = f(x) - S_N(x),$$

which represents the error between f and the partial sum S_N.

Theorem 6.6.3 (Lagrange's Remainder Theorem). *Let f be differentiable $N+1$ times on $(-R, R)$, define $a_n = f^{(n)}(0)/n!$ for $n = 0, 1, \ldots, N$, and let*

$$S_N(x) = a_0 + a_1 x + a_2 x^2 + \cdots + a_N x^N.$$

Given $x \neq 0$ in $(-R, R)$, there exists a point c satisfying $|c| < |x|$ where the error function $E_N(x) = f(x) - S_N(x)$ satisfies

$$E_N(x) = \frac{f^{(N+1)}(c)}{(N+1)!} x^{N+1}.$$

Before embarking on a proof, let's examine the significance of this result. Proving $S_N(x) \to f(x)$ is equivalent to showing $E_N(x) \to 0$. There are three components to the expression for $E_N(x)$. In the denominator, we have $(N+1)!$, which helps to make E_N small as N tends to infinity. In the numerator, we have x^{N+1}, which potentially grows depending on the size of x. Thus, we should expect that a Taylor series is less likely to converge the farther x is chosen from the origin. Finally, we have $f^{(N+1)}(c)$, which is a bit of a mystery. For functions with straightforward derivatives, this term can often be handled using a suitable upper bound.

Example 6.6.4. Consider the Taylor series for $\sin(x)$ generated earlier. How well does

$$S_5(x) = x - \frac{1}{3!}x^3 + \frac{1}{5!}x^5$$

approximate $\sin(x)$ on the interval $[-2, 2]$? Lagrange's Remainder Theorem asserts that the difference between these two functions is

$$E_5(x) = \sin(x) - S_5(x) = \frac{-\sin(c)}{6!}x^6$$

for some c in the interval $(-|x|, |x|)$. Not knowing the value of c, we can still be quite certain that $|\sin(c)| \leq 1$. Because $x \in [-2, 2]$, we have

$$|E_5(x)| \leq \frac{2^6}{6!} \approx .089.$$

To prove that $S_N(x)$ converges uniformly to $\sin(x)$ on $[-2, 2]$, we observe that the $f^{(N+1)}(c)$ term in the Lagrange formula will never exceed 1 in absolute value. Thus,

$$|E_N(x)| = \left| \frac{f^{(N+1)}(c)}{(N+1)!}x^{N+1} \right| \leq \frac{1}{(N+1)!}2^{N+1}$$

for $x \in [-2, 2]$. Because factorials grow significantly faster than exponentials, it follows that $E_N(x) \to 0$ uniformly on $[-2, 2]$.

Replacing the constant 2 with an arbitrary constant R has no effect on the validity of the argument, and so the Taylor series converges uniformly to $\sin(x)$ on every interval of the form $[-R, R]$.

Proof of Lagrange's Remainder Theorem: The Taylor coefficients are chosen so that the function f and the polynomial S_N have the same derivatives at zero, at least up through the Nth derivative, after which S_N becomes the zero function. In other words, $f^{(n)}(0) = S_N^{(n)}(0)$ for all $0 \leq n \leq N$, which implies the error function $E_N(x) = f(x) - S_N(x)$ satisfies

$$E_N^{(n)}(0) = 0 \quad \text{for all } n = 0, 1, 2, \ldots, N.$$

The key ingredient in this argument is the Generalized Mean Value Theorem (Theorem 5.3.5) from Chapter 5. To simplify notation, let's assume $x > 0$ and apply the Generalized Mean Value Theorem to the functions $E_N(x)$ and x^{N+1} on the interval $[0, x]$. Thus, there exists a point $x_1 \in (0, x)$ such that

$$\frac{E_N(x)}{x^{N+1}} = \frac{E'_N(x_1)}{(N+1)x_1^N}.$$

Now apply the Generalized Mean Value Theorem to the functions $E'_N(x)$ and $(N+1)x^N$ on the interval $[0, x_1]$ to get that there exists a point $x_2 \in (0, x_1)$ where

$$\frac{E_N(x)}{x^{N+1}} = \frac{E'_N(x_1)}{(N+1)x_1^N} = \frac{E''_N(x_2)}{(N+1)Nx_2^{N-1}}.$$

Continuing in this manner we find

$$\frac{E_N(x)}{x^{N+1}} = \frac{E_N^{(N+1)}(x_{N+1})}{(N+1)!}$$

where $x_{N+1} \in (0, x_N) \subseteq \cdots \subseteq (0, x)$. Now set $c = x_{N+1}$. Because $S_N^{(N+1)}(x) = 0$, we have $E_N^{(N+1)}(x) = f^{(N+1)}(x)$ and it follows that

$$E_N(x) = \frac{f^{(N+1)}(c)}{(N+1)!}x^{N+1}$$

as desired.   □

Taylor Series Centered at $a \neq 0$.

Throughout this chapter we have focused our attention on series expansions centered at zero, but there is nothing special about zero other than notational simplicity. If f is defined in some neighborhood of $a \in \mathbf{R}$ and infinitely differentiable at a, then the Taylor series expansion around a takes the form

$$\sum_{n=0}^{\infty} c_n(x-a)^n \quad \text{where} \quad c_n = \frac{f^{(n)}(a)}{n!}.$$

Setting $E_N(x) = f(x) - S_N(x)$ as usual, Lagrange's Remainder Theorem in this case says that there exists a value c between a and x where

$$E_N(x) = \frac{f^{(N+1)}(c)}{(N+1)!}(x-a)^{N+1}.$$

In Exercise 6.6.9, we derive an alternate remainder formula due to Cauchy that requires these more general expansions for its derivation.

A Counterexample

Lagrange's Remainder Theorem is extremely useful for determining how well the partial sums of the Taylor series approximate the original function, but it leaves unresolved the central question of whether or not the Taylor series necessarily converges to the function that generated it. The appearance of $f^{(N+1)}(c)$ in the error formula makes any general statement impossible. The Cauchy form of the remainder just mentioned provides another way to represent the error between the partial sum $S_N(x)$ and the function $f(x)$, and there are others still, but none lend themselves to a proof that $S_N \to f$. This is because no such proof exists! Let

$$g(x) = \begin{cases} e^{-1/x^2} & \text{for } x \neq 0, \\ 0 & \text{for } x = 0. \end{cases}$$

Computing the Taylor coefficients for this function, it's clear that $a_0 = g(0) = 0$. To compute a_1 we write

$$a_1 = g'(0) = \lim_{x \to 0} \frac{g(x) - g(0)}{x - 0} = \lim_{x \to 0} \frac{e^{-1/x^2}}{x} = \lim_{x \to 0} \frac{1/x}{e^{1/x^2}}$$

where both numerator and denominator tend to ∞ as x approaches zero. Applying the ∞/∞ version of L'Hospital's Rule (Theorem 5.3.8) we see

$$a_1 = \lim_{x \to 0} \frac{-1/x^2}{e^{1/x^2}(-2/x^3)} = \lim_{x \to 0} \frac{x}{2e^{1/x^2}} = 0.$$

This tells us that g is flat at the origin. In Exercise 6.6.6, we outline the rest of the proof showing that $g^{(n)}(0) = 0$ for all $n \in \mathbf{N}$; in other words, g is *extremely flat* at the origin.

The implications of this example are highly significant. The function g is infinitely differentiable, and every one of its Taylor coefficients is equal to zero. By default, then, its Taylor series converges uniformly on all of $\mathbf{R}$ to the zero function. But other than at $x = 0$, $g(x)$ is never equal to zero. *The Taylor series for $g(x)$ converges, but it does not converge to $g(x)$ except at the center point $x = 0$.* The unmistakable conclusion is that not every infinitely differentiable function can be represented by its Taylor series.

Exercises

Exercise 6.6.1. The derivation in Example 6.6.1 shows the Taylor series for $\arctan(x)$ is valid for all $x \in (-1, 1)$. Notice, however, that the series also converges when $x = 1$. Assuming that $\arctan(x)$ is continuous, explain why the value of the series at $x = 1$ must necessarily be $\arctan(1)$. What interesting identity do we get in this case?

Exercise 6.6.2. Starting from one of the previously generated series in this section, use manipulations similar to those in Example 6.6.1 to find Taylor series representations for each of the following functions. For precisely what values of x is each series representation valid?

(a) $x \cos(x^2)$

(b) $x/(1 + 4x^2)^2$

(c) $\log(1 + x^2)$

Exercise 6.6.3. Derive the formula for the Taylor coefficients given in Theorem 6.6.2.

Exercise 6.6.4. Explain how Lagrange's Remainder Theorem can be modified to prove

$$1 - \frac{1}{2} + \frac{1}{3} - \frac{1}{4} + \frac{1}{5} - \frac{1}{6} + \cdots = \log(2).$$

Exercise 6.6.5.   (a) Generate the Taylor coefficients for the exponential function $f(x) = e^x$, and then prove that the corresponding Taylor series converges uniformly to e^x on any interval of the form $[-R, R]$.

(b) Verify the formula $f'(x) = e^x$.

(c) Use a substitution to generate the series for e^{-x}, and then informally calculate $e^x \cdot e^{-x}$ by multiplying together the two series and collecting common powers of x.

Exercise 6.6.6. Review the proof that $g'(0) = 0$ for the function

$$g(x) = \begin{cases} e^{-1/x^2} & \text{for } x \neq 0, \\ 0 & \text{for } x = 0. \end{cases}$$

introduced at the end of this section.

(a) Compute $g'(x)$ for $x \neq 0$. Then use the definition of the derivative to find $g''(0)$.

(b) Compute $g''(x)$ and $g'''(x)$ for $x \neq 0$. Use these observations and invent whatever notation is needed to give a general description for the nth derivative $g^{(n)}(x)$ at points different from zero.

(c) Construct a general argument for why $g^{(n)}(0) = 0$ for all $n \in \mathbf{N}$.

Exercise 6.6.7. Find an example of each of the following or explain why no such function exists.

(a) An infinitely differentiable function $g(x)$ on all of $\mathbf{R}$ with a Taylor series that converges to $g(x)$ only for $x \in (-1, 1)$.

(b) An infinitely differentiable function $h(x)$ with the same Taylor series as $\sin(x)$ but such that $h(x) \neq \sin(x)$ for all $x \neq 0$.

(c) An infinitely differentiable function $f(x)$ on all of $\mathbf{R}$ with a Taylor series that converges to $f(x)$ if and only if $x \leq 0$.

Exercise 6.6.8. Here is a weaker form of Lagrange's Remainder Theorem whose proof is arguably more illuminating than the one for the stronger result.

(a) First establish a lemma: If g and h are differentiable on $[0, x]$ with $g(0) = h(0)$ and $g'(t) \leq h'(t)$ for all $t \in [0, x]$, then $g(t) \leq h(t)$ for all $t \in [0, x]$.

(b) Let f, S_N, and E_N be as Theorem 6.6.3, and take $0 < x < R$. If $|f^{(N+1)}(t)| \leq M$ for all $t \in [0, x]$, show
$$|E_N(x)| \leq \frac{Mx^{N+1}}{(N+1)!}.$$

Exercise 6.6.9 (Cauchy's Remainder Theorem). Let f be differentiable $N + 1$ times on $(-R, R)$. For each $a \in (-R, R)$, let $S_N(x, a)$ be the partial sum of the Taylor series for f centered at a; in other words, define
$$S_N(x, a) = \sum_{n=0}^{N} c_n(x - a)^n \quad \text{where} \quad c_n = \frac{f^{(n)}(a)}{n!}.$$

Let $E_N(x, a) = f(x) - S_N(x, a)$. Now fix $x \neq 0$ in $(-R, R)$ and consider $E_N(x, a)$ as a function of a.

(a) Find $E_N(x, x)$.

(b) Explain why $E_N(x, a)$ is differentiable with respect to a, and show
$$E_N'(x, a) = \frac{-f^{(N+1)}(a)}{N!}(x - a)^N.$$

(c) Show
$$E_N(x) = E_N(x, 0) = \frac{f^{(N+1)}(c)}{N!}(x - c)^N x$$
for some c between 0 and x. This is Cauchy's form of the remainder for Taylor series centered at the origin.

Exercise 6.6.10. Consider $f(x) = 1/\sqrt{1 - x}$.

(a) Generate the Taylor series for f centered at zero, and use Lagrange's Remainder Theorem to show the series converges to f on $[0, 1/2]$. (The case $x < 1/2$ is more straightforward while $x = 1/2$ requires some extra care.) What happens when we attempt this with $x > 1/2$?

(b) Use Cauchy's Remainder Theorem proved in Exercise 6.6.9 to show the series representation for f holds on $[0, 1)$.

6.7 The Weierstrass Approximation Theorem

Karl Weierstrass's name is attached to a number of significant results discussed already. The Bolzano-Weierstrass Theorem was fundamental to understanding the relationship between convergence, completeness, and compactness worked out in the early chapters. In this chapter, the Weierstrass M-Test emerged as the primary tool for demonstrating uniform convergence of infinite series.

As discussed in Section 5.4, Weierstrass was also responsible for one of the earliest examples of a continuous, nowhere differentiable function, making this discovery in 1872.

In 1885, Weierstrass proved a result that served as an interesting counterpoint to his nowhere differentiable function. This theorem, which also bears his name, would become the catalyst for a new branch of analysis called approximation theory.

Theorem 6.7.1 (Weierstrass Approximation Theorem). *Let $f : [a, b] \to$* **R** *be continuous. Given $\epsilon > 0$, there exists a polynomial $p(x)$ satisfying*

$$|f(x) - p(x)| < \epsilon$$

for all $x \in [a, b]$.

A restatement of the Weierstrass Approximation Theorem (WAT) without all the symbols is that every continuous function on a closed interval can be uniformly approximated by a polynomial.

Exercise 6.7.1. Assuming WAT, show that if f is continuous on $[a, b]$, then there exists a sequence (p_n) of polynomials such that $p_n \to f$ uniformly on $[a, b]$.

Our work in the previous section provides a nice starting point for understanding what WAT is saying. Given a function such as $\sin(x)$, we saw in Example 6.6.4 that the resulting Taylor series converges uniformly on compact sets back to $\sin(x)$. Because the partial sums of a Taylor series are polynomials, this example constitutes a proof of WAT in the very special case of $f(x) = \sin(x)$. It should be clear, however, that Taylor series won't work in general. To construct a Taylor series, we need f to be an infinitely differentiable function (and even then the Taylor series might fail to approximate f), while WAT requires only that f be continuous.

So should we be surprised that such a theorem is true? This is hard to say. On a purely intuitive level, if we consider a smooth curve like $f(x) = \sqrt{1-x}$ on $[-1, 1]$, then it doesn't take too much imagination to believe that a polynomial might exist that tracks closely with $\sqrt{1-x}$ as x moves over the domain. But one of the lessons of Section 5.4 is that a continuous function does not have to be smooth. Although it is not Weierstrass's original example, a careful look at the nowhere differentiable function shown in Figure 5.7 makes the point just as well. Despite the unimaginably jagged nature of the graph, according to WAT, it is still possible to find a polynomial that uniformly approximates this unruly function to any prescribed degree of accuracy.

Interpolation

Weierstrass's theorem deals with approximating polynomials, but a good way to get a feel for the content of this result is to temporarily replace the polynomials in WAT with the collection of all continuous, piecewise-linear functions.

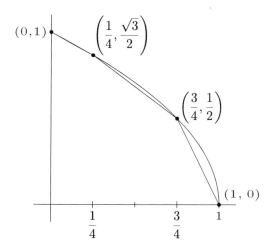

Figure 6.6: POLYGONAL APPROXIMATION OF $f(x) = \sqrt{1-x}$.

Definition 6.7.2. A continuous function $\phi : [a, b] \to \mathbf{R}$ is *polygonal* if there is a partition

$$a = x_0 < x_1 < x_2 < \cdots < x_n = b$$

of $[a, b]$ such that ϕ is linear on each subinterval $[x_{i-1}, x_i]$, where $i = 1, \ldots n$.

The term "interpolation" refers to the process of finding a function whose graph passes through a given set of points. If, for example, we take the points

$$(0, 1), \left(\frac{1}{4}, \frac{\sqrt{3}}{2}\right), \left(\frac{3}{4}, \frac{1}{2}\right), (1, 0)$$

then there is an obvious polygonal function that interpolates these points: it is just the function we get by connecting the points with line segments. Now these four points all lie on the graph of $f(x) = \sqrt{1-x}$, and notice that the resulting polygonal interpolation does a reasonable job of imitating the graph of f (Fig. 6.6). This is not an accident.

Theorem 6.7.3. *Let $f : [a, b] \to \mathbf{R}$ be continuous. Given $\epsilon > 0$, there exists a polygonal function ϕ satisfying*

$$|f(x) - \phi(x)| < \epsilon$$

for all $x \in [a, b]$.

Exercise 6.7.2. Prove Theorem 6.7.3.

Notice how similar Theorem 6.7.3 is to WAT, the only difference being that we have substituted a polygonal function in place of the polynomial.

The strategy for the proof of Theorem 6.7.3 is to first choose an appropriate numbers of points on the graph of f, and then show that the resulting polygonal interpolation of these points does the trick. It's not unreasonable to suspect that a similar strategy might lead to a proof of the Weierstrass Approximation Theorem. Can we prove WAT by constructing a polynomial interpolation of points on the graph of f? Well, no as it turns out, but this is not so easy to see.

Exercise 6.7.3.   (a) Find the second degree polynomial $p(x) = q_0 + q_1 x + q_2 x^2$ that interpolates the three points $(-1, 1), (0, 0)$, and $(1, 1)$ on the graph of $g(x) = |x|$. Sketch $g(x)$ and $p(x)$ over $[-1, 1]$ on the same set of axes.

   (b) Find the fourth degree polynomial that interpolates $g(x) = |x|$ at the points $x = -1, -1/2, 0, 1/2$, and 1. Add a sketch of this polynomial to the graph from (a).

   The previous exercise may still give the impression that a polynomial interpolation approach is going to lead to a proof of WAT, but that isn't the case. Continuing on with larger and larger numbers of equally spaced points yields high degree polynomials that oscillate very rapidly and actually do a poor job of approximating g *between* the interpolating points. In fact, it turns out that the resulting sequence of polynomials only converges to $g(x)$ when $x = -1, 0$, or 1.

Approximating the Absolute Value Function

Having reached a temporary dead end, we need to back up a bit and take a different turn. Let's return to Theorem 6.7.3 which asserts that every continuous function can be uniformly approximated by a polygonal function. This should feel like a promising first step toward a proof of WAT and indeed it is. If we can find a way to approximate an arbitrary polygonal function with polynomials, then a triangle inequality argument would finish the proof.

   Before we get too excited about this line of attack, keep in mind that the absolute value function from Exercise 6.7.3 is an example of a polygonal function and we are currently unsure how to produce polynomials to approximate it. What has changed, however, is our motivation for doing so. A moment's thought reveals that handling the absolute value function might be the key to solving the whole problem. Why is this? Every polygonal function is made up of line segments that meet at corners. If we can find polynomials that uniformly approximate $g(x) = |x|$ with its right angled corner at the origin, then with a little cleverness we ought to be able to handle more general polygonal functions and prove WAT using Theorem 6.7.3.

Cauchy's Remainder Formula for Taylor Series

One elegant way to show $g(x) = |x|$ is the uniform limit of polynomials is via Taylor series, which is a bit surprising given that $|x|$ is not differentiable. The trick, as we will see, is to start by computing the Taylor series for the infinitely differentiable function $\sqrt{1 - x}$.

Exercise 6.7.4. Show that $f(x) = \sqrt{1-x}$ has Taylor series coefficients a_n where $a_0 = 1$ and

$$a_n = \frac{-1 \cdot 3 \cdot 5 \cdots (2n-3)}{2 \cdot 4 \cdot 6 \cdots 2n}$$

for $n \geq 1$.

Our goal is to show

(1)
$$\sqrt{1-x} = \sum_{n=0}^{\infty} a_n x^n$$

for all $x \in [-1, 1]$ by showing that the error function

$$E_N(x) = \sqrt{1-x} - \sum_{n=0}^{N} a_n x^n$$

tends to 0 as $N \to \infty$. To this point, Lagrange's Remainder Theorem has been the featured tool for jobs like this, but it comes up short in this case. To see exactly why, fix $x \in (0, 1]$. Then Theorem 6.6.3 asserts that there exists a $c \in (0, x)$ (dependent on N) such that

$$E_N(x) = \frac{f^{(N+1)}(c)}{(N+1)!} x^{N+1}$$

$$= \frac{1}{(N+1)!} \left(\frac{-1 \cdot 3 \cdot 5 \cdots (2N-1)}{2^{N+1}(1-c)^{N+1/2}} \right) x^{N+1}$$

$$= \left(\frac{-1 \cdot 3 \cdot 5 \cdots (2N-1)}{2 \cdot 4 \cdot 6 \cdots (2N+2)} \right) \left(\frac{x}{1-c} \right)^{N+1/2} x^{1/2}.$$

The problem is that $x/(1-c)$ is largest when $c = x$, and $(x/(1-x))^{N+1/2}$ goes exponentially to infinity when x is bigger than $1/2$. This doesn't mean our Taylor series is only valid on $[0, 1/2]$; it just means we are using the wrong remainder formula.

Exercise 6.7.5. (a) Follow the advice in Exercise 6.6.9 to prove the Cauchy form of the remainder:

$$E_N(x) = \frac{f^{(N+1)}(c)}{N!} (x-c)^N x$$

for some c between 0 and x.

(b) Use this result to prove equation (1) is valid for all $x \in (-1, 1)$.

Although Cauchy's Remainder Theorem doesn't tell us so, equation (1) is also valid at $x = \pm 1$.

Exercise 6.7.6.   (a) Let

$$c_n = \frac{1 \cdot 3 \cdot 5 \cdots (2n-1)}{2 \cdot 4 \cdot 6 \cdots 2n}$$

for $n \geq 1$. Show $c_n < \frac{2}{\sqrt{2n+1}}$.

(b) Use (a) to show that $\sum_{n=0}^{\infty} a_n$ converges (absolutely, in fact) where a_n is the sequence of Taylor coefficients generated in Exercise 6.7.4.

(c) Carefully explain how this verifies that equation (1) holds for all $x \in [-1, 1]$.

Recall that our goal is to find polynomials that uniformly approximate the absolute value function on an interval containing the non-differentiable point at the origin. Our Taylor series for $\sqrt{1-x}$ provides a clever shortcut for handling this task.

Exercise 6.7.7.   (a) Use the fact that $|a| = \sqrt{a^2}$ to prove that, given $\epsilon > 0$, there exists a polynomial $q(x)$ satisfying

$$||x| - q(x)| < \epsilon$$

for all $x \in [-1, 1]$.

(b) Generalize this conclusion to an arbitrary interval $[a, b]$.

Proving WAT

Earlier we suggested that proving WAT for the special case of the absolute value function was the key to the whole proof. Now it is time to fill in the details.

Exercise 6.7.8.   (a) Fix $a \in [-1, 1]$ and sketch

$$h_a(x) = \frac{1}{2}(|x - a| + (x - a))$$

over $[-1, 1]$. Note that h_a is polygonal and satisfies $h_a(x) = 0$ for all $x \in [-1, a]$.

(b) Explain why we know $h_a(x)$ can be uniformly approximated with a polynomial on $[-1, 1]$.

(c) Let ϕ be a polygonal function that is linear on each subinterval of the partition

$$-1 = a_0 < a_1 < a_2 < \cdots < a_n = 1 .$$

Show there exist constants $b_0, b_1, \ldots, b_{n-1}$ so that

$$\phi(x) = \phi(-1) + b_0 h_{a_0}(x) + b_1 h_{a_1}(x) + \cdots + b_{n-1} h_{a_{n-1}}(x)$$

for all $x \in [-1, 1]$.

(d) Complete the proof of WAT for the interval $[-1, 1]$, and then generalize to an arbitrary interval $[a, b]$.

Exercise 6.7.9. (a) Find a counterexample which shows that WAT is not true if we replace the closed interval $[a, b]$ with the open interval (a, b).

(b) What happens if we replace $[a, b]$ with the closed set $[a, \infty)$. Does the theorem still hold?

Exercise 6.7.10. Is there a countable subset of polynomials $\mathcal{C}$ with the property that every continuous function on $[a, b]$ can be uniformly approximated by polynomials from $\mathcal{C}$?

Exercise 6.7.11. Assume that f has a continuous derivative on $[a, b]$. Show that there exists a polynomial $p(x)$ such that

$$|f(x) - p(x)| < \epsilon \quad \text{and} \quad |f'(x) - p'(x)| < \epsilon$$

for all $x \in [a, b]$.

6.8 Epilogue

The argument sketched out here for the Weierstrass Approximation Theorem is due to Henri Lebesque, who published his proof in 1898. Its greatest virtue is its relative simplicity. Starting from a single special case—the absolute value function—we managed to bootstrap our way up to an arbitrary continuous function. A downside of this approach is that by the time we reach the case of a general continuous function, there is no practical way to explicitly write down a formula for the polynomial that approximates it.

There are a number of other proofs for WAT that don't have this drawback. A particularly popular one was provided by Sergei Bernstein. Bernstein employs a family of polynomials—now called Bernstein polynomials—that have become important in their own right. Weierstrass's original approach was also quite elegant. His proof has much in common with the proof of Fejér's Theorem in Section 8.5 on Fourier series. Not coincidentally, it is possible to derive yet another proof of WAT as a corollary to Fejér's Theorem. (See Exercise 8.5.11.)

The Weierstrass Approximation Theorem is set on a closed interval $[a, b]$. Exercise 6.7.9 is included to emphasize the importance of the closed and bounded nature of the domain, but it should not be too surprising that the theorem will remain true if we replace $[a, b]$ with an arbitrary compact set. What about replacing the set of polynomials? Are there other collections of relatively simple continuous functions that can be used to approximate an arbitrary continuous function? Sure there are. In Theorem 6.7.3 we saw that polygonal functions have this property, and there are other examples as well. In the late 1930s, Marshall Stone proved a far-reaching generalization of the Weierstrass Approximation Theorem. Stone's version of WAT starts with an arbitrary compact set K and a collection $\mathcal{C}$ of continuous functions on K with the following three properties:

(i) the constant function $k(x) = 1$ is in $\mathcal{C}$,

(ii) if $p, q \in \mathcal{C}$ and $c \in \mathbf{R}$ then $p + q, pq,$ and cp are all in $\mathcal{C}$,

(iii) if $x \neq y$ in K, then there exists $p \in \mathcal{C}$ with $p(x) \neq p(y)$.

Under these conditions, Stone showed that any continuous function on K could be uniformly approximated by functions in $\mathcal{C}$. This result, referred to as the Stone–Weierstrass Theorem, has a slightly more involved proof that tracks very closely with Lebesgue's proof of WAT outlined in the previous section. In particular, both arguments depend fundamentally on being able to approximate the absolute value function with polynomials.

A collection of functions that possesses property (ii) of the Stone–Weierstrass Theorem is called an *algebra*. An algebra that possesses property (iii) is said to *separate points*. Having the constant function $k(x) = 1$ in the algebra ensures we don't have some $x_0 \in K$ where $p(x_0) = 0$ for all functions in our algebra. (Why would this be problematic?) It is straightforward to check that the set of polynomials as well as the set of polygonal functions form algebras that separate points, and so both WAT and Theorem 6.7.3 become special cases of Stone's general result. For a new example, consider the collection of polynomials with only even powers on the interval $[0, 1]$. The Stone–Weierstrass Theorem tells us that this subset of polynomials can still uniformly approximate an arbitrary continuous function, although if we were to switch our domain to $[-1, 1]$ then this algebra would no longer separate points. As a final example, consider the set

$$\mathcal{C} = \{a_0 + a_1 \cos(x) + \cdots + a_n \cos(nx) : a_0, a_1, \ldots, a_n \in \mathbf{R}\}.$$

In Section 8.5 we take up the theory of Fourier series which explores when a function has a representation as an infinite series of trigonometric functions. As a precursor to that conversation, notice that the Stone–Weierstrass Theorem tells us at the outset that at least every continuous function on $[0, \pi]$ is the uniform limit of functions from $\mathcal{C}$.

The story from Section 6.6 surrounding Taylor series expansions also deserves a final word. The ingenuity with which Euler and others found and exploited power series representations for the cast of familiar functions from calculus understandably led to speculation that every function could be represented in such a fashion. (The term "function" at this time implicitly referred to functions that were infinitely differentiable.) This point of view effectively ended with Cauchy's discovery in 1821 of the counterexample presented at the end of Section 6.6. So under what conditions does the Taylor series necessarily converge to the generating function? Lagrange's Remainder Theorem states that the difference between the Taylor polynomial $S_N(x)$ and the function $f(x)$ is given by

$$E_N(x) = \frac{f^{(N+1)}(c)}{(N+1)!} x^{N+1}.$$

The $(N+1)!$ term in the denominator grows more rapidly than the x^{N+1} term in the numerator. Thus, if we knew for instance that

$$|f^{(N+1)}(c)| \leq M$$

for all $c \in (-R, R)$ and $N \in \mathbf{N}$, we could be sure that $E_N(x) \to 0$ and hence that $S_N(x) \to f(x)$. This is the case for $\sin(x)$, $\cos(x)$, and e^x, whose derivatives do not grow at all as $N \to \infty$. It is also possible to formulate weaker conditions on the rate of growth of $f^{(N+1)}$ that guarantee convergence.

It is not altogether clear whether Cauchy's counterexample should come as a surprise. The fact that every previous search for a Taylor series ended in success certainly gives the impression that a power series representation is an intrinsic property of infinitely differentiable functions. But notice what we are saying here. A Taylor series for a function f is constructed from the values of f and its derivatives at the origin. If the Taylor series converges to f on some interval $(-R, R)$, then the behavior of f near zero completely determines its behavior at every point in $(-R, R)$. One implication of this would be that if two functions with Taylor series agree on some small neighborhood $(-\epsilon, \epsilon)$, then these two functions would have to be the same everywhere. When it is put this way, we probably should not expect a Taylor series to always converge back to the function from which it was derived. As we have seen, this is not the case for real-valued functions. What is fascinating, however, is that results of this nature *do* hold for functions of a complex variable. The definition of the derivative looks symbolically the same when the real numbers are replaced by complex numbers, but the implications are profoundly different. In this setting, a function that is differentiable at every point in some open disc must necessarily be infinitely differentiable on this set. This supplies the ingredients to construct the Taylor series that in every instance converges uniformly on compact sets to the function that generated it.

Chapter 7

The Riemann Integral

7.1 Discussion: How Should Integration be Defined?

The Fundamental Theorem of Calculus is a statement about the inverse relationship between differentiation and integration. It comes in two parts, depending on whether we are differentiating an integral or integrating a derivative. Under suitable hypotheses on the functions f and F, the Fundamental Theorem of Calculus states that

(i) $\displaystyle\int_a^b F'(x)\,dx = F(b) - F(a)$ and

(ii) if $\displaystyle G(x) = \int_a^x f(t)\,dt$, then $G'(x) = f(x)$.

Before we can undertake any type of rigorous investigation of these statements, we need to settle on a definition for $\int_a^b f$. Historically, the concept of integration was defined as the inverse process of differentiation. In other words, the integral of a function f was understood to be a function F that satisfied $F' = f$. Newton, Leibniz, Fermat, and the other founders of calculus then went on to explore the relationship between antiderivatives and the problem of computing areas. This approach is ultimately unsatisfying from the point of view of analysis because it results in a very limited number of functions that can be integrated. Recall that every derivative satisfies the intermediate value property (Darboux's Theorem, Theorem 5.2.7). This means that any function with a jump discontinuity cannot be a derivative. If we want to define integration via antidifferentiation, then we must accept the consequence that a function as simple as

© Springer Science+Business Media New York 2015
S. Abbott, *Understanding Analysis*, Undergraduate Texts in Mathematics, DOI 10.1007/978-1-4939-2712-8_7

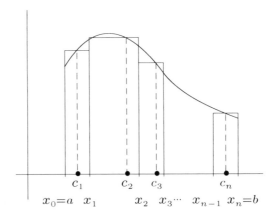

Figure 7.1: A Riemann Sum.

$$h(x) = \begin{cases} 1 & \text{for } 0 \leq x < 1 \\ 2 & \text{for } 1 \leq x \leq 2 \end{cases}$$

is not integrable on the interval $[0, 2]$.

A very interesting shift in emphasis occurred around 1850 in the work of Cauchy, and soon after in the work of Bernhard Riemann. The idea was to completely divorce integration from the derivative and instead use the notion of "area under the curve" as a starting point for building a rigorous definition of the integral. The reasons for this were complicated. As we have mentioned earlier (Section 1.2), the concept of *function* was undergoing a transformation. The traditional understanding of a function as a holistic formula such as $f(x) = x^2$ was being replaced with a more liberal interpretation, which included such bizarre constructions as Dirichlet's function discussed in Section 4.1. Serving as a catalyst to this evolution was the budding theory of Fourier series (discussed in Section 8.5), which required, among other things, the need to be able to integrate these more unruly objects.

The Riemann integral, as it is called today, is the one usually discussed in introductory calculus. Starting with a function f on $[a, b]$, we partition the domain into small subintervals. On each subinterval $[x_{k-1}, x_k]$, we pick some point $c_k \in [x_{k-1}, x_k]$ and use the y-value $f(c_k)$ as an approximation for f on $[x_{k-1}, x_k]$. Graphically speaking, the result is a row of thin rectangles constructed to approximate the area between f and the x-axis. The area of each rectangle is $f(c_k)(x_k - x_{k-1})$, and so the total area of all of the rectangles is given by the *Riemann sum* (Fig. 7.1)

$$\sum_{k=1}^{n} f(c_k)(x_k - x_{k-1}).$$

Note that "area" here comes with the understanding that areas below the x-axis are assigned a negative value.

What should be evident from the graph is that the accuracy of the Riemann-sum approximation seems to improve as the rectangles get thinner. In some sense, we take the *limit* of these approximating Riemann sums as the width of the individual subintervals of the partitions tends to zero. This limit, if it exists, is Riemann's definition of $\int_a^b f$.

This brings us to a handful of questions. Creating a rigorous meaning for the limit just referred to is not too difficult. What will be of most interest to us—and was also to Riemann—is deciding what types of functions can be integrated using this procedure. Specifically, what conditions on f guarantee that this limit exists?

The theory of the Riemann integral turns on the observation that smaller subintervals produce better approximations to the function f. On each subinterval $[x_{k-1}, x_k]$, the function f is approximated by its value at some point $c_k \in [x_{k-1}, x_k]$. The quality of the approximation is directly related to the difference

$$|f(x) - f(c_k)|$$

as x ranges over the subinterval. Because the subintervals can be chosen to have arbitrarily small width, this means that we want $f(x)$ to be close to $f(c_k)$ whenever x is close to c_k. But this sounds like a discussion of continuity! We will soon see that the continuity of f is intimately related to the existence of the Riemann integral $\int_a^b f$.

Is continuity sufficient to prove that the Riemann sums converge to a well-defined limit? Is it necessary, or can the Riemann integral handle a discontinuous function such as $h(x)$ mentioned earlier? Relying on the intuitive notion of area, it would seem that $\int_0^2 h = 3$, but does the Riemann integral reach this conclusion? If so, how discontinuous can a function be before it fails to be integrable? Can the Riemann integral make sense out of something as pathological as Dirichlet's function on the interval $[0, 1]$?

A function such as

$$g(x) = \begin{cases} x^2 \sin(\frac{1}{x}) & \text{for } x \neq 0 \\ 0 & \text{for } x = 0 \end{cases}$$

raises another interesting question. Here is an example of a differentiable function, studied in Section 5.1, where the derivative $g'(x)$ is *not* continuous. As we explore the class of integrable functions, some attempt must be made to reunite the integral with the derivative. Having defined integration independently of differentiation, we would like to come back and investigate the conditions under which equations (i) and (ii) from the Fundamental Theorem of Calculus stated earlier hold. If we are making a wish list for the types of functions that we want to be integrable, then in light of equation (i) it seems desirable to expect this set to at least contain the set of derivatives. The fact that derivatives are not always continuous is further motivation not to content ourselves with an integral that cannot handle some discontinuities.

7.2   The Definition of the Riemann Integral

Although it has the benefit of some polish due to Darboux, the development of the integral presented in this chapter is closely related to the procedure just discussed. In place of Riemann sums, we will construct *upper sums* and *lower sums* (Fig. 7.2), and in place of a limit we will use a supremum and an infimum.

Throughout this section, it is assumed that we are working with a *bounded* function f on a closed interval $[a, b]$, meaning that there exists an $M > 0$ such that $|f(x)| \le M$ for all $x \in [a, b]$.

Partitions, Upper Sums, and Lower Sums

Definition 7.2.1. A *partition* P of $[a, b]$ is a finite set of points from $[a, b]$ that includes both a and b. The notational convention is to always list the points of a partition $P = \{x_0, x_1, x_2, \ldots, x_n\}$ in increasing order; thus,

$$a = x_0 < x_1 < x_2 < \cdots < x_n = b.$$

For each subinterval $[x_{k-1}, x_k]$ of P, let

$$m_k = \inf\{f(x) : x \in [x_{k-1}, x_k]\} \quad \text{and} \quad M_k = \sup\{f(x) : x \in [x_{k-1}, x_k]\}.$$

The *lower sum* of f with respect to P is given by

$$L(f, P) = \sum_{k=1}^{n} m_k(x_k - x_{k-1}).$$

Likewise, we define the *upper sum* of f with respect to P by

$$U(f, P) = \sum_{k=1}^{n} M_k(x_k - x_{k-1}).$$

For a particular partition P, it is clear that $U(f, P) \ge L(f, P)$. The fact that this same inequality holds if the upper and lower sums are computed with respect to different partitions is the content of the next two lemmas.

Definition 7.2.2. A partition Q is a *refinement* of a partition P if Q contains all of the points of P; that is, if $P \subseteq Q$.

Lemma 7.2.3. *If $P \subseteq Q$, then $L(f, P) \le L(f, Q)$, and $U(f, P) \ge U(f, Q)$.*

Proof. Consider what happens when we refine P by adding a single point z to some subinterval $[x_{k-1}, x_k]$ of P.

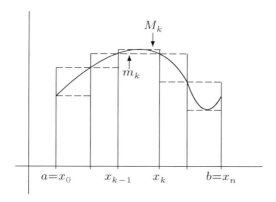

Figure 7.2:   Upper and Lower Sums.

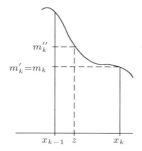

Focusing on the lower sum for a moment, we have

$$m_k(x_k - x_{k-1}) \quad = \quad m_k(x_k - z) + m_k(z - x_{k-1})$$
$$\leq \quad m_k'(x_k - z) + m_k''(z - x_{k-1}),$$

where

$$m_k' = \inf \{f(x) : x \in [z, x_k]\} \quad \text{and} \quad m_k'' = \inf \{f(x) : x \in [x_{k-1}, z]\}$$

are each necessarily as large or larger than m_k.

By induction, we have $L(f, P) \leq L(f, Q)$, and an analogous argument holds for the upper sums.   □

Lemma 7.2.4. *If P_1 and P_2 are any two partitions of $[a, b]$, then $L(f, P_1) \leq U(f, P_2)$.*

Proof. Let $Q = P_1 \cup P_2$ be the so-called *common refinement* of P_1 and P_2. Because $P_1 \subseteq Q$ and $P_2 \subseteq Q$, it follows that

$$L(f, P_1) \leq L(f, Q) \leq U(f, Q) \leq U(f, P_2).$$ □

Integrability

Intuitively, it helps to visualize a particular upper sum as an overestimate for the value of the integral and a lower sum as an underestimate. As the partitions get more refined, the upper sums get potentially smaller while the lower sums get potentially larger. A function is *integrable* if the upper and lower sums "meet" at some common value in the middle.

Rather than taking a limit of these sums, we will instead make use of the Axiom of Completeness and consider the *infimum* of the upper sums and the *supremum* of the lower sums.

Definition 7.2.5. Let $\mathcal{P}$ be the collection of all possible partitions of the interval $[a, b]$. The *upper integral* of f is defined to be

$$U(f) = \inf\{U(f, P) : P \in \mathcal{P}\}.$$

In a similar way, define the *lower integral* of f by

$$L(f) = \sup\{L(f, P) : P \in \mathcal{P}\}.$$

The following fact is not surprising.

Lemma 7.2.6. *For any bounded function f on $[a, b]$, it is always the case that $U(f) \geq L(f)$.*

Proof. Exercise 7.2.1.   □

Definition 7.2.7 (Riemann Integrability). A bounded function f defined on the interval $[a, b]$ is *Riemann-integrable* if $U(f) = L(f)$. In this case, we define $\int_a^b f$ or $\int_a^b f(x)\, dx$ to be this common value; namely,

$$\int_a^b f = U(f) = L(f).$$

The modifier "Riemann" in front of "integrable" accurately suggests that there are other ways to define the integral. In fact, our work in this chapter will expose the need for a different approach, one of which is discussed in Section 8.1. In this chapter, the Riemann integral is the only method under consideration, so it will usually be convenient to drop the modifier "Riemann" and simply refer to a function as being "integrable."

Criteria for Integrability

To summarize the situation thus far, it is always the case for a bounded function f on $[a, b]$ that

$$\sup\{L(f, P) : P \in \mathcal{P}\} = L(f) \leq U(f) = \inf\{U(f, P) : P \in \mathcal{P}\}.$$

The function f is integrable if the inequality is an equality. The major thrust of our investigation of the integral is to describe, as best we can, the class of integrable functions. The preceding inequality reveals that integrability is really equivalent to the existence of partitions whose upper and lower sums are arbitrarily close together.

Theorem 7.2.8 (Integrability Criterion). *A bounded function f is integrable on $[a, b]$ if and only if, for every $\epsilon > 0$, there exists a partition P_ϵ of $[a, b]$ such that*

$$U(f, P_\epsilon) - L(f, P_\epsilon) < \epsilon.$$

Proof. Let $\epsilon > 0$. If such a partition P_ϵ exists, then

$$U(f) - L(f) \leq U(f, P_\epsilon) - L(f, P_\epsilon) < \epsilon.$$

Because ϵ is arbitrary, it must be that $U(f) = L(f)$, so f is integrable. (To be absolutely precise here, we could throw in a reference to Theorem 1.2.6.)

The proof of the converse statement is a familiar triangle inequality argument with parentheses in place of absolute value bars because, in each case, we know which quantity is larger. Because $U(f)$ is the greatest lower bound of the upper sums, we know that, given some $\epsilon > 0$, there must exist a partition P_1 such that

$$U(f, P_1) < U(f) + \frac{\epsilon}{2}.$$

Likewise, there exists a partition P_2 satisfying

$$L(f, P_2) > L(f) - \frac{\epsilon}{2}.$$

Now, let $P_\epsilon = P_1 \cup P_2$ be the common refinement. Keeping in mind that the integrability of f means $U(f) = L(f)$, we can write

$$
\begin{aligned}
U(f, P_\epsilon) - L(f, P_\epsilon) &\leq U(f, P_1) - L(f, P_2) \\
&< \left(U(f) + \frac{\epsilon}{2} \right) - \left(L(f) - \frac{\epsilon}{2} \right) \\
&= \frac{\epsilon}{2} + \frac{\epsilon}{2} = \epsilon.
\end{aligned}
$$

$\square$

In the discussion at the beginning of this chapter, it became clear that integrability is closely tied to the concept of continuity. To make this observation more precise, let $P = \{x_0, x_1, x_2, \ldots, x_n\}$ be an arbitrary partition of $[a, b]$, and define $\Delta x_k = x_k - x_{k-1}$. Then,

$$U(f, P) - L(f, P) = \sum_{k=1}^{n} (M_k - m_k) \Delta x_k,$$

where M_k and m_k are the supremum and infimum of the function on the interval $[x_{k-1}, x_k]$, respectively. Our ability to control the size of $U(f, P) - L(f, P)$ hinges on the differences $M_k - m_k$, which we can interpret as the variation in the range of the function over the interval $[x_{k-1}, x_k]$. Restricting the variation of f over arbitrarily small intervals in $[a, b]$ is *precisely* what it means to say that f is uniformly continuous on this set.

Theorem 7.2.9. *If f is continuous on $[a, b]$, then it is integrable.*

Proof. Because f is continuous on a compact set, it must be bounded. It is also uniformly continuous for the same reason. This means that, given $\epsilon > 0$, there exists a $\delta > 0$ so that $|x - y| < \delta$ guarantees

$$|f(x) - f(y)| < \frac{\epsilon}{b - a}.$$

Now, let P be a partition of $[a, b]$ where $\Delta x_k = x_k - x_{k-1}$ is less than δ for every subinterval of P.

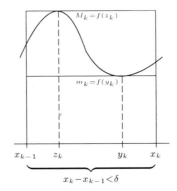

Given a particular subinterval $[x_{k-1}, x_k]$ of P, we know from the Extreme Value Theorem (Theorem 4.4.2) that the supremum $M_k = f(z_k)$ for some $z_k \in [x_{k-1}, x_k]$. In addition, the infimum m_k is attained at some point y_k also in the interval $[x_{k-1}, x_k]$. But this means $|z_k - y_k| < \delta$, so

$$M_k - m_k = f(z_k) - f(y_k) < \frac{\epsilon}{b - a}.$$

Finally,

$$U(f, P) - L(f, P) = \sum_{k=1}^{n} (M_k - m_k)\Delta x_k < \frac{\epsilon}{b - a} \sum_{k=1}^{n} \Delta x_k = \epsilon,$$

and f is integrable by the criterion given in Theorem 7.2.8.   □

Exercises

Exercise 7.2.1. Let f be a bounded function on $[a, b]$, and let P be an arbitrary partition of $[a, b]$. First, explain why $U(f) \geq L(f, P)$. Now, prove Lemma 7.2.6.

Exercise 7.2.2. Consider $f(x) = 1/x$ over the interval $[1, 4]$. Let P be the partition consisting of the points $\{1, 3/2, 2, 4\}$.

(a) Compute $L(f, P)$, $U(f, P)$, and $U(f, P) - L(f, P)$.

(b) What happens to the value of $U(f, P) - L(f, P)$ when we add the point 3 to the partition?

(c) Find a partition P' of $[1, 4]$ for which $U(f, P') - L(f, P') < 2/5$.

Exercise 7.2.3 (Sequential Criterion for Integrability). (a) Prove that a bounded function f is integrable on $[a, b]$ if and only if there exists a sequence of partitions $(P_n)_{n=1}^{\infty}$ satisfying

$$\lim_{n \to \infty} [U(f, P_n) - L(f, P_n)] = 0,$$

and in this case $\int_a^b f = \lim_{n \to \infty} U(f, P_n) = \lim_{n \to \infty} L(f, P_n)$.

(b) For each n, let P_n be the partition of $[0, 1]$ into n equal subintervals. Find formulas for $U(f, P_n)$ and $L(f, P_n)$ if $f(x) = x$. The formula $1 + 2 + 3 + \cdots + n = n(n+1)/2$ will be useful.

(c) Use the sequential criterion for integrability from (a) to show directly that $f(x) = x$ is integrable on $[0, 1]$ and compute $\int_0^1 f$.

Exercise 7.2.4. Let g be bounded on $[a, b]$ and assume there exists a partition P with $L(g, P) = U(g, P)$. Describe g. Is it integrable? If so, what is the value of $\int_a^b g$?

Exercise 7.2.5. Assume that, for each n, f_n is an integrable function on $[a, b]$. If $(f_n) \to f$ uniformly on $[a, b]$, prove that f is also integrable on this set. (We will see that this conclusion does not necessarily follow if the convergence is pointwise.)

Exercise 7.2.6. A *tagged partition* $(P, \{c_k\})$ is one where in addition to a partition P we choose a sampling point c_k in each of the subintervals $[x_{k-1}, x_k]$. The corresponding *Riemann sum*,

$$R(f, P) = \sum_{k=1}^{n} f(c_k) \Delta x_k,$$

is discussed in Section 7.1, where the following definition is alluded to.
Riemann's Original Definition of the Integral: A bounded function f is *integrable* on $[a, b]$ with $\int_a^b f = A$ if for all $\epsilon > 0$ there exists a $\delta > 0$ such that for any tagged partition $(P, \{c_k\})$ satisfying $\Delta x_k < \delta$ for all k, it follows that

$$|R(f, P) - A| < \epsilon.$$

Show that if f satisfies Riemann's definition above, then f is integrable in the sense of Definition 7.2.7. (The full equivalence of these two characterizations of integrability is proved in Section 8.1.)

Exercise 7.2.7. Let $f : [a, b] \to \mathbf{R}$ be increasing on the set $[a, b]$ (i.e., $f(x) \le f(y)$ whenever $x < y$). Show that f is integrable on $[a, b]$.

7.3   Integrating Functions with Discontinuities

The fact that continuous functions are integrable is not so much a fortunate discovery as it is evidence for a well-designed integral. Riemann's integral is a modification of Cauchy's definition of the integral, and Cauchy's definition was crafted specifically to work on continuous functions. The interesting issue is discovering just how dependent the Riemann integral is on the continuity of the integrand.

Example 7.3.1. Consider the function

$$f(x) = \begin{cases} 1 & \text{for } x \neq 1 \\ 0 & \text{for } x = 1 \end{cases}$$

on the interval $[0, 2]$. If P is any partition of $[0, 2]$, a quick calculation reveals that $U(f, P) = 2$. The lower sum $L(f, P)$ will be less than 2 because any subinterval of P that contains $x = 1$ will contribute zero to the value of the lower sum. The way to show that f is integrable is to construct a partition that minimizes the effect of the discontinuity by embedding $x = 1$ into a very small subinterval.

Let $\epsilon > 0$, and consider the partition $P_\epsilon = \{0, 1 - \epsilon/3, 1 + \epsilon/3, 2\}$. Then,

$$\begin{aligned} L(f, P_\epsilon) &= 1\left(1 - \frac{\epsilon}{3}\right) + 0(\epsilon) + 1\left(1 - \frac{\epsilon}{3}\right) \\ &= 2 - \frac{2}{3}\epsilon. \end{aligned}$$

Because $U(f, P_\epsilon) = 2$, we have

$$U(f, P_\epsilon) - L(f, P_\epsilon) = \frac{2}{3}\epsilon < \epsilon.$$

We can now use Theorem 7.2.8 to conclude that f is integrable.

Although the function in Example 7.3.1 is extremely simple, the method used to show it is integrable is really the same one used to prove that any bounded function with a single discontinuity is integrable. The notation in the following proof is more cumbersome, but the essence of the argument is that the misbehavior of the function at its discontinuity is isolated inside a particularly small subinterval of the partition.

Theorem 7.3.2. *If $f : [a, b] \to \mathbf{R}$ is bounded, and f is integrable on $[c, b]$ for all $c \in (a, b)$, then f is integrable on $[a, b]$. An analogous result holds at the other endpoint.*

Proof. Let $\epsilon > 0$. As usual, our task is to produce a partition P such that $U(f, P) - L(f, P) < \epsilon$. For any partition, we can always write

$$\begin{aligned} U(f, P) - L(f, P) &= \sum_{k=1}^{n} (M_k - m_k)\Delta x_k \\ &= (M_1 - m_1)(x_1 - a) + \sum_{k=2}^{n} (M_k - m_k)\Delta x_k, \end{aligned}$$

so the first step is to choose x_1 close enough to a so that

$$(M_1 - m_1)(x_1 - a) < \frac{\epsilon}{2}.$$

This is not too difficult. Because f is bounded, we know there exists $M > 0$ satisfying $|f(x)| \leq M$ for all $x \in [a, b]$. Noting that $M_1 - m_1 \leq 2M$, let's pick x_1 so that

$$x_1 - a < \frac{\epsilon}{4M}.$$

Now, by hypothesis, f is integrable on $[x_1, b]$, so there exists a partition P_1 of $[x_1, b]$ for which

$$U(f, P_1) - L(f, P_1) < \frac{\epsilon}{2}.$$

Finally, we let $P = \{a\} \cup P_1$ be a partition of $[a, b]$, from which it follows that

$$
\begin{aligned}
U(f, P) - L(f, P) &\leq (2M)(x_1 - a) + (U(f, P_1) - L(f, P_1)) \\
&< \frac{\epsilon}{2} + \frac{\epsilon}{2} = \epsilon.
\end{aligned}
$$

$\square$

Theorem 7.3.2 enables us to prove that a bounded function on a closed interval with a single discontinuity at an endpoint is still integrable. In the next section, we will prove that integrability on the intervals $[a, b]$ and $[b, d]$ is equivalent to integrability on $[a, d]$. This property, together with an induction argument, leads to the conclusion that any function with a *finite* number of discontinuities is still integrable. What if the number of discontinuities is infinite?

Example 7.3.3. Recall Dirichlet's function

$$g(x) = \begin{cases} 1 & \text{for } x \text{ rational} \\ 0 & \text{for } x \text{ irrational} \end{cases}$$

from Section 4.1. If P is some partition of $[0, 1]$, then the density of the rationals in $\mathbf{R}$ implies that every subinterval of P will contain a point where $g(x) = 1$. It follows that $U(g, P) = 1$. On the other hand, $L(g, P) = 0$ because the irrationals are also dense in $\mathbf{R}$. Because this is the case for every partition P, we see that the upper integral $U(f) = 1$ and the lower integral $L(f) = 0$. The two are not equal, so we conclude that Dirichlet's function is *not* integrable.

How discontinuous can a function be before it fails to be integrable? Before jumping to the hasty (and incorrect) conclusion that the Riemann integral fails for functions with more than a finite number of discontinuities, we should realize that Dirichlet's function is discontinuous at *every* point in $[0, 1]$. It would be useful to investigate a function where the discontinuities are infinite in number but do not necessarily make up all of $[0, 1]$. Thomae's function, also defined in Section 4.1, is one such example. The discontinuous points of this function are precisely the rational numbers in $[0, 1]$. In the exercises to follow we will see that Thomae's function *is* Riemann-integrable, raising the bar for allowable discontinuous points to include potentially infinite sets.

The conclusion of this story is contained in the doctoral dissertation of Henri Lebesgue, who presented his work in 1901. Lebesgue's elegant criterion for Riemann integrability is explored in great detail in Section 7.6. For the moment, though, we will take a short detour from questions of integrability and construct a proof of the celebrated Fundamental Theorem of Calculus.

Exercises

Exercise 7.3.1. Consider the function

$$h(x) = \begin{cases} 1 & \text{for } 0 \le x < 1 \\ 2 & \text{for } x = 1 \end{cases}$$

over the interval $[0, 1]$.

(a) Show that $L(f, P) = 1$ for every partition P of $[0, 1]$.

(b) Construct a partition P for which $U(f, P) < 1 + 1/10$.

(c) Given $\epsilon > 0$, construct a partition P_ϵ for which $U(f, P_\epsilon) < 1 + \epsilon$.

Exercise 7.3.2. Recall that Thomae's function

$$t(x) = \begin{cases} 1 & \text{if } x = 0 \\ 1/n & \text{if } x = m/n \in \mathbf{Q}\backslash\{0\} \text{ is in lowest terms with } n > 0 \\ 0 & \text{if } x \notin \mathbf{Q} \end{cases}$$

has a countable set of discontinuities occurring at precisely every rational number. Follow these steps to prove $t(x)$ is integrable on $[0, 1]$ with $\int_0^1 t = 0$.

(a) First argue that $L(t, P) = 0$ for any partition P of $[0, 1]$.

(b) Let $\epsilon > 0$, and consider the set of points $D_{\epsilon/2} = \{x \in [0, 1] : t(x) \ge \epsilon/2\}$. How big is $D_{\epsilon/2}$?

(c) To complete the argument, explain how to construct a partition P_ϵ of $[0, 1]$ so that $U(t, P_\epsilon) < \epsilon$.

Exercise 7.3.3. Let

$$f(x) = \begin{cases} 1 & \text{if } x = 1/n \text{ for some } n \in \mathbf{N} \\ 0 & \text{otherwise.} \end{cases}$$

Show that f is integrable on $[0, 1]$ and compute $\int_0^1 f$.

Exercise 7.3.4. Let f and g be functions defined on (possibly different) closed intervals, and assume the range of f is contained in the domain of g so that the composition $g \circ f$ is properly defined.

(a) Show, by example, that it is not the case that if f and g are integrable, then $g \circ f$ is integrable.

Now decide on the validity of each of the following conjectures, supplying a proof or counterexample as appropriate.

(b) If f is increasing and g is integrable, then $g \circ f$ is integrable.

(c) If f is integrable and g is increasing, then $g \circ f$ is integrable.

Exercise 7.3.5. Provide an example or give a reason why the request is impossible.

(a) A sequence $(f_n) \to f$ pointwise, where each f_n has at most a finite number of discontinuities but f is not integrable.

(b) A sequence $(g_n) \to g$ uniformly where each g_n has at most a finite number of discontinuities and g is not integrable.

(c) A sequence $(h_n) \to h$ uniformly where each h_n is not integrable but h is integrable.

Exercise 7.3.6. Let $\{r_1, r_2, r_3, \ldots\}$ be an enumeration of all the rationals in $[0, 1]$, and define

$$g_n(x) = \begin{cases} 1 & \text{if } x = r_n \\ 0 & \text{otherwise.} \end{cases}$$

(a) Is $G(x) = \sum_{n=1}^{\infty} g_n(x)$ integrable on $[0, 1]$?

(b) Is $F(x) = \sum_{n=1}^{\infty} g_n(x)/n$ integrable on $[0, 1]$?

Exercise 7.3.7. Assume $f : [a, b] \to \mathbf{R}$ is integrable.

(a) Show that if g satisfies $g(x) = f(x)$ for all but a finite number of points in $[a, b]$, then g is integrable as well.

(b) Find an example to show that g may fail to be integrable if it differs from f at a countable number of points.

Exercise 7.3.8. As in Exercise 7.3.6, let $\{r_1, r_2, r_3, \ldots\}$ be an enumeration of the rationals in $[0, 1]$, but this time define

$$h_n(x) = \begin{cases} 1 & \text{if } r_n < x \leq 1 \\ 0 & \text{if } 0 \leq x \leq r_n. \end{cases}$$

Show $H(x) = \sum_{n=1}^{\infty} h_n(x)/2^n$ is integrable on $[0, 1]$ even though it has discontinuities at every rational point.

Exercise 7.3.9 (Content Zero). A set $A \subseteq [a, b]$ has *content zero* if for every $\epsilon > 0$ there exists a finite collection of open intervals $\{O_1, O_2, \ldots, O_N\}$ that contain A in their union and whose lengths sum to ϵ or less. Using $|O_n|$ to refer to the length of each interval, we have

$$A \subseteq \bigcup_{n=1}^{N} O_n \quad \text{and} \quad \sum_{n=1}^{N} |O_n| \leq \epsilon.$$

(a) Let f be bounded on $[a, b]$. Show that if the set of discontinuous points of f has content zero, then f is integrable.

(b) Show that any finite set has content zero.

(c) Content zero sets do not have to be finite. They do not have to be count-able. Show that the Cantor set C defined in Section 3.1 has content zero.

(d) Prove that

$$h(x) = \begin{cases} 1 & \text{if } x \in C \\ 0 & \text{if } x \notin C. \end{cases}$$

is integrable, and find the value of the integral.

7.4   Properties of the Integral

Before embarking on the proof of the Fundamental Theorem of Calculus, we need to verify what are probably some very familiar properties of the integral. The discussion in the previous section has already made use of the following fact.

Theorem 7.4.1. *Assume $f : [a, b] \to \mathbf{R}$ is bounded, and let $c \in (a, b)$. Then, f is integrable on $[a, b]$ if and only if f is integrable on $[a, c]$ and $[c, b]$. In this case, we have*

$$\int_a^b f = \int_a^c f + \int_c^b f.$$

Proof. If f is integrable on $[a, b]$, then for $\epsilon > 0$ there exists a partition P such that $U(f, P) - L(f, P) < \epsilon$. Because refining a partition can only potentially bring the upper and lower sums closer together, we can simply add c to P if it is not already there. Then, let $P_1 = P \cap [a, c]$ be a partition of $[a, c]$, and $P_2 = P \cap [c, b]$ be a partition of $[c, b]$. It follows that

$$U(f, P_1) - L(f, P_1) < \epsilon \quad \text{and} \quad U(f, P_2) - L(f, P_2) < \epsilon,$$

implying that f is integrable on $[a, c]$ and $[c, b]$.

Conversely, if we are given that f is integrable on the two smaller intervals $[a, c]$ and $[c, b]$, then given an $\epsilon > 0$ we can produce partitions P_1 and P_2 of $[a, c]$ and $[c, b]$, respectively, such that

$$U(f, P_1) - L(f, P_1) < \frac{\epsilon}{2} \quad \text{and} \quad U(f, P_2) - L(f, P_2) < \frac{\epsilon}{2}.$$

Letting $P = P_1 \cup P_2$ produces a partition of $[a, b]$ for which

$$U(f, P) - L(f, P) < \epsilon.$$

Thus, f is integrable on $[a, b]$.

Continuing to let $P = P_1 \cup P_2$ as earlier, we have

$$
\begin{aligned}
\int_a^b f \leq U(f, P) \quad &< \quad L(f, P) + \epsilon \\
&= \quad L(f, P_1) + L(f, P_2) + \epsilon \\
&\leq \quad \int_a^c f + \int_c^b f + \epsilon,
\end{aligned}
$$

which implies $\int_a^b f \leq \int_a^c f + \int_c^b f$. To get the other inequality, observe that

$$
\begin{aligned}
\int_a^c f + \int_c^b f \quad &\leq \quad U(f, P_1) + U(f, P_2) \\
&< \quad L(f, P_1) + L(f, P_2) + \epsilon \\
&= \quad L(f, P) + \epsilon \\
&\leq \quad \int_a^b f + \epsilon.
\end{aligned}
$$

Because $\epsilon > 0$ is arbitrary, we must have $\int_a^c f + \int_c^b f \leq \int_a^b f$, so

$$\int_a^c f + \int_c^b f = \int_a^b f,$$

as desired. ☐

The proof of Theorem 7.4.1 demonstrates some of the standard techniques involved for proving facts about the Riemann integral. The next result catalogs the remainder of the basic properties of the integral that we will need in our upcoming arguments.

Theorem 7.4.2. *Assume f and g are integrable functions on the interval $[a, b]$.*

(i) *The function $f + g$ is integrable on $[a, b]$ with $\int_a^b (f + g) = \int_a^b f + \int_a^b g$.*

(ii) *For $k \in \mathbf{R}$, the function kf is integrable with $\int_a^b kf = k \int_a^b f$.*

(iii) *If $m \leq f(x) \leq M$ on $[a, b]$, then $m(b - a) \leq \int_a^b f \leq M(b - a)$.*

(iv) *If $f(x) \leq g(x)$ on $[a, b]$, then $\int_a^b f \leq \int_a^b g$.*

(v) *The function $|f|$ is integrable and $|\int_a^b f| \leq \int_a^b |f|$.*

Proof. Properties (i) and (ii) are reminiscent of the Algebraic Limit Theorem and its many descendants (Theorems 2.3.3, 2.7.1, 4.2.4, and 5.2.4). In fact, there is a way to use the Algebraic Limit Theorem for this argument as well. An immediate corollary to Theorem 7.2.8 is that a function f is integrable on $[a, b]$ if and only if there exists a sequence of partitions (P_n) satisfying

$$(1) \qquad \lim_{n \to \infty} [U(f, P_n) - L(f, P_n)] = 0,$$

and in this case $\int_a^b f = \lim U(f, P_n) = \lim L(f, P_n)$. (A proof for this was requested as Exercise 7.2.3.)

To prove (ii) for the case $k \geq 0$, first verify that for any partition P we have

$$U(kf, P) = kU(f, P) \quad \text{and} \quad L(kf, P) = kL(f, P).$$

Exercise 1.3.5 is used here. Because f is integrable, there exist partitions (P_n) satisfying (1). Turning our attention to the function (kf), we see that

$$\lim_{n \to \infty} [U(kf, P_n) - L(kf, P_n)] = \lim_{n \to \infty} k[U(f, P_n) - L(f, P_n)] = 0,$$

and the formula in (ii) follows. The case where $k < 0$ is similar except that we have

$$U(kf, P_n) = kL(f, P_n) \quad \text{and} \quad L(kf, P_n) = kU(f, P_n).$$

A proof for (i) can be constructed using similar methods and is requested in Exercise 7.4.5.

To prove (iii), observe that

$$U(f, P) \geq \int_a^b f \geq L(f, P)$$

for any partition P. Statement (iii) follows if we take P to be the trivial partition consisting of only the endpoints a and b.

For (iv), let $h = g - f$ and use (i), (ii), and (iii).

Because $-|f(x)| \leq f(x) \leq |f(x)|$ on $[a, b]$, statement (v) will follow from (iv) provided that we can show that $|f|$ is actually integrable. The proof of this fact is outlined in Exercise 7.4.1. $\qquad \square$

To this point, the quantity $\int_a^b f$ is only defined in the case where $a < b$.

Definition 7.4.3. If f is integrable on the interval $[a, b]$, define

$$\int_b^a f = - \int_a^b f.$$

Also, for $c \in [a, b]$ define

$$\int_c^c f = 0.$$

Definition 7.4.3 is a natural convention to simplify the algebra of integrals. If f is an integrable function on some interval I, then it is straightforward to verify that the equation

$$\int_a^b f = \int_a^c f + \int_c^b f$$

from Theorem 7.4.1 remains valid for *any* three points a, b, and c chosen in any order from I.

Uniform Convergence and Integration

If (f_n) is a sequence of integrable functions on $[a, b]$, and if $f_n \to f$, then we are inevitably going to want to know whether

(2) $$\int_a^b f_n \to \int_a^b f.$$

This is an archetypical instance of one of the major themes of analysis: When does a mathematical manipulation such as integration respect the limiting process?

If the convergence is pointwise, then any number of things can go wrong. It is possible for each f_n to be integrable but for the limit f not to be integrable (Exercise 7.3.5). Even if the limit function f is integrable, equation (2) may fail to hold. As an example of this, let

$$f_n(x) = \begin{cases} n & \text{if } 0 < x < 1/n \\ 0 & \text{if } x = 0 \text{ or } x \geq 1/n. \end{cases}$$

Each f_n has two discontinuities on $[0, 1]$ and so is integrable with $\int_0^1 f_n = 1$. For each $x \in [0, 1]$, we have $\lim f_n(x) = 0$ so that $f_n \to 0$ pointwise on $[0, 1]$. But now observe that the limit function $f = 0$ certainly integrates to 0, and

$$0 \neq \lim_{n \to \infty} \int_0^1 f_n.$$

As a final remark on what can go wrong in (2), we should point out that it is possible to modify this example to produce a situation where $\lim \int_0^1 f_n$ does not even exist.

One way to resolve all of these problems is to add the assumption of uniform convergence.

Theorem 7.4.4 (Integrable Limit Theorem). *Assume that $f_n \to f$ uniformly on $[a, b]$ and that each f_n is integrable. Then, f is integrable and*

$$\lim_{n \to \infty} \int_a^b f_n = \int_a^b f.$$

Proof. The proof that f is integrable was requested as Exercise 7.2.5. The properties of the integral listed in Theorem 7.4.2 allow us to assert that for any f_n.

$$\left| \int_a^b f_n - \int_a^b f \right| = \left| \int_a^b (f_n - f) \right| \leq \int_a^b |f_n - f|.$$

Let $\epsilon > 0$ be arbitrary. Because $f_n \to f$ uniformly, there exists an N such that

$$|f_n(x) - f(x)| < \epsilon/(b-a) \quad \text{for all } n \geq N \text{ and } x \in [a, b].$$

Thus, for $n \geq N$ we see that

$$\left| \int_a^b f_n - \int_a^b f \right| \leq \int_a^b |f_n - f|$$

$$\leq \int_a^b \frac{\epsilon}{b-a} = \epsilon,$$

and the result follows.   □

Exercises

Exercise 7.4.1. Let f be a bounded function on a set A, and set

$$M = \sup\{f(x) : x \in A\}, \quad m = \inf\{f(x) : x \in A\},$$

$$M' = \sup\{|f(x)| : x \in A\}, \quad \text{and} \quad m' = \inf\{|f(x)| : x \in A\}.$$

(a) Show that $M - m \geq M' - m'$.

(b) Show that if f is integrable on the interval $[a, b]$, then $|f|$ is also integrable on this interval.

(c) Provide the details for the argument that in this case we have $|\int_a^b f| \leq \int_a^b |f|$.

Exercise 7.4.2. (a) Let $g(x) = x^3$, and classify each of the following as positive, negative, or zero.

(i) $\int_0^{-1} g + \int_0^1 g$ (ii) $\int_1^0 g + \int_0^1 g$ (iii) $\int_1^{-2} g + \int_0^1 g$.

(b) Show that if $b \leq a \leq c$ and f is integrable on the interval $[b, c]$, then it is still the case that $\int_a^b f = \int_a^c f + \int_c^b f$.

Exercise 7.4.3. Decide which of the following conjectures is true and supply a short proof. For those that are not true, give a counterexample.

(a) If $|f|$ is integrable on $[a, b]$, then f is also integrable on this set.

(b) Assume g is integrable and $g(x) \geq 0$ on $[a, b]$. If $g(x) > 0$ for an infinite number of points $x \in [a, b]$, then $\int_a^b g > 0$.

(c) If g is continuous on $[a, b]$ and $g(x) \geq 0$ with $g(y_0) > 0$ for at least one point $y_0 \in [a, b]$, then $\int_a^b g > 0$.

Exercise 7.4.4. Show that if $f(x) > 0$ for all $x \in [a, b]$ and f is integrable, then $\int_a^b f > 0$.

Exercise 7.4.5. Let f and g be integrable functions on $[a, b]$.

(a) Show that if P is any partition of $[a, b]$, then

$$U(f + g, P) \leq U(f, P) + U(g, P).$$

Provide a specific example where the inequality is strict. What does the corresponding inequality for lower sums look like?

(b) Review the proof of Theorem 7.4.2 (ii), and provide an argument for part (i) of this theorem.

Exercise 7.4.6. Although not part of Theorem 7.4.2, it is true that the product of integrable functions is integrable. Provide the details for each step in the following proof of this fact:

(a) If f satisfies $|f(x)| \leq M$ on $[a, b]$, show

$$|(f(x))^2 - (f(y))^2| \leq 2M|f(x) - f(y)|.$$

(b) Prove that if f is integrable on $[a, b]$, then so is f^2.

(c) Now show that if f and g are integrable, then fg is integrable. (Consider $(f + g)^2$.)

Exercise 7.4.7. Review the discussion immediately preceding Theorem 7.4.4.

(a) Produce an example of a sequence $f_n \to 0$ pointwise on $[0, 1]$ where $\lim_{n \to \infty} \int_0^1 f_n$ does not exist.

(b) Produce an example of a sequence g_n with $\int_0^1 g_n \to 0$ but $g_n(x)$ does not converge to zero for any $x \in [0, 1]$. To make it more interesting, let's insist that $g_n(x) \geq 0$ for all x and n.

Exercise 7.4.8. For each $n \in \mathbf{N}$, let

$$h_n(x) = \begin{cases} 1/2^n & \text{if } 1/2^n < x \leq 1 \\ 0 & \text{if } 0 \leq x \leq 1/2^n \end{cases},$$

and set $H(x) = \sum_{n=1}^{\infty} h_n(x)$. Show H is integrable and compute $\int_0^1 H$.

Exercise 7.4.9. Let g_n and g be uniformly bounded on $[0, 1]$, meaning that there exists a single $M > 0$ satisfying $|g(x)| \leq M$ and $|g_n(x)| \leq M$ for all $n \in \mathbf{N}$ and $x \in [0, 1]$. Assume $g_n \to g$ pointwise on $[0, 1]$ and uniformly on any set of the form $[0, \alpha]$, where $0 < \alpha < 1$.

If all the functions are integrable, show that $\lim_{n \to \infty} \int_0^1 g_n = \int_0^1 g$.

Exercise 7.4.10. Assume g is integrable on $[0, 1]$ and continuous at 0. Show

$$\lim_{n \to \infty} \int_0^1 g(x^n) dx = g(0).$$

Exercise 7.4.11. Review the original definition of integrability in Section 7.2, and in particular the definition of the upper integral $U(f)$. One reasonable suggestion might be to bypass the complications introduced in Definition 7.2.7 and simply define the integral to be the value of $U(f)$. Then *every* bounded function is integrable! Although tempting, proceeding in this way has some significant drawbacks. Show by example that several of the properties in Theorem 7.4.2 no longer hold if we replace our current definition of integrability with the proposal that $\int_a^b f = U(f)$ for every bounded function f.

7.5   The Fundamental Theorem of Calculus

The derivative and the integral have been independently defined, each in its own rigorous mathematical terms. The definition of the derivative is motivated by the problem of finding slopes of tangent lines and is given in terms of functional limits of difference quotients. The definition of the integral grows out of the desire to calculate areas under nonconstant functions and is given in terms of supremums and infimums of finite sums. The Fundamental Theorem of Calculus reveals the remarkable inverse relationship between the two processes.

The result is stated in two parts. The first is a computational statement that describes how an antiderivative can be used to evaluate an integral over a particular interval. The second statement is more theoretical in nature, expressing the fact that every continuous function is the derivative of its indefinite integral.

Theorem 7.5.1 (Fundamental Theorem of Calculus).   (i) *If $f : [a, b] \to$*
   R *is integrable, and $F : [a, b] \to \mathbf{R}$ satisfies $F'(x) = f(x)$ for all $x \in [a, b]$,*
   then

$$\int_a^b f = F(b) - F(a).$$

   (ii) *Let $g : [a, b] \to \mathbf{R}$ be integrable, and for $x \in [a, b]$, define*

$$G(x) = \int_a^x g.$$

   Then G is continuous on $[a, b]$. If g is continuous at some point $c \in [a, b]$,
   then G is differentiable at c and $G'(c) = g(c)$.

Proof. (i) Let P be a partition of $[a, b]$ and apply the Mean Value Theorem to F on a typical subinterval $[x_{k-1}, x_k]$ of P. This yields a point $t_k \in (x_{k-1}, x_k)$ where

$$
\begin{aligned}
F(x_k) - F(x_{k-1}) &= F'(t_k)(x_k - x_{k-1}) \\
&= f(t_k)(x_k - x_{k-1}).
\end{aligned}
$$

Now, consider the upper and lower sums $U(f, P)$ and $L(f, P)$. Because $m_k \leq f(t_k) \leq M_k$ (where m_k is the infimum on $[x_{k-1}, x_k]$ and M_k is the supremum), it follows that

$$
L(f, P) \leq \sum_{k=1}^{n} [F(x_k) - F(x_{k-1})] \leq U(f, P).
$$

But notice that the sum in the middle telescopes so that

$$
\sum_{k=1}^{n} [F(x_k) - F(x_{k-1})] = F(b) - F(a),
$$

which is *independent* of the partition P. Thus we have

$$
L(f) \leq F(b) - F(a) \leq U(f).
$$

Because $L(f) = U(f) = \int_a^b f$, we conclude that $\int_a^b f = F(b) - F(a)$.

(ii) To prove the second statement, take $x > y$ in $[a, b]$ and observe that

$$
\begin{aligned}
|G(x) - G(y)| = \left| \int_a^x g - \int_a^y g \right| &= \left| \int_y^x g \right| \\
&\leq \int_y^x |g| \\
&\leq M(x - y),
\end{aligned}
$$

where $M > 0$ is a bound on $|g|$. This shows that G is Lipschitz and so is uniformly continuous on $[a, b]$ (Exercise 4.4.9).

Now, let's assume that g is continuous at $c \in [a, b]$. In order to show that $G'(c) = g(c)$, we rewrite the limit for $G'(c)$ as

$$
\begin{aligned}
\lim_{x \to c} \frac{G(x) - G(c)}{x - c} &= \lim_{x \to c} \frac{1}{x - c} \left(\int_a^x g(t)\, dt - \int_a^c g(t)\, dt \right) \\
&= \lim_{x \to c} \frac{1}{x - c} \left(\int_c^x g(t)\, dt \right).
\end{aligned}
$$

We would like to show that this limit equals $g(c)$. Thus, given an $\epsilon > 0$, we must produce a $\delta > 0$ such that if $|x - c| < \delta$, then

(1)
$$
\left| \frac{1}{x - c} \left(\int_c^x g(t)\, dt \right) - g(c) \right| < \epsilon.
$$

The assumption of continuity of g gives us control over the difference $|g(t)-g(c)|$. In particular, we know that there exists a $\delta > 0$ such that

$$|t - c| < \delta \quad \text{implies} \quad |g(t) - g(c)| < \epsilon.$$

To take advantage of this, we cleverly write the constant $g(c)$ as

$$g(c) = \frac{1}{x - c} \int_c^x g(c)\, dt$$

and combine the two terms in equation (1) into a single integral. Keeping in mind that $|x - c| \geq |t - c|$, we have that for all $|x - c| < \delta$,

$$\left| \frac{1}{x - c} \left(\int_c^x g(t)\, dt \right) - g(c) \right| = \left| \frac{1}{x - c} \int_c^x (g(t) - g(c))\, dt \right|$$

$$\leq \frac{1}{(x - c)} \int_c^x |g(t) - g(c)|\, dt$$

$$< \frac{1}{(x - c)} \int_c^x \epsilon\, dt = \epsilon. \qquad \square$$

Exercises

Exercise 7.5.1. (a) Let $f(x) = |x|$ and define $F(x) = \int_{-1}^x f$. Find a piece-wise algebraic formula for $F(x)$ for all x. Where is F continuous? Where is F differentiable? Where does $F'(x) = f(x)$?

(b) Repeat part (a) for the function

$$f(x) = \begin{cases} 1 & \text{if } x < 0 \\ 2 & \text{if } x \geq 0. \end{cases}$$

Exercise 7.5.2. Decide whether each statement is true or false, providing a short justification for each conclusion.

(a) If $g = h'$ for some h on $[a, b]$, then g is continuous on $[a, b]$.

(b) If g is continuous on $[a, b]$, then $g = h'$ for some h on $[a, b]$.

(c) If $H(x) = \int_a^x h$ is differentiable at $c \in [a, b]$, then h is continuous at c.

Exercise 7.5.3. The hypothesis in Theorem 7.5.1 (i) that $F'(x) = f(x)$ for all $x \in [a, b]$ is slightly stronger than it needs to be. Carefully read the proof and state exactly what needs to be assumed with regard to the relationship between f and F for the proof to be valid.

Exercise 7.5.4. Show that if $f : [a, b] \to \mathbf{R}$ is continuous and $\int_a^x f = 0$ for all $x \in [a, b]$, then $f(x) = 0$ everywhere on $[a, b]$. Provide an example to show that this conclusion does not follow if f is not continuous.

Exercise 7.5.5. The Fundamental Theorem of Calculus can be used to supply a shorter argument for Theorem 6.3.1 under the additional assumption that the sequence of derivatives is continuous.

Assume $f_n \to f$ pointwise and $f'_n \to g$ uniformly on $[a, b]$. Assuming each f'_n is continuous, we can apply Theorem 7.5.1 (i) to get

$$\int_a^x f'_n = f_n(x) - f_n(a)$$

for all $x \in [a, b]$. Show that $g(x) = f'(x)$.

Exercise 7.5.6 (Integration-by-parts). (a) Assume $h(x)$ and $k(x)$ have continuous derivatives on $[a, b]$ and derive the familiar integration-by-parts formula

$$\int_a^b h(t)k'(t)dt = h(b)k(b) - h(a)k(a) - \int_a^b h'(t)k(t)dt \ .$$

(b) Explain how the result in Exercise 7.4.6 can be used to slightly weaken the hypothesis in part (a).

Exercise 7.5.7. Use part (ii) of Theorem 7.5.1 to construct another proof of part (i) of Theorem 7.5.1 under the stronger hypothesis that f is continuous. (To get started, set $G(x) = \int_a^x f$.)

Exercise 7.5.8 (Natural Logarithm and Euler's Constant). Let

$$L(x) = \int_1^x \frac{1}{t} dt,$$

where we consider only $x > 0$.

(a) What is $L(1)$? Explain why L is differentiable and find $L'(x)$.

(b) Show that $L(xy) = L(x) + L(y)$. (Think of y as a constant and differentiate $g(x) = L(xy)$.)

(c) Show $L(x/y) = L(x) - L(y)$.

(d) Let

$$\gamma_n = \left(1 + \frac{1}{2} + \frac{1}{3} + \cdots + \frac{1}{n}\right) - L(n).$$

Prove that (γ_n) converges. The constant $\gamma = \lim \gamma_n$ is called Euler's constant.

(e) Show how consideration of the sequence $\gamma_{2n} - \gamma_n$ leads to the interesting identity

$$L(2) = 1 - \frac{1}{2} + \frac{1}{3} - \frac{1}{4} + \frac{1}{5} - \frac{1}{6} + \cdots .$$

Exercise 7.5.9. Given a function f on $[a, b]$, define the *total variation* of f to be

$$Vf = \sup \left\{ \sum_{k=1}^{n} |f(x_k) - f(x_{k-1})| \right\},$$

where the supremum is taken over all partitions P of $[a, b]$.

(a) If f is continuously differentiable (f' exists as a continuous function), use the Fundamental Theorem of Calculus to show $Vf \leq \int_a^b |f'|$.

(b) Use the Mean Value Theorem to establish the reverse inequality and conclude that $Vf = \int_a^b |f'|$.

Exercise 7.5.10 (Change-of-variable Formula). Let $g : [a, b] \to \mathbf{R}$ be differentiable and assume g' is continuous. Let $f : [c, d] \to \mathbf{R}$ be continuous, and assume that the range of g is contained in $[c, d]$ so that the composition $f \circ g$ is properly defined.

(a) Why are we sure f is the derivative of some function? How about $(f \circ g)g'$?

(b) Prove the change-of-variable formula

$$\int_a^b f(g(x))g'(x)dx = \int_{g(a)}^{g(b)} f(t)dt.$$

Exercise 7.5.11. Assume f is integrable on $[a, b]$ and has a "jump discontinuity" at $c \in (a, b)$. This means that both one-sided limits exist as x approaches c from the left and from the right, but that

$$\lim_{x \to c^-} f(x) \neq \lim_{x \to c^+} f(x).$$

(This phenomenon is discussed in more detail in Section 4.6.)

(a) Show that, in this case, $F(x) = \int_a^x f$ is not differentiable at $x = c$.

(b) The discussion in Section 5.5 mentions the existence of a continuous monotone function that fails to be differentiable on a dense subset of $\mathbf{R}$. Combine the results of part (a) with Exercise 6.4.10 to show how to construct such a function.

7.6 Lebesgue's Criterion for Riemann Integrability

We now return to our investigation of the relationship between continuity and the Riemann integral. We have proved that continuous functions are integrable and that the integral also exists for functions with only a finite number of discontinuities. At the opposite end of the spectrum, we saw that Dirichlet's function, which is discontinuous at every point on $[0, 1]$, fails to be Riemann-integrable. The next examples show that the set of discontinuities of an integrable function can be infinite and even uncountable. (These also appear as exercises in Section 7.3.)

Riemann-integrable Functions with Infinite Discontinuities

Recall from Section 4.1 that Thomae's function

$$t(x) = \begin{cases} 1 & \text{if } x = 0 \\ 1/n & \text{if } x = m/n \in \mathbf{Q}\backslash\{0\} \text{ is in lowest terms with } n > 0 \\ 0 & \text{if } x \notin \mathbf{Q} \end{cases}$$

is continuous on the set of irrationals and has discontinuities at every rational point. Let's prove that Thomae's function is integrable on $[0, 1]$ with $\int_0^1 t = 0$.

Let $\epsilon > 0$. The strategy, as usual, is to construct a partition P_ϵ of $[0, 1]$ for which $U(t, P_\epsilon) - L(t, P_\epsilon) < \epsilon$.

Exercise 7.6.1. (a) First, argue that $L(t, P) = 0$ for any partition P of $[0, 1]$.

(b) Consider the set of points $D_{\epsilon/2} = \{x : t(x) \geq \epsilon/2\}$. How big is $D_{\epsilon/2}$?

(c) To complete the argument, explain how to construct a partition P_ϵ of $[0, 1]$ so that $U(t, P_\epsilon) < \epsilon$.

We first met the Cantor set C in Section 3.1. We have since learned that C is a compact, uncountable subset of the interval $[0, 1]$.

Exercise 7.6.2. Define

$$h(x) = \begin{cases} 1 & \text{if } x \in C \\ 0 & \text{if } x \notin C \end{cases}.$$

(a) Show h has discontinuities at each point of C and is continuous at every point of the complement of C. Thus, h is not continuous on an uncountably infinite set.

(b) Now prove that h is integrable on $[0, 1]$.

Sets of Measure Zero

Thomae's function fails to be continuous at each rational number in $[0, 1]$. Although this set is infinite, we have seen that any infinite subset of $\mathbf{Q}$ is countable. Countably infinite sets are the smallest type of infinite set. The Cantor set is uncountable, but it is also small in a sense that we are now ready to make precise. In the introduction to Chapter 3, we presented an argument that the Cantor set has zero "length." The term "length" is awkward here because it really should only be applied to intervals or finite unions of intervals, which the Cantor set is not. There is a generalization of the concept of length to more general sets called the *measure* of a set. Of interest to our discussion are subsets that have *measure zero*.

Definition 7.6.1. A set $A \subseteq \mathbf{R}$ has *measure zero* if, for all $\epsilon > 0$, there exists a countable collection of open intervals O_n with the property that A is contained in the union of all of the intervals O_n and the sum of the lengths of all of the intervals is less than or equal to ϵ. More precisely, if $|O_n|$ refers to the length of the interval O_n, then we have

$$A \subseteq \bigcup_{n=1}^{\infty} O_n \quad \text{and} \quad \sum_{n=1}^{\infty} |O_n| \leq \epsilon.$$

Example 7.6.2. Consider a finite set $A = \{a_1, a_2, \ldots, a_N\}$. To show that A has measure zero, let $\epsilon > 0$ be arbitrary. For each $1 \leq n \leq N$, construct the interval

$$G_n = \left(a_n - \frac{\epsilon}{2N}, a_n + \frac{\epsilon}{2N} \right).$$

Clearly, A is contained in the union of these intervals, and

$$\sum_{n=1}^{N} |G_n| = \sum_{n=1}^{N} \frac{\epsilon}{N} = \epsilon.$$

Exercise 7.6.3. Show that any countable set has measure zero.

Exercise 7.6.4. Prove that the Cantor set has measure zero.

Exercise 7.6.5. Show that if two sets A and B each have measure zero, then $A \cup B$ has measure zero as well. In addition, discuss the proof of the stronger statement that the countable union of sets of measure zero also has measure zero. (This second statement is true, but a completely rigorous proof requires a result about double summations discussed in Section 2.8.)

α-Continuity

Definition 7.6.3. Let f be defined on $[a, b]$, and let $\alpha > 0$. The function f is *α-continuous at $x \in [a, b]$* if there exists $\delta > 0$ such that for all $y, z \in (x - \delta, x + \delta)$ it follows that $|f(y) - f(z)| < \alpha$.

Let f be a bounded function on $[a, b]$. For each $\alpha > 0$, define D^α to be the set of points in $[a, b]$ where the function f fails to be α-continuous; that is,

(1) $D^\alpha = \{x \in [a, b] : f \text{ is not } \alpha\text{-continuous at } x.\}$

The concept of α-continuity was previously introduced in Section 4.6. Several of the ensuing exercises appeared as exercises in this section as well.

Exercise 7.6.6. If $\alpha < \alpha'$, show that $D^{\alpha'} \subseteq D^\alpha$.

Now, let

(2) $D = \{x \in [a, b] : f \text{ is not continuous at } x \}.$

Exercise 7.6.7.   (a) Let $\alpha > 0$ be given. Show that if f is continuous at $x \in [a, b]$, then it is α-continuous at x as well. Explain how it follows that $D^\alpha \subseteq D$.

(b) Show that if f is not continuous at x, then f is not α-continuous for some $\alpha > 0$. Now, explain why this guarantees that

$$D = \bigcup_{n=1}^{\infty} D^{\alpha_n} \quad \text{where } \alpha_n = 1/n.$$

Exercise 7.6.8. Prove that for a fixed $\alpha > 0$, the set D^α is closed.

Just as with continuity, α-continuity is defined pointwise, and just as with continuity, uniformity is going to play an important role.

For a fixed $\alpha > 0$, a function $f : A \to \mathbf{R}$ is *uniformly α-continuous on A* if there exists a $\delta > 0$ such that whenever x and y are points in A satisfying $|x - y| < \delta$, it follows that $|f(x) - f(y)| < \alpha$. By imitating the proof of Theorem 4.4.7, it is completely straightforward to show that if f is α-continuous at every point on some compact set K, then f is uniformly α-continuous on K.

Compactness Revisited

Compactness of subsets of the real line can be described in three equivalent ways. The following theorem appears toward the end of Section 3.3.

Theorem 7.6.4. *Let $K \subseteq \mathbf{R}$. The following three statements are all equivalent, in the sense that if any one is true, then so are the two others.*

(i) *Every sequence contained in K has a convergent subsequence that converges to a limit in K.*

(ii) *K is closed and bounded.*

(iii) *Given a collection of open intervals $\{G_\lambda : \lambda \in \Lambda\}$ that covers K (that is, $K \subseteq \bigcup_{\lambda \in \Lambda} G_\lambda$) there exists a finite subcollection $\{G_{\lambda_1}, G_{\lambda_2}, G_{\lambda_3}, \ldots, G_{\lambda_N}\}$ of the original set that also covers K.*

The equivalence of (i) and (ii) has been used throughout the core material in the text. Characterization (iii) has been less central but is essential to the upcoming argument. If the characterization of compactness in terms of open covers is not familiar, take a moment to review the second half of Section 3.3 and complete the proof that (i) and (ii) imply (iii) outlined in Exercise 3.3.9.

Lebesgue's Theorem

We are now prepared to completely categorize the collection of Riemann-integrable functions in terms of continuity.

Theorem 7.6.5 (Lebesgue's Theorem). *Let f be a bounded function defined on the interval $[a, b]$. Then, f is Riemann-integrable if and only if the set of points where f is not continuous has measure zero.*

Proof. Let $M > 0$ satisfy $|f(x)| \leq M$ for all $x \in [a, b]$, and let D and D^{α} be defined as in the preceding equations (1) and (2). Let's first assume that D has measure zero and prove that our function is integrable.

    ($\Leftarrow$) Let $\epsilon > 0$ and set

$$\alpha = \frac{\epsilon}{2(b-a)}.$$

Exercise 7.6.9. Show that there exists a *finite* collection of disjoint open intervals $\{G_1, G_2, \ldots, G_N\}$ whose union contains D^{α} and that satisfies

$$\sum_{n=1}^{N} |G_n| < \frac{\epsilon}{4M}.$$

Exercise 7.6.10. Let K be what remains of the interval $[a, b]$ after the open intervals G_n are all removed; that is, $K = [a, b] \backslash \bigcup_{n=1}^{N} G_n$. Argue that f is uniformly α-continuous on K.

Exercise 7.6.11. Finish the proof in this direction by explaining how to construct a partition P_{ϵ} of $[a, b]$ such that $U(f, P_{\epsilon}) - L(f, P_{\epsilon}) \leq \epsilon$. It will be helpful to break the sum

$$U(f, P_{\epsilon}) - L(f, P_{\epsilon}) = \sum_{k=1}^{n} (M_k - m_k) \Delta x_k$$

into two parts—one over those subintervals that contain points of D^{α} and the other over subintervals that do not.

    ($\Rightarrow$) For the other direction, assume f is Riemann-integrable. We must argue that the set D of discontinuities of f has measure zero.

    Let $\epsilon > 0$ be arbitrary, and fix $\alpha > 0$. Because f is Riemann-integrable, there exists a partition P_{ϵ} of $[a, b]$ such that $U(f, P_{\epsilon}) - L(f, P_{\epsilon}) < \alpha \epsilon$.

Exercise 7.6.12.   (a) Prove that D^{α} has measure zero. Point out that it is possible to choose a cover for D^{α} that consists of a finite number of open intervals.

    (b) Show how this implies that D has measure zero.   □

Our main agenda in the remainder of this section is to employ Lebesgue's Theorem in our pursuit of a non-integrable derivative, but this elegant result has a number of other applications.

Exercise 7.6.13.   (a) Show that if f and g are integrable on $[a, b]$, then so is the product fg. (This result was requested in Exercise 7.4.6, but notice how much easier the argument is now.)

(b) Show that if g is integrable on $[a, b]$ and f is continuous on the range of g, then the composition $f \circ g$ is integrable on $[a, b]$.

If we instead assume that f is integrable and g is continuous, it actually doesn't follow that the composition $f \circ g$ is an integrable function. Producing a counterexample, however, requires a few more ingredients.

A Nonintegrable Derivative

To this point, our one example of a nonintegrable function is Dirichlet's nowhere-continuous function. We close this section with another example that has special significance. The content of the Fundamental Theorem of Calculus is that integration and differentiation are inverse processes of each other. If a function f is differentiable on $[a, b]$, then part (i) of the Fundamental Theorem tells us that

$$(3) \qquad \int_a^b f' = f(b) - f(a),$$

provided f' is integrable. But shouldn't f' be integrable just by virtue of being a derivative? A curious side-effect of staring at equation (3) for any length of time is that it starts to feel as though *every* derivative should be integrable because we have an obvious candidate for what the value of the integral ought to be. Alas, for the Riemann integral at least, reality comes up short of our expectations. What follows is the construction of a differentiable function f for which equation (3) fails because $\int_a^b f'$ does not exist.

We will once again be interested in the Cantor set

$$C = \bigcap_{n=0}^{\infty} C_n,$$

defined in Section 3.1. As an initial step, let's create a function $f(x)$ that is differentiable on $[0, 1]$ and whose derivative $f'(x)$ has discontinuities at every point of C. The key ingredient for this construction is the function

$$g(x) = \begin{cases} x^2 \sin(1/x) & \text{if } x > 0 \\ 0 & \text{if } x \leq 0. \end{cases}$$

Exercise 7.6.14. (a) Find $g'(0)$.

(b) Use the standard rules of differentiation to compute $g'(x)$ for $x \neq 0$.

(c) Explain why, for every $\delta > 0$, $g'(x)$ attains every value between 1 and -1 as x ranges over the set $(-\delta, \delta)$. Conclude that g' is not continuous at $x = 0$.

Now, we want to transport the behavior of g around zero to each of the endpoints of the closed intervals that make up the sets C_n used in the definition of

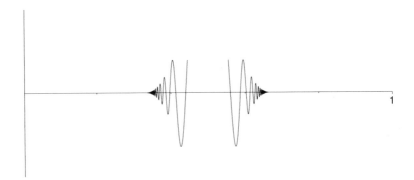

Figure 7.3:   A PRELIMINARY SKETCH OF $f_1(x)$.

the Cantor set. The formulas are awkward but the basic idea is straightforward. Start by setting

$$f_0(x) = 0 \quad \text{on} \quad C_0 = [0, 1].$$

To define f_1 on $[0, 1]$, first assign

$$f_1(x) = 0 \quad \text{for all} \quad x \in C_1 = \left[0, \frac{1}{3}\right] \cup \left[\frac{2}{3}, 1\right].$$

In the remaining open middle third, put translated "copies" of g oscillating toward the two endpoints (Fig. 7.3). In terms of a formula, we have

$$f_1(x) = \begin{cases} 0 & \text{if } x \in [0, 1/3] \\ g(x - 1/3) & \text{if } x \text{ is just to the right of } 1/3 \\ g(-x + 2/3) & \text{if } x \text{ is just to the left of } 2/3 \\ 0 & \text{if } x \in [2/3, 1]. \end{cases}$$

Finally, we splice the two oscillating pieces of f_1 together in a way that makes f_1 differentiable and such that

$$|f_1(x)| \le (x - 1/3)^2 \quad \text{and} \quad |f_1(x)| \le (-x + 2/3)^2.$$

This splicing is no great feat, and we will skip the details so as to keep our attention focused on the two endpoints $1/3$ and $2/3$. These are the points where $f_1'(x)$ fails to be continuous.

To define $f_2(x)$, we start with $f_1(x)$ and do the same trick as before, this time in the two open intervals $(1/9, 2/9)$ and $(7/9, 8/9)$. The result (Fig. 7.4) is a differentiable function that is zero on C_2 and has a derivative that is not continuous on the set

$$\left\{ \frac{1}{9}, \frac{2}{9}, \frac{1}{3}, \frac{2}{3}, \frac{7}{9}, \frac{8}{9} \right\}.$$

Continuing in this fashion yields a sequence of functions $f_0, f_1, f_2, \ldots$ defined on $[0, 1]$.

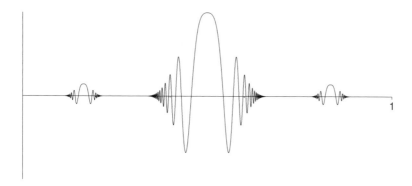

Figure 7.4: A GRAPH OF $f_2(x)$.

Exercise 7.6.15. (a) If $c \in C$, what is $\lim_{n\to\infty} f_n(c)$?

(b) Why does $\lim_{n\to\infty} f_n(x)$ exist for $x \notin C$?

Now, set

$$f(x) = \lim_{n\to\infty} f_n(x).$$

Exercise 7.6.16. (a) Explain why $f'(x)$ exists for all $x \notin C$.

(b) If $c \in C$, argue that $|f(x)| \leq (x - c)^2$ for all $x \in [0, 1]$. Show how this implies $f'(c) = 0$.

(c) Give a careful argument for why $f'(x)$ fails to be continuous on C. Remember that C contains many points besides the endpoints of the intervals that make up $C_1, C_2, C_3, \ldots$.

Let's take inventory of the situation. Our goal is to create a nonintegrable derivative. Our function $f(x)$ is differentiable, and f' fails to be continuous on C. We are not quite done.

Exercise 7.6.17. Why is $f'(x)$ Riemann-integrable on $[0, 1]$?

The reason the Cantor set has measure zero is that, at each stage, 2^{n-1} open intervals of length $1/3^n$ are removed from C_{n-1}. The resulting sum

$$\sum_{n=1}^{\infty} 2^{n-1}\left(\frac{1}{3^n}\right)$$

converges to one, which means that the approximating sets $C_1, C_2, C_3, \ldots$ have total lengths tending to zero. Instead of removing open intervals of length $1/3^n$ at each stage, let's see what happens when we remove intervals of length $1/3^{n+1}$.

Exercise 7.6.18. Show that, under these circumstances, the sum of the lengths of the intervals making up each C_n no longer tends to zero as $n \to \infty$. What is this limit?

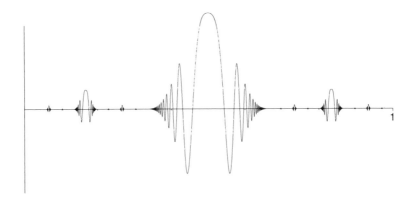

Figure 7.5:   A DIFFERENTIABLE FUNCTION WITH A NON-INTEGRABLE DERIVATIVE.

If we again take the intersection $\bigcap_{n=0}^{\infty} C_n$, the result is a Cantor-type set with the same topological properties—it is closed, compact, perfect, and contains no intervals. But a consequence of the previous exercise is that it no longer has measure zero. This is just what we need to define our desired function. By repeating the preceding construction of $f(x)$ on this new Cantor-type set of *strictly positive* measure, we get a differentiable function whose derivative has too many points of discontinuity (Fig. 7.5). By Lebesgue's Theorem, this derivative cannot be integrated using the Riemann integral.

Exercise 7.6.19. As a final gesture, provide the example advertised in Exercise 7.6.13 of an integrable function f and a continuous function g where the composition $f \circ g$ is properly defined but not integrable. Exercise 4.3.12 may be useful.

7.7   Epilogue

Riemann's definition of the integral was a modification of Cauchy's integral, which was originally designed for the purpose of integrating continuous functions. In this goal, the Riemann integral was a complete success. For continuous functions at least, the process of integration now stood on its own rigorous footing, defined independently of differentiation. As analysis progressed, however, the dependence of integrability on continuity became problematic. The last example of Section 7.6 highlights one type of weakness: not every derivative can be integrated. Another limitation of the Riemann integral arises in association with limits of sequences of functions. To get a sense of this, let's once again consider Dirichlet's function $g(x)$ introduced in Section 4.1. Recall that $g(x) = 1$ whenever x is rational, and $g(x) = 0$ at every irrational point. Focusing on the interval $[0, 1]$ for a moment, let

$$\{r_1, r_2, r_3, r_4 \ldots\}$$

be an enumeration of the countable number of rational points in this interval. Now, let $g_1(x) = 1$ if $x = r_1$ and define $g_1(x) = 0$ otherwise. Next, define $g_2(x) = 1$ if x is either r_1 or r_2, and let $g_2(x) = 0$ at all other points. In general, for each $n \in \mathbf{N}$, define

$$g_n(x) = \begin{cases} 1 & \text{if } x \in \{r_1, r_2, \dots, r_n\} \\ 0 & \text{otherwise.} \end{cases}$$

Notice that each g_n has only a finite number of discontinuities and so is Riemann-integrable with $\int_0^1 g_n = 0$. But we also have $g_n \to g$ pointwise on the interval $[0, 1]$. The problem arises when we remember that Dirichlet's nowhere-continuous function is not Riemann-integrable. Thus, the equation

(1)
$$\lim_{n \to \infty} \int_0^1 g_n = \int_0^1 g$$

fails to hold, not because the values on each side of the equal sign are different but because the value on the right-hand side does not exist. The content of Theorem 7.4.4 is that this equation does hold whenever we have $g_n \to g$ *uniformly*. This is a reasonable way to resolve the situation, but it is a bit unsatisfying because the deficiency in this case is not entirely with the type of convergence but lies in the strength of the Riemann integral. If we could make sense of the right-hand side via some other definition of integration, then maybe equation (1) would actually be true.

Such a definition was introduced by Henri Lebesgue in 1901. Generally speaking, Lebesgue's integral is constructed using a generalization of length called the *measure* of a set. In the previous section, we studied sets of *measure zero*. In particular, we showed that the rational numbers in $[0,1]$ (because they are countable) have measure zero. The irrational numbers in $[0,1]$ have measure one. This should not be too surprising because we now have that the measures of these two disjoint sets add up to the length of the interval $[0, 1]$. Rather than chopping up the x-axis to approximate the area under the curve, Lebesgue suggested partitioning the y-axis. In the case of Dirichlet's function g, there are only two range values—zero and one. The integral, according to Lebesgue, could be defined via

$$\int_0^1 g = 1 \cdot [\text{measure of set where } g = 1] + 0 \cdot [\text{measure of set where } g = 0]$$
$$= 1 \cdot 0 + 0 \cdot 1 = 0.$$

With this interpretation of $\int_0^1 g$, equation (1) is now valid!

The Lebesgue integral is presently the standard integral in advanced mathematics. The theory is taught to all graduate students, as well as to many undergraduates, and it is the integral used in most research papers where integration is required. The Lebesgue integral generalizes the Riemann integral in the sense that any function that is Riemann-integrable is Lebesgue-integrable and integrates to the same value. The real strength of the Lebesgue integral is that the class of integrable functions is much larger. Most importantly, this class includes the limits of different types of Cauchy sequences of integrable functions. This leads to a group of extremely important convergence theorems related to equation (1) with hypotheses much weaker than the uniform convergence assumed in Theorem 7.4.4.

Despite its prevalence, the Lebesgue integral does have a few drawbacks. There are functions whose *improper* Riemann integrals exist but that are not Lebesgue-integrable. Another disappointment arises from the relationship between integration and differentiation. Even with the Lebesgue integral, it is still not possible to prove

$$\int_a^b f' = f(b) - f(a)$$

without some additional assumptions on f. Around 1960, a new integral was proposed that can integrate a larger class of functions than either the Riemann integral or the Lebesgue integral and suffers from neither of the preceding weaknesses. Remarkably, this integral is actually a return to Riemann's original technique for defining integration, with some small modifications in how we describe the "fineness" of the partitions. An introduction to the generalized Riemann integral is the topic of Section 8.1.

Chapter 8

Additional Topics

The foundation in analysis provided by the first seven chapters is sufficient background for the exploration of some advanced and historically important topics. The writing in this chapter is similar to that in the concluding project sections of each individual chapter. Exercises are included within the exposition and are designed to make each section a narrative investigation into a significant achievement in the field of analysis.

8.1 The Generalized Riemann Integral

Chapter 7 concluded with Henri Lebesgue's elegant result that a bounded function is Riemann-integrable if and only if its points of discontinuity form a set of measure zero. To eliminate the dependence of integrability on continuity, Lebesgue proposed a new method of integration that has become the standard integral in mathematics. In the Epilogue to Chapter 7, we briefly outlined some of the strengths and weaknesses of the Lebesgue integral, concluding with a look back to the Fundamental Theorem of Calculus (Theorem 7.5.1). (Lebesgue's measure-zero criterion is not a prerequisite for understanding the material in this section, but the discussion in Section 7.7 provides some useful context for what follows.)

If F is a differentiable function on $[a, b]$, then in a perfect world we might hope to prove that

$$(1) \qquad \int_a^b F' = F(b) - F(a).$$

Notice that although this is the conclusion of part (i) of Theorem 7.5.1, there we needed the additional requirement that F' be Riemann-integrable. To drive this point home, Section 7.6 concluded with an example of a function that has

© Springer Science+Business Media New York 2015
S. Abbott, *Understanding Analysis*, Undergraduate Texts
in Mathematics, DOI 10.1007/978-1-4939-2712-8_8

a derivative that the Riemann integral cannot handle. The Lebesgue integral alluded to earlier is a significant improvement. It can integrate our example from Section 7.6, but ultimately it too suffers from the same setback. Not every derivative is integrable, no matter which integral is used.

What follows is a short introduction to the generalized Riemann integral, discovered independently around 1960 by Jaroslav Kurzweil and Ralph Henstock. As mentioned in Section 7.7, this lesser-known modification of the Riemann integral can actually integrate a larger class of functions than Lebesgue's ubiquitous integral and yields a surprisingly simple proof of equation (1) above with no additional hypotheses.

The Riemann Integral as a Limit

Let

$$P = \{x_0, x_1, x_2, \ldots, x_n\}$$

be a partition of $[a, b]$. A *tagged partition* is one where in addition to P we have chosen points c_k in each of the subintervals $[x_{k-1}, x_k]$. This sets the stage for the concept of a Riemann sum. Given a function $f : [a, b] \to \mathbf{R}$, and a tagged partition $(P, \{c_k\}_{k=1}^n)$, the *Riemann sum* generated by this partition is given by

$$R(f, P) = \sum_{k=1}^n f(c_k)(x_k - x_{k-1}).$$

Looking back at the definition of the upper sum

$$U(f, P) = \sum_{k=1}^n M_k(x_k - x_{k-1}) \quad \text{where} \quad M_k = \sup\{f(x) : x \in [x_{k-1}, x_k]\},$$

and the lower sum

$$L(f, P) = \sum_{k=1}^n m_k(x_k - x_{k-1}) \quad \text{where} \quad m_k = \inf\{f(x) : x \in [x_{k-1}, x_k]\},$$

it should be clear that

$$L(f, P) \leq R(f, P) \leq U(f, P)$$

for any bounded function f. In Definition 7.2.7, we characterized integrability by insisting that the infimum of the upper sums equal the supremum of the lower sums. Any Riemann sum is going to fall between a particular upper and lower sum. If the upper and lower sums are converging to some common value, then the Riemann sums are also eventually close to this value as well. The next theorem shows that it is possible to characterize Riemann integrability in a way equivalent to Definition 7.2.7 using an ϵ–δ-type definition applied to Riemann sums.

Definition 8.1.1. Let $\delta > 0$. A partition P is δ-*fine* if every subinterval $[x_{k-1}, x_k]$ satisfies $x_k - x_{k-1} < \delta$. In other words, every subinterval has width less than δ.

Theorem 8.1.2 (Limit Criterion for Riemann Integrability). *A bounded function $f : [a, b] \to \mathbf{R}$ is Riemann-integrable with*

$$\int_a^b f = A$$

if and only if, for every $\epsilon > 0$, there exists a $\delta > 0$ such that, for any tagged partition $(P, \{c_k\})$ that is δ-fine, it follows that

$$|R(f, P) - A| < \epsilon.$$

Before attempting the proof, we should point out that, in some treatments, the criterion in Theorem 8.1.2 is actually taken as the *definition* of Riemann integrability. In fact, this is how Riemann originally defined the concept. The spirit of this theorem is close to what is taught in most introductory calculus courses. To approximate the area under the curve, Riemann sums are constructed. The hope is that as the partitions become finer, the corresponding approximations get closer to the value of the integral. The content of Theorem 8.1.2 is that if the function is integrable, then these approximations do indeed converge to the value of the integral, regardless of how the tags are chosen. Conversely, if the approximating Riemann sums for finer and finer partitions collect around some value A, then the function is integrable and integrates to A.

Proof. ($\Rightarrow$) For the forward direction, we begin with the assumption that f is integrable on $[a, b]$. Given an $\epsilon > 0$, we must produce a $\delta > 0$ such that if $(P, \{c_k\})$ is any tagged partition that is δ-fine, then $|R(f, P) - \int_a^b f| < \epsilon$.

Because f is integrable, we know there exists a partition P_ϵ such that

$$U(f, P_\epsilon) - L(f, P_\epsilon) < \frac{\epsilon}{3}.$$

Let $M > 0$ be a bound on $|f|$, and let n be the number of subintervals of P_ϵ (so that P_ϵ really consists of $n + 1$ points in $[a, b]$). We will argue that choosing

$$\delta = \epsilon/9nM$$

has the desired property.

Here is the idea. Let $(P, \{c_k\})$ be an arbitrary tagged partition of $[a, b]$ that is δ-fine, and let $P' = P \cup P_\epsilon$. The key is to establish the string of inequalities

$$L(f, P') - \frac{\epsilon}{3} < L(f, P) \le U(f, P) < U(f, P') + \frac{\epsilon}{3}.$$

Exercise 8.1.1. (a) Explain why both the Riemann sum $R(f, P)$ and $\int_a^b f$ fall between $L(f, P)$ and $U(f, P)$.

(b) Explain why $U(f, P') - L(f, P') < \epsilon/3$.

By the previous exercise, if we can show $U(f, P) < U(f, P') + \epsilon/3$ (and similarly $L(f, P') - \epsilon/3 < L(f, P)$), then it will follow that

$$\left| R(f, P) - \int_a^b f \right| < \epsilon$$

and the proof will be done. Thus, we turn our attention toward estimating the distance between $U(f, P)$ and $U(f, P')$.

Exercise 8.1.2. Explain why $U(f, P) - U(f, P') \geq 0$.

A typical term in either $U(f, P)$ or $U(f, P')$ has the form $M_k(x_k - x_{k-1})$, where M_k is the supremum of f over $[x_{k-1}, x_k]$. A good number of these terms appear in both upper sums and so cancel out.

Exercise 8.1.3.   (a) In terms of n, what is the largest number of terms of the form $M_k(x_k - x_{k-1})$ that could appear in one of $U(f, P)$ or $U(f, P')$ but not the other?

(b) Finish the proof in this direction by arguing that

$$U(f, P) - U(f, P') < \epsilon/3.$$

($\Leftarrow$) For this direction, we assume that the ϵ-δ criterion in Theorem 8.1.2 holds and argue that f is integrable. Integrability, as we have defined it, depends on our ability to choose partitions for which the upper sums are close to the lower sums. We have remarked that given any partition P, it is always the case that

$$L(f, P) \leq R(f, P) \leq U(f, P)$$

no matter which tags are chosen to compute $R(f, P)$.

Exercise 8.1.4.   (a) Show that if f is continuous, then it is possible to pick tags $\{c_k\}_{k=1}^n$ so that
$$R(f, P) = U(f, P).$$

Similarly, there are tags for which $R(f, P) = L(f, P)$ as well.

(b) If f is not continuous, it may not be possible to find tags for which $R(f, P) = U(f, P)$. Show, however, that given an arbitrary $\epsilon > 0$, it is possible to pick tags for P so that

$$U(f, P) - R(f, P) < \epsilon.$$

The analogous statement holds for lower sums.

Exercise 8.1.5. Use the results of the previous exercise to finish the proof of Theorem 8.1.2.   □

Gauges and $\delta(x)$-fine Partitions

The key to the generalized Riemann integral is to allow the δ in Theorem 8.1.2 to be a *function of x*.

Definition 8.1.3. A function $\delta : [a, b] \to \mathbf{R}$ is called a *gauge* on $[a, b]$ if $\delta(x) > 0$ for all $x \in [a, b]$.

Definition 8.1.4. Given a particular gauge $\delta(x)$, a tagged partition $(P, \{c_k\}_{k=1}^n)$ is $\delta(x)$-*fine* if every subinterval $[x_{k-1}, x_k]$ satisfies $x_k - x_{k-1} < \delta(c_k)$. In other words, each subinterval $[x_{k-1}, x_k]$ has width less than $\delta(c_k)$.

It is important to see that if $\delta(x)$ is a constant function, then Definition 8.1.4 says precisely the same thing as Definition 8.1.1. In the case where $\delta(x)$ is not a constant, Definition 8.1.4 describes a way of measuring the fineness of partitions that is quite different.

Exercise 8.1.6. Consider the interval $[0, 1]$.

(a) If $\delta(x) = 1/9$, find a $\delta(x)$-fine tagged partition of $[0, 1]$. Does the choice of tags matter in this case?

(b) Let

$$\delta(x) = \begin{cases} 1/4 & \text{if } x = 0 \\ x/3 & \text{if } 0 < x \leq 1. \end{cases}$$

Construct a $\delta(x)$-fine tagged partition of $[0,1]$.

The tinkering required in Exercise 8.1.6 (b) may cast doubt on whether an arbitrary gauge always admits a $\delta(x)$-fine partition. However, it is not too difficult to show that this is indeed the case.

Theorem 8.1.5. *Given a gauge $\delta(x)$ on an interval $[a, b]$, there exists a tagged partition $(P, \{c_k\}_{k=1}^n)$ that is $\delta(x)$-fine.*

Proof. Let $I_0 = [a, b]$. It may be possible to find a tag such that the trivial partition $P = \{a, b\}$ works. Specifically, if $b - a < \delta(x)$ for some $x \in [a, b]$, then we can set c_1 equal to such an x and notice that $(P, \{c_1\})$ is $\delta(x)$-fine. If no such x exists, then bisect $[a, b]$ into two equal halves.

Exercise 8.1.7. Finish the proof of Theorem 8.1.5. $\qquad\qquad\square$

Generalized Riemann Integrability

Keeping in mind that Theorem 8.1.2 offers an equivalent way to define Riemann integrability, we now propose a new method for defining the value of the integral.

Definition 8.1.6. A function f on $[a, b]$ has *generalized Riemann integral A* if, for every $\epsilon > 0$, there exists a gauge $\delta(x)$ on $[a, b]$ such that for each tagged partition $(P, \{c_k\}_{k=1}^n)$ that is $\delta(x)$-fine, it is true that

$$|R(f, P) - A| < \epsilon.$$

In this case, we write $A = \int_a^b f$.

Theorem 8.1.7. *If a function has a generalized Riemann integral, then the value of the integral is unique.*

Proof. Assume that a function f has generalized Riemann integral A_1 and that it also has generalized Riemann integral A_2. We must prove $A_1 = A_2$.

$\square$

Exercise 8.1.8. Finish the argument.

The implications of Definition 8.1.6 on the resulting class of integrable functions are far reaching. This is somewhat surprising given that the criteria for integrability in Definition 8.1.6 and Theorem 8.1.2 differ in such a small way. One observation that should be immediately evident is the following.

Exercise 8.1.9. Explain why every function that is Riemann-integrable with $\int_a^b f = A$ must also have generalized Riemann integral A.

The converse statement is not true, and that is the important point. One example that we have of a non-Riemann-integrable function is Dirichlet's function

$$g(x) = \begin{cases} 1 & \text{if } x \in \mathbf{Q} \\ 0 & \text{if } x \notin \mathbf{Q} \end{cases}$$

which has discontinuities at every point of $\mathbf{R}$.

Theorem 8.1.8. *Dirichlet's function $g(x)$ is generalized Riemann-integrable on $[0, 1]$ with $\int_0^1 g = 0$.*

Proof. Let $\epsilon > 0$. By Definition 8.1.6, we must construct a gauge $\delta(x)$ on $[0, 1]$ such that whenever $(P, \{c_k\}_{k=1}^n)$ is a $\delta(x)$-fine tagged partition, it follows that

$$0 \leq \sum_{k=1}^n g(c_k)(x_k - x_{k-1}) < \epsilon.$$

The gauge represents a restriction on the size of $\Delta x_k = x_k - x_{k-1}$ in the sense that $\Delta x_k < \delta(c_k)$. The Riemann sum consists of products of the form $g(c_k)\Delta x_k$. Thus, for irrational tags, there is nothing to worry about because $g(c_k) = 0$ in this case. Our task is to make sure that any time a tag c_k is rational, it comes from a suitably thin subinterval.

Let $\{r_1, r_2, r_3, \ldots\}$ be an enumeration of the countable set of rational numbers contained in $[0, 1]$. For each r_k, set $\delta(r_k) = \epsilon/2^{k+1}$. For x irrational, set $\delta(x) = 1$.

Exercise 8.1.10. Show that if $(P, \{c_k\}_{k=1}^n)$ is a $\delta(x)$-fine tagged partition, then $R(g, P) < \epsilon$. □

Dirichlet's function fails to be Riemann-integrable because, given any (un-tagged) partition, it is possible to make $R(g, P) = 1$ or $R(g, P) = 0$ by choosing the tags to be either all rational or all irrational. For the generalized Riemann integral, choosing all rational tags results in a tagged partition that is not $\delta(x)$-fine (when $\delta(x)$ is small on rational points) and so does not have to be considered. In general, allowing for nonconstant gauges allows us to be more discriminating about which tagged partitions qualify as $\delta(x)$-fine. The result, as we have just seen, is that it may be easier to achieve the inequality

$$|R(f, P) - A| < \epsilon$$

for the often smaller and more carefully selected set of tagged partitions that remain.

The Fundamental Theorem of Calculus

We conclude this brief introduction to the generalized Riemann integral with a proof of the Fundamental Theorem of Calculus. As was alluded to earlier, the most notable distinction between the following theorem and part (i) of Theorem 7.5.1 is that here we do not need to assume that the derivative function is integrable. Using the generalized Riemann integral, every derivative is integrable, and the integral can be evaluated using the antiderivative in the familiar way. It is also interesting to note that in Theorem 7.5.1 the Mean Value Theorem played the crucial role in the argument, but it is not needed here.

Theorem 8.1.9. *Assume $F : [a, b] \to \mathbf{R}$ is differentiable at each point in $[a, b]$ and set $f(x) = F'(x)$. Then, f has the generalized Riemann integral*

$$\int_a^b f = F(b) - F(a).$$

Proof. Let $P = \{x_0, x_1, x_2, \ldots, x_n\}$ be a partition of $[a, b]$. Both this proof and the proof of Theorem 7.5.1 make use of the following fact.

Exercise 8.1.11. Show that

$$F(b) - F(a) = \sum_{k=1}^n [F(x_k) - F(x_{k-1})].$$

If $\{c_k\}_{k=1}^{n}$ is a set of tags for P, then we can estimate the difference between the Riemann sum $R(f, P)$ and $F(b) - F(a)$ by

$$|F(b) - F(a) - R(f, P)| \;=\; \left| \sum_{k=1}^{n} [F(x_k) - F(x_{k-1}) - f(c_k)(x_k - x_{k-1})] \right|$$

$$\leq \;\; \sum_{k=1}^{n} |F(x_k) - F(x_{k-1}) - f(c_k)(x_k - x_{k-1})| .$$

Let $\epsilon > 0$. To prove the theorem, we must construct a gauge $\delta(c)$ such that

(2) $$|F(b) - F(a) - R(f, P)| < \epsilon$$

for all $(P, \{c_k\})$ that are $\delta(c)$-fine. (Using the variable c in the gauge function is more convenient than x in this case.)

Exercise 8.1.12. For each $c \in [a, b]$, explain why there exists a $\delta(c) > 0$ (a $\delta > 0$ depending on c) such that

$$\left| \frac{F(x) - F(c)}{x - c} - f(c) \right| < \epsilon$$

for all $0 < |x - c| < \delta(c)$.

This $\delta(c)$ is the desired gauge on $[a, b]$. Let $(P, \{c_k\}_{k=1}^{n})$ be a $\delta(c)$-fine partition of $[a, b]$. It just remains to show that equation (2) is satisfied for this tagged partition.

Exercise 8.1.13.   (a) For a particular $c_k \in [x_{k-1}, x_k]$ of P, show that

$$|F(x_k) - F(c_k) - f(c_k)(x_k - c_k)| < \epsilon(x_k - c_k)$$

and

$$|F(c_k) - F(x_{k-1}) - f(c_k)(c_k - x_{k-1})| < \epsilon(c_k - x_{k-1}).$$

(b) Now, argue that

$$|F(x_k) - F(x_{k-1}) - f(c_k)(x_k - x_{k-1})| < \epsilon(x_k - x_{k-1}),$$

and use this fact to complete the proof of the theorem.   □

If we consider the function

$$F(x) = \begin{cases} x^{3/2} \sin(1/x) & \text{if } x \neq 0 \\ 0 & \text{if } x = 0 \end{cases}$$

then it is not too difficult to show that F is differentiable everywhere, including $x = 0$, with

$$F'(x) = \begin{cases} (3/2)\sqrt{x}\sin(1/x) - (1/\sqrt{x})\cos(1/x) & \text{if } x \neq 0 \\ 0 & \text{if } x = 0. \end{cases}$$

What is notable here is that the derivative is unbounded near the origin. The theory of the ordinary Riemann integral begins with the assumption that we only consider bounded functions on closed intervals, but there is no such restriction for the generalized Riemann integral. Theorem 8.1.9 proves that F' has a generalized integral. Now, *improper* Riemann integrals have been created to extend Riemann integration to some unbounded functions, but it is another interesting fact about the generalized Riemann integral that any function having an improper integral must already be integrable in the sense described in Definition 8.1.6.

As a parting gesture, let's show how Theorem 8.1.9 yields a short verification of the substitution technique from calculus.

Theorem 8.1.10 (Change-of-variable Formula). *Let $g : [a, b] \to \mathbf{R}$ be differentiable at each point of $[a, b]$, and assume F is differentiable on the set $g([a, b])$. If $f(x) = F'(x)$ for all $x \in g([a, b])$, then*

$$\int_a^b (f \circ g) \cdot g' = \int_{g(a)}^{g(b)} f.$$

Proof. The hypothesis of the theorem guarantees that the function $(F \circ g)(x)$ is differentiable for all $x \in [a, b]$.

Exercise 8.1.14. (a) Why are we sure that f and $(F \circ g)'$ have generalized Riemann integrals?

(b) Use Theorem 8.1.9 to finish the proof. $\square$

The impressive properties of the generalized Riemann integral do not end here. The central source for the material in this section is Robert Bartle's award winning article "Return to the Riemann Integral," which appeared in the *American Mathematical Monthly*, October, 1996. The article goes on to discuss convergence theorems for this new integral in the spirit of Theorem 7.4.4, and outlines the argument that the collection of integrable functions is strictly larger when the Lebesgue integral is replaced by the generalized Riemann integral. In light of this, the author boldly declares that "*the time has come to discard the Lebesgue integral as the primary integral.*" (Italics in the original.)

That this revolution has not come to pass may simply be due to a case of overwhelming inertia, but a contributing factor is very likely the geometrically satisfying intuition of Lebesgue's theory. At the heart of Lebesgue's approach to integration is the desire to generalize the concepts of length and area. Although one can certainly use a properly developed integral to give a rigorous definition for the length—or measure—of a general set, there is a compelling argument that this puts the ideas in the wrong pedagogical order. Rather than using a sophisticated integral to generalize a primitive notion such as length, Lebesgue found an effective way to talk about the length of a very wide class of sets, and used that to build his definition of the integral. The very elegant result of his endeavor is likely to be the industry standard for a long time to come.

8.2   Metric Spaces and the Baire Category Theorem

A natural question to ask is whether the theorems we have proved about sequences, series, and functions in $\mathbf{R}$ have analogues in the plane $\mathbf{R}^2$ or in even higher dimensions. Looking back over the proofs, one crucial observation is that most of the arguments depend on just a few basic properties of the absolute value function. Interpreting the statement "$|x - y|$" to mean the "distance from x to y in $\mathbf{R}$," our aim is to experiment with other ways of measuring distance on other sets such as $\mathbf{R}^2$ and $C[0, 1]$, the space of continuous functions on $[0, 1]$.

Definition 8.2.1. Given a set X, a function $d : X \times X \to \mathbf{R}$ is a *metric* on X if for all $x, y \in X$:

(i) $d(x, y) \geq 0$ with $d(x, y) = 0$ if and only if $x = y$,

(ii) $d(x, y) = d(y, x)$, and

(iii) for all $z \in X$, $d(x, y) \leq d(x, z) + d(z, y)$.

A *metric space* is a set X together with a metric d.

Property (iii) in the previous definition is the "triangle inequality." The next two exercises illustrate the point that the same set X can be home to several different metrics. When referring to a metric space, we must specify the set *and* the particular distance function d.

Exercise 8.2.1. Decide which of the following are metrics on $X = \mathbf{R}^2$. For each, we let $x = (x_1, x_2)$ and $y = (y_1, y_2)$ be points in the plane.

(a) $d(x, y) = \sqrt{(x_1 - y_1)^2 + (x_2 - y_2)^2}$.

(b) $d(x, y) = \max\{|x_1 - y_1|, |x_2 - y_2|\}$.

(c) $d(x, y) = |x_1 x_2 + y_1 y_2|$.

The metric in part (a) of the previous exercise is the familiar Euclidean distance between two points in the plane. This is often referred to as the "usual" or "standard" metric on $\mathbf{R}^2$. The usual metric on $\mathbf{R}$ is our old friend $d(x, y) = |x - y|$.

Exercise 8.2.2. Let $C[0, 1]$ be the collection of continuous functions on the closed interval $[0, 1]$. Decide which of the following are metrics on $C[0, 1]$.

(a) $d(f, g) = \sup\{|f(x) - g(x)| : x \in [0, 1]\}$.

(b) $d(f, g) = |f(1) - g(1)|$.

(c) $d(f, g) = \int_0^1 |f - g|$.

The following distance function is called the *discrete metric* and can be defined on any set X. For any $x, y \in X$, let

$$\rho(x, y) = \begin{cases} 1 & \text{if } x \neq y \\ 0 & \text{if } x = y. \end{cases}$$

Exercise 8.2.3. Verify that the discrete metric is actually a metric.

Basic Definitions

Definition 8.2.2. Let (X, d) be a metric space. A sequence $(x_n) \subseteq X$ *converges* to an element $x \in X$ if for all $\epsilon > 0$ there exists an $N \in \mathbf{N}$ such that $d(x_n, x) < \epsilon$ whenever $n \geq N$.

Definition 8.2.3. A sequence (x_n) in a metric space (X, d) is a *Cauchy sequence* if for all $\epsilon > 0$ there exists an $N \in \mathbf{N}$ such that $d(x_m, x_n) < \epsilon$ whenever $m, n \geq N$.

Exercise 8.2.4. Show that a convergent sequence is Cauchy.

The Cauchy Criterion, as it is called in $\mathbf{R}$, was an "if and only if" statement. In the general metric space setting, however, the converse statement does not always hold. Recall that, in $\mathbf{R}$, the assertion that "Cauchy sequences converge" was shown to be equivalent to the Axiom of Completeness. In order to transport the Axiom of Completeness into a metric space, we would need to have an ordering on our space so that we could discuss such things as upper bounds. It is an interesting observation that not every set can be ordered in a satisfying way (the points in $\mathbf{R}^2$ for example). Even without an ordering, we are still going to want completeness. For metric spaces, the convergence of Cauchy sequences is taken to be the definition of completeness.

Definition 8.2.4. A metric space (X, d) is *complete* if every Cauchy sequence in X converges to an element of X.

Exercise 8.2.5. (a) Consider $\mathbf{R}^2$ with the discrete metric $\rho(x, y)$ examined in Exercise 8.2.3. What do Cauchy sequences look like in this space? Is $\mathbf{R}^2$ complete with respect to this metric?

(b) Show that $C[0, 1]$ is complete with respect to the metric in Exercise 8.2.2 (a).

(c) Define $C^1[0, 1]$ to be the collection of differentiable functions on $[0,1]$ whose derivatives are also continuous. Is $C^1[0, 1]$ complete with respect to the metric defined in Exercise 8.2.2 (a)?

Because completeness is a prerequisite for doing anything significant in the way of analysis, the metric in Exercise 8.2.2 (a) is the most natural metric to consider when working with $C[0, 1]$. The notation

$$\|f - g\|_\infty = d(f, g) = \sup\{|f(x) - g(x)| : x \in [0, 1]\}$$

is standard, and setting $g = 0$ gives the so-called "sup norm"

$$\|f\|_\infty = d(f, 0) = \sup\{|f(x)| : x \in [0, 1]\}.$$

In all upcoming discussions, it is assumed that the space $C[0, 1]$ is endowed with this metric unless otherwise specified.

Definition 8.2.5. Let (X, d_1) and (Y, d_2) be metric spaces. A function $f : X \to Y$ is *continuous at* $x \in X$ if for all $\epsilon > 0$ there exists a $\delta > 0$ such that $d_2(f(x), f(y)) < \epsilon$ whenever $d_1(x, y) < \delta$.

Exercise 8.2.6. Which of these functions from $C[0, 1]$ to $\mathbf{R}$ (with the usual metric) are continuous?

(a) $g(f) = \int_0^1 fk$, where k is some fixed function in $C[0, 1]$.

(b) $g(f) = f(1/2)$.

(c) $g(f) = f(1/2)$, but this time with respect to the metric on $C[0, 1]$ from Exercise 8.2.2 (c).

Topology on Metric Spaces

Definition 8.2.6. Given $\epsilon > 0$ and an element x in the metric space (X, d), the ϵ-*neighborhood of* x is the set $V_\epsilon(x) = \{y \in X : d(x, y) < \epsilon\}$.

Exercise 8.2.7. Describe the ϵ-neighborhoods in $\mathbf{R}^2$ for each of the different metrics described in Exercise 8.2.1. How about for the discrete metric?

With the definition of an ϵ-neighborhood, we can now define *open sets*, *limit points*, and *closed sets* exactly as we did before. A set $O \subseteq X$ is *open* if for every $x \in O$ we can find a neighborhood $V_\epsilon(x) \subseteq O$. A point x is a *limit point* of a set A if every $V_\epsilon(x)$ intersects A in some point other than x. A set C is *closed* if it contains its limit points.

Exercise 8.2.8. Let (X, d) be a metric space.

(a) Verify that a typical ϵ-neighborhood $V_\epsilon(x)$ is an open set. Is the set

$$C_\epsilon(x) = \{y \in X : d(x, y) \le \epsilon\}$$

a closed set?

(b) Show that a set $E \subseteq X$ is open if and only if its complement is closed.

Exercise 8.2.9. (a) Show that the set $Y = \{f \in C[0, 1] : \|f\|_\infty \le 1\}$ is closed in $C[0, 1]$.

(b) Is the set $T = \{f \in C[0, 1] : f(0) = 0\}$ open, closed, or neither in $C[0, 1]$?

We define compactness in metric spaces just as we did for $\mathbf{R}$.

Definition 8.2.7. A subset K of a metric space (X, d) is *compact* if every sequence in K has a convergent subsequence that converges to a limit in K.

An extremely useful characterization of compactness in $\mathbf{R}$ is the proposition that a set is compact if and only if it is closed and bounded. For abstract metric spaces, this proposition only holds in the forward direction.

Exercise 8.2.10. (a) Supply a definition for *bounded* subsets of a metric space (X, d).

(b) Show that if K is a compact subset of the metric space (X, d), then K is closed and bounded.

(c) Show that $Y \subseteq C[0, 1]$ from Exercise 8.2.9 (a) is closed and bounded but not compact.

A good hint for part (c) of the previous exercise can be found in Exercise 6.2.14 from Chapter 6. This exercise defines the concept of an *equicontinuous* family of functions, which is a key ingredient in the Arzela–Ascoli Theorem (Exercise 6.2.15). The Arzela–Ascoli Theorem states that any bounded, *equicontinuous* collection of functions in $C[0, 1]$ must have a uniformly convergent subsequence. One way to summarize this famous result—which we did not have the language for in Chapter 6—is as a statement describing a particular class of compact subsets in $C[0, 1]$. Looking at the definition of compactness, and remembering that the uniform limit of continuous functions is continuous, the Arzela–Ascoli Theorem states that any closed, bounded, equicontinuous collection of functions is a compact subset of $C[0, 1]$.

Definition 8.2.8. Given a subset E of a metric space (X, d), the *closure* $\overline{E}$ is the union of E together with its limit points. The *interior* of E is denoted by E° and is defined as

$$E^\circ = \{x \in E : \text{there exists } V_\epsilon(x) \subseteq E\}.$$

Closure and interior are dual concepts. Results about these concepts come in pairs and exhibit an elegant and useful symmetry.

Exercise 8.2.11. (a) Show that E is closed if and only if $\overline{E} = E$. Show that E is open if and only if $E^\circ = E$.

(b) Show that $\overline{E}^c = (E^c)^\circ$, and similarly that $(E^\circ)^c = \overline{E^c}$.

A good hint for this exercise is to review the proofs from Chapter 3, where closure at least is discussed. Thinking of all of these concepts as they relate to $\mathbf{R}$ or $\mathbf{R}^2$ with the usual metric is not a bad idea. However, it is important to remember also that rigorous proofs must be constructed purely from the relevant definitions.

Exercise 8.2.12. (a) Show

$$\overline{V_\epsilon(x)} \subseteq \{y \in X : d(x, y) \le \epsilon\},$$

in an arbitrary metric space (X, d).

(b) To keep things from sounding too familiar, find an example of a specific metric space where

$$\overline{V_\epsilon(x)} \neq \{y \in X : d(x, y) \leq \epsilon\}.$$

We are on our way toward the Baire Category Theorem. The next definitions provide the final bit of vocabulary needed to state the result.

Definition 8.2.9. A set $A \subseteq X$ is *dense* in the metric space (X, d) if $\overline{A} = X$. A subset E of a metric space (X, d) is *nowhere-dense* in X if $\overline{E}^{\circ}$ is empty.

Exercise 8.2.13. If E is a subset of a metric space (X, d), show that E is nowhere-dense in X if and only if $\overline{E}^{c}$ is dense in X.

The Baire Category Theorem

In Section 3.5, we proved Baire's Theorem, which states that it is impossible to write the real numbers $\mathbf{R}$ as the countable union of nowhere-dense sets. Previous to this, we knew that $\mathbf{R}$ was too big to be written as the countable union of single points ($\mathbf{R}$ is uncountable), but Baire's Theorem improves on this by asserting that the only way to make $\mathbf{R}$ from a countable union of arbitrary sets is for the closure of at least one of these sets to contain an interval. The keystone to the proof of Baire's Theorem is the completeness of $\mathbf{R}$. The idea now is to replace $\mathbf{R}$ with an arbitrary complete metric space and prove the theorem in this more general setting. This leads to a statement that can be used to discuss the size and structure of other spaces such as $\mathbf{R}^2$ and $C[0, 1]$. At the end of Chapter 3, we mentioned one particularly fascinating implication of this result for $C[0, 1]$, which is that—despite the substantial difficulty required to produce an example of one—most continuous functions are nowhere-differentiable. It would be a good idea at this point to reread Sections 3.6 and 5.5. We are now equipped to carry out the details promised in these discussions.

Theorem 8.2.10. *Let (X, d) be a complete metric space, and let $\{O_n\}$ be a countable collection of dense, open subsets of X. Then, $\bigcap_{n=1}^{\infty} O_n$ is not empty.*

Proof. When we proved this theorem on $\mathbf{R}$, completeness manifested itself in the form of the Nested Interval Property. We could derive something akin to NIP in the metric space setting, but instead let's take an approach that uses the convergence of Cauchy sequences (because this is how we have defined completeness).

Pick $x_1 \in O_1$. Because O_1 is open, there exists an $\epsilon_1 > 0$ such that $V_{\epsilon_1}(x_1) \subseteq O_1$.

Exercise 8.2.14.   (a) Give the details for why we know there exists a point $x_2 \in V_{\epsilon_1}(x_1) \cap O_2$ and an $\epsilon_2 > 0$ satisfying $\epsilon_2 < \epsilon_1/2$ with $V_{\epsilon_2}(x_2)$ contained in O_2 and

$$\overline{V_{\epsilon_2}(x_2)} \subseteq V_{\epsilon_1}(x_1).$$

(b) Proceed along this line and use the completeness of (X, d) to produce a single point $x \in O_n$ for every $n \in \mathbf{N}$. $\qquad\square$

Theorem 8.2.11 (Baire Category Theorem). *A complete metric space is not the union of a countable collection of nowhere-dense sets.*

Exercise 8.2.15. Complete the proof of the theorem.

This result is called the Baire Category Theorem because it creates two categories of size for subsets in a metric space. A set of "first category" is one that *can* be written as a countable union of nowhere-dense sets. These are the small, intuitively thin subsets of a metric space. We now see that if our metric space is complete, then it is necessarily of "second category," meaning it cannot be written as a countable union of nowhere-dense sets. Given a subset A of a complete metric space X, showing that A is of first category is a mathematically precise way of demonstrating that A constitutes a very minor portion of the set X. The term "meager" is often used to mean a set of first category.

With the stage set, we now outline the argument that continuous functions that are differentiable at even one point of [0,1] form a meager subset of the metric space $C[0, 1]$.

Theorem 8.2.12. *The set*

$$D = \{f \in C[0, 1] : f'(x) \text{ exists for some } x \in [0, 1]\}$$

is a set of first category in $C[0, 1]$.

Proof. For each pair of natural numbers m, n, define

$$A_{m,n} = \left\{ f \in C[0, 1] : \text{ there exists } x \in [0, 1] \text{ where} \right.$$

$$\left. \left| \frac{f(x) - f(t)}{x - t} \right| \le n \text{ whenever } 0 < |x - t| < \frac{1}{m} \right\}.$$

This definition takes some time to digest. Think of $1/m$ as defining a δ-neighborhood around the point x, and view n as an upper bound on the magnitude of the slopes of lines through the two points $(x, f(x))$ and $(t, f(t))$. The set $A_{m,n}$ contains any function in $C[0, 1]$ for which it is possible to find at least one point x where the slopes through $(x, f(x))$ and points on the function nearby—within $1/m$ to be precise—are bounded by n.

Exercise 8.2.16. Show that if $f \in C[0, 1]$ is differentiable at a point $x \in [0, 1]$, then $f \in A_{m,n}$ for some pair $m, n \in \mathbf{N}$.

The collection of subsets $\{A_{m,n} : m, n \in \mathbf{N}\}$ is countable, and we have just seen that the union of these sets contains our set D. Because it is not difficult to see that a subset of a set of first category is first category, the final hurdle in the argument is to prove that each $A_{m,n}$ is nowhere-dense in $C[0, 1]$.

Fix m and n. The first order of business is to prove that $A_{m,n}$ is a closed set. To this end, let (f_k) be a sequence in $A_{m,n}$ and assume $f_k \to f$ in $C[0,1]$. We need to show $f \in A_{m,n}$.

Because $f_k \in A_{m,n}$, then for each $k \in \mathbf{N}$ there exists a point $x_k \in [0,1]$ where

$$\left| \frac{f_k(x_k) - f_k(t)}{x_k - t} \right| \leq n \quad \text{for all} \quad 0 < |x_k - t| < 1/m.$$

Exercise 8.2.17.   (a) The sequence (x_k) does not necessarily converge, but explain why there exists a subsequence (x_{k_l}) that is convergent. Let $x = \lim(x_{k_l})$.

(b) Prove that $f_{k_l}(x_{k_l}) \to f(x)$.

(c) Now finish the proof that $A_{m,n}$ is closed.

Because $A_{m,n}$ is closed, $\overline{A_{m,n}} = A_{m,n}$. In order to prove that $A_{m,n}$ is nowhere-dense, we just have to show that it contains no ϵ-neighborhoods, so pick an arbitrary $f \in A_{m,n}$, let $\epsilon > 0$, and consider the ϵ-neighborhood $V_\epsilon(f)$ in $C[0,1]$. To show that this set is not contained in $A_{m,n}$, we must produce a function $g \in C[0,1]$ that satisfies $\|f - g\|_\infty < \epsilon$ and has the property that there is no point $x \in [0,1]$ where

$$\left| \frac{g(x) - g(t)}{x - t} \right| \leq n \quad \text{for all} \quad 0 < |x - t| < 1/m.$$

Exercise 8.2.18. A continuous function is called *polygonal* if its graph consists of a finite number of line segments.

(a) Show that there exists a polygonal function $p \in C[0,1]$ satisfying $\|f - p\|_\infty < \epsilon/2$.

(b) Show that if h is any function in $C[0,1]$ that is bounded by 1, then the function

$$g(x) = p(x) + \frac{\epsilon}{2} h(x)$$

satisfies $g \in V_\epsilon(f)$.

(c) Construct a polygonal function $h(x)$ in $C[0,1]$ that is bounded by 1 and leads to the conclusion $g \notin A_{m,n}$, where g is defined as in (b). Explain how this completes the argument for Theorem 8.2.12.   □

8.3   Euler's Sum

In Section 6.1 we saw Euler's first and most famous derivation of the formula

$$1 + \frac{1}{4} + \frac{1}{9} + \frac{1}{16} + \frac{1}{25} + \cdots = \frac{\pi^2}{6}.$$

At the crux of this argument are two representations for the function $\sin(x)$. The first is the standard Taylor series representation

(1) $$\sin(x) = x - \frac{x^3}{3!} + \frac{x^5}{5!} - \frac{x^7}{7!} + \cdots,$$

and the second is an infinite product representation

(2) $$\sin(x) = x\left(1 - \frac{x}{\pi}\right)\left(1 + \frac{x}{\pi}\right)\left(1 - \frac{x}{2\pi}\right)\left(1 + \frac{x}{2\pi}\right)\cdots.$$

Although we have since made rigorous sense of the first equation (Example 6.6.4), proving the validity of equation (2) is still beyond our means.

The news is not all bad, however. In the time since Euler first made this discovery, dozens of different proofs for this result have been published, starting with several by Euler himself and continuing right up to the present. The machinery required in these arguments runs the gamut from multi-variable calculus to Fourier series to complex integration, but one in particular due to Boo Rim Choe relies mainly on Taylor series expansions and properties of uniformly convergent series. Choe's argument was published in 1987 but actually has much in common with one of Euler's original attempts. The proof outlined in this section follows Choe's argument with some simplifications due to Peter Duren.[1]

Wallis's Product

Even though we don't currently have the tools to prove the infinite product formula for $\sin(x)$ in equation (2), we can prove a special case.

Exercise 8.3.1. Supply the details to show that when $x = \pi/2$ the product formula in (2) is equivalent to

(3) $$\frac{\pi}{2} = \lim_{n\to\infty}\left(\frac{2\cdot2}{1\cdot3}\right)\left(\frac{4\cdot4}{3\cdot5}\right)\left(\frac{6\cdot6}{5\cdot7}\right)\cdots\left(\frac{2n\cdot2n}{(2n-1)(2n+1)}\right),$$

where the infinite product in (2) is interpreted to be a limit of partial products. (Although it is not necessary for what follows, it might be useful to review the treatment of infinite products in Exercises 2.4.10 and 2.7.10.)

The goal of the next few exercises is to supply a proper proof for equation (3). This curious formula involving π was first discovered by John Wallis (1616–1703) and will provide some key ingredients for our proof of Euler's sum. It resurfaces again in Section 8.4 where the factorial function is defined.

Set

$$b_n = \int_0^{\frac{\pi}{2}} \sin^n(x)\,dx, \quad \text{for } n = 0, 1, 2, \ldots.$$

The first few terms are easy enough to calculate; in particular,

$$b_0 = \int_0^{\frac{\pi}{2}} 1\,dx = \frac{\pi}{2} \quad \text{and} \quad b_1 = \int_0^{\frac{\pi}{2}} \sin(x)\,dx = 1.$$

---

[1] [13], p. 92–95

Exercise 8.3.2. Assume $h(x)$ and $k(x)$ have continuous derivatives on $[a, b]$ and derive the integration-by-parts formula

$$\int_a^b h(t)k'(t)\,dt = h(b)k(b) - h(a)k(a) - \int_a^b h'(t)k(t)\,dt \ .$$

Exercise 8.3.3.   (a) Using the simple identity $\sin^n(x) = \sin^{n-1}(x)\sin(x)$ and the previous exercise, derive the recurrence relation

$$b_n = \frac{n-1}{n} b_{n-2} \quad \text{for all } n \geq 2.$$

(b) Use this relation to generate the first three even terms and the first three odd terms of the sequence (b_n).

(c) Write a general expression for b_{2n} and b_{2n+1}.

Because $0 \leq \sin^{n+1}(x) \leq \sin^n(x)$ on $[0, \pi/2]$, it follows that $b_{n+1} \leq b_n$ and (b_n) is decreasing. It turns out that $(b_n) \to 0$ but that isn't the limit we are interested in at the moment.

Exercise 8.3.4. Show

$$\lim_{n\to\infty} \frac{b_{2n}}{b_{2n+1}} = 1,$$

and use this fact to finish the proof of Wallis's product formula in (3).

There are some standard techniques for working with the notation of equation (3). For instance,
$$2 \cdot 4 \cdot 6 \cdots (2n) = 2^n n!$$

and
$$1 \cdot 3 \cdot 5 \cdots (2n+1) = \frac{(2n+1)!}{2 \cdot 4 \cdot 6 \cdots (2n)} = \frac{(2n+1)!}{2^n n!}.$$

Exercise 8.3.5. Derive the following alternative form of Wallis's product formula:
$$\sqrt{\pi} = \lim_{n\to\infty} \frac{2^{2n}(n!)^2}{(2n)!\sqrt{n}} \ .$$

Taylor Series

The next step in the argument is to generate the Taylor series for $\arcsin(x)$. This is not really possible to do directly from Taylor's formula for the coefficients, but keeping in mind that

$$(\arcsin(x))' = \frac{1}{\sqrt{1 - x^2}},$$

we can get where we want to go by first finding the expansion for $1/\sqrt{1 - x}$.

Exercise 8.3.6. Show that $1/\sqrt{1-x}$ has Taylor expansion $\sum_{n=0}^{\infty} c_n x^n$, where $c_0 = 1$ and

$$c_n = \frac{(2n)!}{2^{2n}(n!)^2} = \frac{1 \cdot 3 \cdot 5 \cdots (2n-1)}{2 \cdot 4 \cdot 6 \cdots 2n}$$

for $n \geq 1$.

The coefficients c_n should look familiar from our work on Wallis's product. Exercise 8.3.5 can be rephrased as

$$\sqrt{\pi} = \lim_{n\to\infty} \frac{1}{c_n \sqrt{n}}.$$

Exercise 8.3.7. Show that $\lim c_n = 0$ but $\sum_{n=0}^{\infty} c_n$ diverges.

The divergence of $\sum_{n=0}^{\infty} c_n$ makes sense when we consider the Taylor series for $1/\sqrt{1-x}$. We want to determine the values of x for which

(4)
$$\frac{1}{\sqrt{1-x}} = \sum_{n=0}^{\infty} c_n x^n,$$

and $x = 1$ is not in the domain of the left side. We do aim to prove (4) for all $x \in (-1,1)$ but the usual word of warning is in order. Having computed the coefficients c_n, it is not enough to simply argue that the series on the right side converges when $|x| < 1$. To properly establish (4) we are going to show that the error function

$$E_N(x) = \frac{1}{\sqrt{1-x}} - \sum_{n=0}^{N} c_n x^n$$

tends to zero as $N \to \infty$. Back in Section 6.6, the primary tool we used for this task was Lagrange's Remainder Theorem (Theorem 6.6.3), but it is not up to this particular challenge

Exercise 8.3.8. Using the expression for $E_N(x)$ from Lagrange's Remainder Theorem, show that equation (4) is valid for all $|x| < 1/2$. What goes wrong when we try to use this method to prove (4) for $x \in (1/2, 1)$?

The Integral Form of the Remainder

The moral of the previous exercise is that we need a different method for estimating $E_N(x)$. The Lagrange form of the remainder grows out of the Mean Value Theorems and yields a formula for the error function in terms of the derivative $f^{(N+1)}$. Now that we are in possession of a proper definition of the integral, we can derive another useful formula for $E_N(x)$.

Theorem 8.3.1 (Integral Remainder Theorem). *Let f be differentiable $N + 1$ times on $(-R, R)$ and assume $f^{(N+1)}$ is continuous. Define $a_n = f^{(n)}(0)/n!$ for $n = 0, 1, \ldots, N$, and let*

$$S_N(x) = a_0 + a_1 x + a_2 x^2 + \cdots + a_N x^N.$$

For all $x \in (-R, R)$, the error function $E_N(x) = f(x) - S_N(x)$ satisfies

$$E_N(x) = \frac{1}{N!} \int_0^x f^{(N+1)}(t)(x-t)^N \, dt \ .$$

Proof. The case $x = 0$ is easy to check, so let's take $x \neq 0$ in $(-R, R)$ and keep in mind that x is a fixed constant in what follows. To avoid a few technical distractions, let's just consider the case $x > 0$.

Exercise 8.3.9.   (a) Show

$$f(x) = f(0) + \int_0^x f'(t) \, dt \ .$$

(b) Now use a previous result from this section to show

$$f(x) = f(0) + f'(0)x + \int_0^x f''(t)(x-t) \, dt \ .$$

(c) Continue in this fashion to complete the proof of the theorem.   □

   To gain a better understanding of this formulation for $E_N(x)$ and simultaneously make some headway on our exploration of equation (4), let's return to the special case $f(x) = 1/\sqrt{1-x}$.

Exercise 8.3.10.   (a) Make a rough sketch of $1/\sqrt{1-x}$ and $S_2(x)$ over the interval $(-1, 1)$, and compute $E_2(x)$ for $x = 1/2, 3/4$, and $8/9$.

(b) For a general x satisfying $-1 < x < 1$, show

$$E_2(x) = \frac{15}{16} \int_0^x \left(\frac{x-t}{1-t}\right)^2 \frac{1}{(1-t)^{3/2}} \, dt \ .$$

(c) Explain why the inequality

$$\left|\frac{x-t}{1-t}\right| \leq |x|$$

is valid, and use this to find an overestimate for $|E_2(x)|$ that no longer involves an integral. Note that this estimate will necessarily depend on x. Confirm that things are going well by checking that this overestimate is in fact larger than $|E_2(x)|$ at the three computed values from part (a).

(d) Finally, show $E_N(x) \to 0$ as $N \to \infty$ for an arbitrary $x \in (-1, 1)$.

Having established that the Taylor series in (4) does indeed converge for all $|x| < 1$, it is now clear sailing to produce a Taylor series representation for $\arcsin(x)$. The first step is to substitute x^2 for x in (4) to get

$$\frac{1}{\sqrt{1-x^2}} = \sum_{n=0}^{\infty} c_n x^{2n} \quad \text{for all } |x| < 1.$$

The next step is to take the term-by-term anti-derivative of this series. Any time we start manipulating infinite series as though they were finite in nature we need to pause and make sure we are on solid footing.

Exercise 8.3.11. Assuming that the derivative of $\arcsin(x)$ is indeed $1/\sqrt{1-x^2}$, supply the justification that allows us to conclude

$$(5) \qquad \arcsin(x) = \sum_{n=0}^{\infty} \frac{c_n}{2n+1} x^{2n+1} \quad \text{for all } |x| < 1 .$$

Exercise 8.3.12. Our work thus far shows that the Taylor series in (5) is valid for all $|x| < 1$, but note that $\arcsin(x)$ is continuous for all $|x| \leq 1$. Carefully explain why the series in (5) converges uniformly to $\arcsin(x)$ on the closed interval $[-1, 1]$.

Summing $\sum_{n=1}^{\infty} 1/n^2$

Every proof of Euler's sum contains a moment of genuine ingenuity at some point, and this is where our proof takes an unanticipated turn.

Let's make the substitution $x = \sin(\theta)$ in (5) where we restrict our attention to $-\pi/2 \leq \theta \leq \pi/2$. The result is

$$\theta = \arcsin(\sin(\theta)) = \sum_{n=0}^{\infty} \frac{c_n}{2n+1} \sin^{2n+1}(\theta)$$

which converges uniformly on $[-\pi/2, \pi/2]$.

Exercise 8.3.13. (a) Show

$$\int_0^{\pi/2} \theta \, d\theta = \sum_{n=0}^{\infty} \frac{c_n}{2n+1} b_{2n+1},$$

being careful to justify each step in the argument. The term b_{2n+1} refers back to our earlier work on Wallis's product.

(b) Deduce

$$\frac{\pi^2}{8} = \sum_{n=0}^{\infty} \frac{1}{(2n+1)^2},$$

and use this to finish the proof that $\pi^2/6 = \sum_{n=1}^{\infty} 1/n^2$.

The Riemann-Zeta Function

Euler's determination of the value of $\sum 1/n^2$ brought him international recognition and represented a significant milestone in what would be a lifelong exploration of series of the form $\sum 1/n^s$. Euler's original argument for summing $\sum 1/n^2$ discussed in Section 6.1 involved equating the coefficient of x^2 in two different series expansions for $\sin(x)/x$. By equating the coefficients of higher powers of x he was also able to sum $\sum 1/n^s$ for $s = 4, 6, 8, 10$ and 12. (Try it for $s = 4$.) Eventually, Euler worked out a general formula for any even natural number, and in the process he shifted his focus to thinking about $\sum 1/n^s$ as a *function* of the variable s. The iconic notation

$$\zeta(s) = \sum_{n=1}^{\infty} \frac{1}{n^s} \quad \text{for all } s > 1,$$

and the name—the Riemann-zeta function—would come one hundred years later, but it was Euler who first unearthed many deep properties of this function. Significant among these is a connection to the prime numbers, evident in the Eulerian formula

$$(6) \qquad \sum_{n=1}^{\infty} \frac{1}{n^s} = \left(\frac{1}{1 - 2^{-s}} \right) \left(\frac{1}{1 - 3^{-s}} \right) \left(\frac{1}{1 - 5^{-s}} \right) \left(\frac{1}{1 - 7^{-s}} \right) \cdots ,$$

where the product is taken over all the primes. The mathematics underlying the Riemann-zeta function gets complicated very quickly, but this particular formula is actually quite accessible. Notice that for each prime p,

$$\frac{1}{1 - p^{-s}} = 1 + \frac{1}{p^s} + \frac{1}{p^{2s}} + \frac{1}{p^{3s}} + \frac{1}{p^{4s}} + \cdots .$$

Multiplying out the product on the right in (6) in this fashion and using the fact that every $n \in \mathbf{N}$ is a unique product of primes leads naturally to the given relationship.

Euler returned to study $\zeta(s)$ many times in his career, in part it seems to tend to the unfinished business of evaluating $\sum 1/n^s$ for the odd integers. Amid his many successes, this was a challenge that eluded Euler, as it has eluded every mathematician since.

8.4   Inventing the Factorial Function

The goal of this section is to produce a function $f(x)$, defined on all of $\mathbf{R}$, with the property that $f(n) = n!$ for each $n \in \mathbf{N}$. With no other restriction on f, this is as easy as it is uninteresting—simply define f piecewise in such a way that it passes through the points $(1,1)$, $(2,2)$, $(3, 6)$, $(4, 24)$, and so on. Letting

$$f(x) = \begin{cases} n! & \text{if } n \le x < n+1, \ n \in \mathbf{N} \\ 1 & \text{if } x < 1 \end{cases}$$

does the trick.

To make this problem meaningful we need to be much more discriminating about what properties we require f to have. Should f be continuous? Differentiable? Twice differentiable? We shall see about this. This problem actually has its origins in a series of 1729 letters between Christian Goldbach (of "Goldbach's Conjecture" fame, although that is a different story) and Leonard Euler. The term "function" in Euler's day implicitly referred to a mapping defined by an analytic expression comprised of the elementary functions and operations of calculus. Logarithms, exponentials, polynomials, and power series were examples of 18th century functions; the piecewise concoction proposed above was not.

Thus, a better statement of our goal—although still a little imprecise—is to find a function defined by a single, organic formula which extends the definition of $n!$ in a meaningful way to non-natural numbers.

Exercise 8.4.1. For $n \in \mathbf{N}$, let

$$n\# = n + (n - 1) + (n - 2) + \cdots + 2 + 1.$$

(a) Without looking ahead, decide if there is a natural way to define $0\#$. How about $(-2)\#$? Conjecture a reasonable value for $\frac{7}{2}\#$.

(b) Now prove $n\# = \frac{1}{2}n(n + 1)$ for all $n \in \mathbf{N}$, and revisit part (a).

The formula in part (b) of the previous exercise not only simplifies the calculation of $n\#$ for large values of n, but also yields a properly defined function on $\mathbf{R}$ when the discrete variable n is replaced with the continuous variable x. Indeed, Euler would be perfectly comfortable with the expression $x\# = \frac{1}{2}x(x+1)$.

We are seeking something similar for $n!$. What is the right definition for $x!$ when $x \in \mathbf{R}$?

The Exponential Function

The idea of extending the definition of a function defined on $\mathbf{N}$ to all of $\mathbf{R}$ may at first sound like a somewhat whimsical enterprise, but it is perfectly analogous to the way we come to understand a function like 2^x. Similar to $n!$, 2^n for $n \in \mathbf{N}$ is unambiguous and meaningful the minute we understand multiplication, but something like $2^{-\pi}$ is another matter. Because it is instructive, and because we are going to presently need functions of the form t^x, let's take a moment to define exponential functions in a rigorous way.

Typically the way a function like 2^x gets defined on $\mathbf{R}$ is through a series of domain expansions. Starting with 2^n, we first expand the domain to $\mathbf{Z}$ using reciprocals, then to $\mathbf{Q}$ using roots, and finally to $\mathbf{R}$ using continuity. Although we could follow this strategy, we are going to take a different approach that has the advantage of yielding the important properties we need more efficiently.

Step one is to properly define the natural exponential function e^x. Back in Chapter 6, we assumed e^x was already defined and showed how it could be represented by its Taylor series. Here we flip this process around. The problem on the table is to rigorously construct a proper definition for e^x, and the theory of power series gives us a bedrock foundation on which to build.

Define

(1) $$E(x) = \sum_{n=0}^{\infty} \frac{x^n}{n!} = 1 + x + \frac{x^2}{2!} + \frac{x^3}{3!} + \dots.$$

Exercise 8.4.2. Verify that the series converges absolutely for all $x \in \mathbf{R}$, that $E(x)$ is differentiable on $\mathbf{R}$, and $E'(x) = E(x)$.

Exercise 8.4.3.   (a) Use the results of Exercise 2.8.7 and the binomial formula to show that $E(x + y) = E(x)E(y)$ for all $x, y \in \mathbf{R}$.

 (b) Show that $E(0) = 1$, $E(-x) = 1/E(x)$, and $E(x) > 0$ for all $x \in \mathbf{R}$.

The takeaway here is that the power series $E(x)$ satisfies all the properties we associate with the exponential function, and we can therefore give ourselves permission to go back to the more familiar notation e^x in place of $E(x)$. What happens if we have a momentary relapse and interpret e^x as the real number $e \approx 2.71828\dots$ raised to the power x rather than $E(x)$? Not to worry—the two interpretations coincide, whenever the former is defined in the usual way.

Exercise 8.4.4. Define $e = E(1)$. Show $E(n) = e^n$ and $E(m/n) = (\sqrt[n]{e})^m$ for all $m, n \in \mathbf{Z}$.

One final property of e^x we need is its behavior as $x \to \pm\infty$.

Definition 8.4.1. Given $f : [a, \infty] \to \mathbf{R}$, we say that $\lim_{x \to \infty} f(x) = L$ if, for all $\epsilon > 0$, there exists $M > a$ such that whenever $x \geq M$ it follows that $|f(x) - L| < \epsilon$.

Exercise 8.4.5. Show $\lim_{x \to \infty} x^n e^{-x} = 0$ for all $n = 0, 1, 2, \dots$.
    To get started notice that when $x \geq 0$, all the terms in (1) are positive.

Other Bases

Having set e^x on solid mathematical footing, we can now do the same for t^x where $t > 0$. This requires use of the natural logarithm.

Exercise 8.4.6.   (a) Explain why we know e^x has an inverse function—let's call it $\log x$—defined on the strictly positive real numbers and satisfying

 (i) $\log(e^y) = y$ for all $y \in \mathbf{R}$ and
 (ii) $e^{\log x} = x$, for all $x > 0$.

 (b) Prove $(\log x)' = 1/x$. (See Exercise 5.2.12.)

 (c) Fix $y > 0$ and differentiate $\log(xy)$ with respect to x. Conclude that

$$\log(xy) = \log x + \log y \quad \text{for all } x, y > 0.$$

(d) For $t > 0$ and $n \in \mathbf{N}$, t^n has the usual interpretation as $t \cdot t \cdots t$ (n times). Show that

(2) $$t^n = e^{n \log t} \quad \text{for all } n \in \mathbf{N}.$$

Part (d) of the previous exercise is the pivotal formula because the expression on the right of the equal sign is meaningful if we replace n with $x \in \mathbf{R}$. This is our cue to use the identity in (2) as a template for the definition of t^x on all of $\mathbf{R}$.

Definition 8.4.2. Given $t > 0$, define the exponential function t^x to be

$$t^x = e^{x \log t} \quad \text{for all } x \in \mathbf{R}.$$

Exercise 8.4.7. (a) Show $t^{m/n} = (\sqrt[n]{t})^m$ for all $m, n \in \mathbf{N}$.

(b) Show $\log(t^x) = x \log t$, for all $t > 0$ and $x \in \mathbf{R}$.

(c) Show t^x is differentiable on $\mathbf{R}$ and find the derivative.

Finding the right definition for $x!$ is harder than defining t^x, but the strategy is essentially the same. We are seeking a formula of the form $n! = g(n)$ where g yields a meaningful formula when n is replaced by x. What might such a function $g(x) = x!$ look like when graphed over $\mathbf{R}$? For $x \geq 0$ it must grow extremely rapidly to keep up with $n!$, but how about on $x < 0$? Using a functional equation for $x!$ we can create a reasonable artist's rendering of the function we are looking for.

The Functional Equation

A defining property of the factorial on $\mathbf{N}$ is that $1! = 1$ and $n! = n(n-1)!$ for all $n \geq 2$. Thus it seems reasonable to require the same from our currently mythic function $x!$ defined on $\mathbf{R}$. Whatever $x!$ means it should satisfy

$$x! = x(x-1)! \quad \text{for all } x \in \mathbf{R}.$$

Setting $n = 1$ in this equation, for example, yields $1 = 0!$.

Exercise 8.4.8. Inspired by the fact that $0! = 1$ and $1! = 1$, let $h(x)$ satisfy

(i) $h(x) = 1$ for all $0 \leq x \leq 1$, and

(ii) $h(x) = xh(x-1)$ for all $x \in \mathbf{R}$.

(a) Find a formula for $h(x)$ on $[1, 2]$, $[2, 3]$, and $[n, n+1]$ for arbitrary $n \in \mathbf{N}$.

(b) Now do the same for $[-1, 0]$, $[-2, -1]$, and $[-n, -n+1]$.

(c) Sketch h over the domain $[-4, 4]$.

Notice that $h(x)$ satisfies $h(n) = n!$ and it is at least continuous for $x \geq 0$, but its piecewise definition and its many non-differentiable corners disqualify it from being our sought after factorial function. One legitimate conclusion that arises out of this exercise is that $x!$, when we find it, will exhibit the same asymptotic behavior as h at $x = -1, -2, -3, \ldots$, and thus won't be defined on the negative integers.

Improper Riemann Integrals

For reasons that will become clear, we need to make rigorous sense of an expression like

$$\int_0^\infty e^{-t}\,dt.$$

Most likely familiar from calculus, integrals over unbounded regions like $[0, \infty)$ are called *improper Riemann integrals* and are defined by taking the limit of "proper" integrals.

Definition 8.4.3. Assume f is defined on $[a, \infty)$ and integrable on every interval of the form $[a, b]$. Then define $\int_a^\infty f$ to be

$$\lim_{b \to \infty} \int_a^b f,$$

provided the limit exists. In this case we say the improper integral $\int_a^\infty f$ *converges*.

Exercise 8.4.9. (a) Show that the improper integral $\int_a^\infty f$ converges if and only if, for all $\epsilon > 0$ there exists $M > a$ such that whenever $d > c \geq M$ it follows that

$$\left| \int_c^d f \right| < \epsilon.$$

(In one direction it will be useful to consider the sequence $a_n = \int_a^{a+n} f$.)

(b) Show that if $0 \leq f \leq g$ and $\int_a^\infty g$ converges than $\int_a^\infty f$ converges.

(c) Part (a) is a Cauchy criterion, and part (b) is a comparison test. State and prove an absolute convergence test for improper integrals.

Exercise 8.4.10. (a) Use the properties of e^t previously discussed to show

$$\int_0^\infty e^{-t}\,dt = 1.$$

(b) Show

(3) $$\frac{1}{\alpha} = \int_0^\infty e^{-\alpha t}\,dt, \quad \text{for all } \alpha > 0.$$

Just for a moment, let's take our analysis gloves off and ask what we think might happen if we differentiate formula (3) with respect to α.

On the left-hand side we certainly get

$$\left[\frac{1}{\alpha}\right]' = -\frac{1}{\alpha^2}.$$

On the right-hand side of (3), let's brazenly crash through the integral sign and take the derivative of the integrand $e^{-\alpha t}$ with respect to α (thinking of t as a constant.) The result is

$$\left[e^{-\alpha t}\right]' = e^{-\alpha t} \cdot (-t).$$

The question, then, is whether this is a valid manipulation. Is it true that

(4)
$$\frac{1}{\alpha^2} = \int_0^\infty te^{-\alpha t}\, dt\,?$$

Well, let's compute the integral and find out.

Exercise 8.4.11. (a) Evaluate $\int_0^b te^{-\alpha t}\, dt$ using the integration-by-parts formula from Exercise 7.5.6. The result will be an expression in α and b.

(b) Now compute $\int_0^\infty te^{-\alpha t}\, dt$ and verify equation (4).

Apparently, our bold differentiation of equation (3) into equation (4) worked out. Now it's time to put our analysis gloves back on and see why this is so.

Differentiating Under the Integral

Let $f(x,t)$ be a function of two variables, defined for all $a \le x \le b$ and $c \le t \le d$. The domain of f is then a *rectangle* D in $\mathbf{R}^2$.

What does it mean to say f is continuous at a point (x_0, t_0) in D? Section 8.2 on metric spaces gives a more thorough explanation, but the only real difference from the single variable setting is that we have to replace our sense of *distance* between points (x_0, t_0) and (x, t) with the familiar Euclidean distance formula

$$\|(x,t) - (x_0, t_0)\| = \sqrt{(x - x_0)^2 + (t - t_0)^2}.$$

Definition 8.4.4. A function $f : D \to \mathbf{R}$ is continuous at (x_0, t_0) if for all $\epsilon > 0$, there exists $\delta > 0$ such that whenever $\|(x, t) - (x_0, t_0)\| < \delta$, it follows that

$$|f(x,t) - f(x_0, t_0)| < \epsilon.$$

Exercise 8.4.12. Assume the function $f(x,t)$ is continuous on the rectangle $D = \{(x,t) : a \le x \le b, c \le t \le d\}$. Explain why the function

$$F(x) = \int_c^d f(x,t)\, dt$$

is properly defined for all $x \in [a, b]$.

It should not be too surprising that Theorem 4.4.7 has an analogue in the $\mathbf{R}^2$ setting. The set D is compact in $\mathbf{R}^2$, and a continuous function on D is uniformly continuous in the sense that the δ in Definition 8.4.4 can be chosen independently of the point (x_0, t_0).

Theorem 8.4.5. *If $f(x, t)$ is continuous on D, then $F(x) = \int_c^d f(x, t)\,dt$ is uniformly continuous on $[a, b]$.*

Exercise 8.4.13. Prove Theorem 8.4.5.

Taking inspiration from equations (3) and (4), let's add the assumption that for each fixed value of t in $[c, d]$, the function $f(x, t)$ is a differentiable function of x; that is,

$$f_x(x, t) = \lim_{z \to x} \frac{f(z, t) - f(x, t)}{z - x}$$

exists for all $(x, t) \in D$. In addition, let's assume that the derivative function $f_x(x, t)$ is continuous.

Theorem 8.4.6. *If $f(x, t)$ and $f_x(x, t)$ are continuous on D, then the function $F(x) = \int_c^d f(x, t)\,dt$ is differentiable and*

$$F'(x) = \int_c^d f_x(x, t)\,dt.$$

Proof. Fix x in $[a, b]$ and let $\epsilon > 0$ be arbitrary. Our task is to find a $\delta > 0$ such that

(5)
$$\left| \frac{F(z) - F(x)}{z - x} - \int_c^d f_x(x, t)\,dt \right| < \epsilon$$

whenever $0 < |z - x| < \delta$.

Exercise 8.4.14. Finish the proof of Theorem 8.4.6   □

Improper Integrals, Revisited

Theorem 8.4.6 is a formal justification for differentiating under the integral sign, but we need to extend this result to the case where the integral is improper. Looking back one more time to our motivating example in equation (3), we see that what we have is a function $f(x, t)$ where the domain of the variable t is the unbounded interval $c \le t < \infty$.

Let's fix x from some set $A \subseteq \mathbf{R}$. For such an x, we define

(6)
$$F(x) = \int_c^\infty f(x, t)\,dt = \lim_{d \to \infty} \int_c^d f(x, t)\,dt,$$

provided the limit exists.

Notice that the formula in (6) is a *pointwise* statement. Given an $x \in A$ and $\epsilon > 0$, we can find an M (perhaps dependent on x) where

$$\left| F(x) - \int_c^d f(x,t)\,dt \right| < \epsilon$$

whenever $d \geq M$. As we have seen on numerous occasions, the elixir required to ensure that good behavior in the finite setting extends to the infinite setting is uniformity.

Definition 8.4.7. Given $f(x,t)$ defined on $D = \{(x,t) : x \in A, c \leq t\}$, assume $F(x) = \int_c^\infty f(x,t)\,dt$ exists for all $x \in A$. We say the improper integral *converges uniformly* to $F(x)$ on A if for all $\epsilon > 0$, there exists $M > c$ such that

$$\left| F(x) - \int_c^d f(x,t)\,dt \right| < \epsilon$$

for all $d \geq M$ and all $x \in A$.

Exercise 8.4.15. (a) Show that the improper integral $\int_0^\infty e^{-xt}\,dt$ converges uniformly to $1/x$ on the set $[1/2, \infty)$.

(b) Is the convergence uniform on $(0, \infty)$?

Exercise 8.4.16. Prove the following analogue of the Weierstrass M-Test for improper integrals: If $f(x,t)$ satisfies $|f(x,t)| \leq g(t)$ for all $x \in A$ and $\int_a^\infty g(t)\,dt$ converges, then $\int_a^\infty f(x,t)\,dt$ converges uniformly on A.

An immediate consequence of Definition 8.4.7 is that if the improper integral converges uniformly then the sequence of functions defined by

$$F_n(x) = \int_c^{c+n} f(x,t)\,dt$$

converges uniformly to $F(x)$ on $[a,b]$. This observation gives us access to the host of useful results we developed in Chapter 6.

Theorem 8.4.8. *If $f(x,t)$ is continuous on $D = \{(x,t) : a \leq x \leq b, c \leq t\}$, then*

$$F(x) = \int_c^\infty f(x,t)\,dt$$

is uniformly continuous on $[a,b]$, provided the integral converges uniformly.

Exercise 8.4.17. Prove Theorem 8.4.8.

Theorem 8.4.9. *Assume the function $f(x,t)$ is continuous on $D = \{(x,t) : a \leq x \leq b, c \leq t\}$ and $F(x) = \int_c^\infty f(x,t)\,dt$ exists for each $x \in [a,b]$. If the derivative function $f_x(x,t)$ exists and is continuous, then*

$$(7) \qquad\qquad F'(x) = \int_c^\infty f_x(x,t)\,dt,$$

provided the integral in (7) converges uniformly.

Exercise 8.4.18. Prove Theorem 8.4.9.

The Factorial Function

It's time to return our attention to equation (3) from earlier in this section:

$$\frac{1}{\alpha} = \int_0^\infty e^{-\alpha t}\, dt, \quad \text{for all } \alpha > 0.$$

Exercise 8.4.19. (a) Although we verified it directly, show how to use the theorems in this section to give a second justification for the formula

$$\frac{1}{\alpha^2} = \int_0^\infty t e^{-\alpha t}\, dt, \quad \text{for all } \alpha > 0.$$

(b) Now derive the formula

$$(8) \qquad \frac{n!}{\alpha^{n+1}} = \int_0^\infty t^n e^{-\alpha t}\, dt, \quad \text{for all } \alpha > 0.$$

If we set $\alpha = 1$ in equation (8) we get

$$n! = \int_0^\infty t^n e^{-t}\, dt.$$

The appearance of $n!$ on the left side of this equation is an exciting development, especially because where n appears on the right it can be meaningfully replaced by a real variable x, at least when $x \geq 0$. This is the equation we have been looking for!

Definition 8.4.10. For $x \geq 0$, define the factorial function

$$x! = \int_0^\infty t^x e^{-t}\, dt.$$

Exercise 8.4.20. (a) Show that $x!$ is an infinitely differentiable function on $(0, \infty)$ and produce a formula for the nth derivative. In particular show that $(x!)'' > 0$.

(b) Use the integration-by-parts formula employed earlier to show that $x!$ satisfies the functional equation

$$(x+1)! = (x+1)x!.$$

The previous exercise is our first piece of evidence that we have found the right definition for $x!$. There is more to come.

A consequence of $(x!)'' > 0$ is that $x!$ is a *convex* function. In calculus this is usually referred to as "concave up" and means that the line segment connecting two points on the graph of $x!$ always sits above the curve. Said another way, there are no inflection points in $x!$ and the slope of the curve steadily increases as the graph passes through the points $(n, n!)$ for $n = 0, 1, 2, \ldots$. We did not

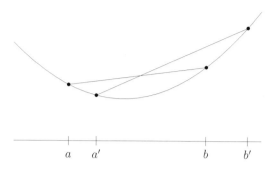

Figure 8.1: Increasing chord slopes on a convex function.

mention this property at the time, but reflecting on our earlier analogy between 2^x and $x!$, convexity is a natural condition to desire in our factorial function.

In fact, not only is $x!$ convex but $\log(x!)$ is also convex. This is a stronger statement. (Consider, for instance, the graphs of $x^2 + 1$ and $\log(x^2 + 1)$.) The proof is a little technical and we won't go through it, but the fact that $\log(x!)$ is convex on $x \geq 0$ is quite significant. Here's why.

Theorem 8.4.11 (Bohr–Mollerup Theorem). *There is a unique positive function f defined on $x \geq 0$ satisfying*

(i) $f(0) = 1$

(ii) $f(x + 1) = (x + 1)f(x)$, *and*

(iii) $\log(f(x))$ *is convex.*

Because $x!$ satisfies properties (i), (ii), and (iii), it follows that $f(x) = x!$.

Proof. We need one more geometrically plausible fact about convex functions. If $[a, b]$ and $[a', b']$ are two intervals in the domain of a convex function ϕ, and $a \leq a'$ and $b \leq b'$, then the slopes of the chords over these intervals satisfy

$$\frac{\phi(b) - \phi(a)}{b - a} \leq \frac{\phi(b') - \phi(a')}{b' - a'}.$$

(See Figure 8.1).

Because f satisfies properties (i) and (ii) we know $f(n) = n!$ for all $n \in \mathbf{N}$. Now fix $n \in \mathbf{N}$ and $x \in (0, 1]$.

Exercise 8.4.21. (a) Use the convexity of $\log(f(x))$ and the three intervals $[n - 1, n]$, $[n, n + x]$, and $[n, n + 1]$ to show

$$x \log(n) \leq \log(f(n + x)) - \log(n!) \leq x \log(n + 1).$$

(b) Show $\log(f(n + x)) = \log(f(x)) + \log((x + 1)(x + 2) \cdots (x + n))$.

(c) Now establish that

$$0 \leq \log(f(x)) - \log\left(\frac{n^x n!}{(x+1)(x+2)\cdots(x+n)}\right) \leq x \log(1 + \frac{1}{n}).$$

(d) Conclude that

$$f(x) = \lim_{n \to \infty} \frac{n^x n!}{(x+1)(x+2)\cdots(x+n)}, \quad \text{for all } x \in (0, 1].$$

(e) Finally, show that the conclusion in (d) holds for all $x \geq 0$.

Because we have arrived at an explicit formula for $f(x)$, the function $f(x)$ must be unique. By virtue of the fact that $x!$ satisfies conditions (i), (ii), and (iii) of the theorem, we can conclude that $x!$ is this unique function; i.e., $f(x) = x!$. Thus, not only have we proved the theorem, but we have also discovered an alternate representation for the factorial function called the *Gauss product formula*:

$$(9) \qquad x! = \int_0^\infty t^x e^{-t} \, dt = \lim_{n \to \infty} \frac{n^x n!}{(x+1)(x+2)\cdots(x+n)},$$

for all $x \geq 0$. $\square$

What happens if $x < 0$? The integral in Definition 8.4.10 becomes improper for a second reason when $x < 0$ because t^x is unbounded and undefined at $t = 0$. If $-1 < x < 0$, it is not hard to show that the integral still converges. On the other hand, the functional equation in Exercise 8.4.20(b) provides a natural way to extend the definition of $x!$ to all of $\mathbf{R}$. Just as in Exercise 8.4.8, the resulting function is never zero, alternating between positive and negative components with vertical asymptotes at $x = -1, -2, -3, \ldots$.

The Gamma Function

The focus of our discussion has been on the ingredients that go into the definition of $x!$—improper integrals, proper definitions of exponential functions, differentiating under the integral sign—but the end result is a function worthy of its own separate chapter. Since its discovery by Euler, the factorial function has become ubiquitous in numerous branches of analysis.

One of the early modifications that occurred was a shift in the domain of $x!$ and a change in the notation. Adrien Marie Legendre introduced the Greek letter Γ (gamma) and set

$$\Gamma(x) = (x - 1)! = \int_0^\infty t^{x-1} e^{-t} \, dt,$$

so that $\Gamma(n+1) = n!$ and $x\Gamma(x) = \Gamma(x+1)$. This convention eventually became the standard, and so it is the gamma function that routinely appears in formulas from number theory, probability, geometry, and beyond.

Philip Davis's article on the history of the gamma function (see [11]) is an excellent place to get a sense of the important role the gamma function has played in the development of analysis.[2] Davis's essay seems to be at least part of the inspiration for a wonderful series of articles by David Fowler that explore the properties of $x!$ in an original and accessible way.[3] Here is one of the anecdotes Fowler offers, which serves as an enticing clue for how intricately the gamma/factorial function is connected to the larger mathematical landscape.

Recall that when $x!$ is extended to all of $\mathbf{R}$ via the functional equation $x! = x(x-1)!$ we get asymptotes at every negative integer. Thus, there is a compelling reason to consider the reciprocal function $1/x!$ which we can take to be zero for $x = -1, -2, -3, \ldots$.

Exercise 8.4.22. (a) Where does $g(x) = \frac{x}{x!(-x)!}$ equal zero? What other familiar function has the same set of roots?

(b) The function e^{-x^2} provides the raw material for the all-important Gaussian bell curve from probability, where it is known that $\int_{-\infty}^{\infty} e^{-x^2}\,dx = \sqrt{\pi}$. Use this fact (and some standard integration techniques) to evaluate $(1/2)!$.

(c) Now use (a) and (b) to conjecture a striking relationship between the factorial function and a well-known function from trigonometry.

Exercise 8.4.23. As a parting shot, use the value for $(1/2)!$ and the Gauss product formula in equation (9) to derive the famous product formula for π discovered by John Wallis in the 1650s:

$$\frac{\pi}{2} = \lim_{n\to\infty}\left(\frac{2\cdot 2}{1\cdot 3}\right)\left(\frac{4\cdot 4}{3\cdot 5}\right)\left(\frac{6\cdot 6}{5\cdot 7}\right)\cdots\left(\frac{2n\cdot 2n}{(2n-1)(2n+1)}\right).$$

8.5 Fourier Series

In his famous treatise, *Théorie Analytique de la Chaleur* (The Analytical Theory of Heat), 1822, Joseph Fourier (1768–1830) boldly asserts, "Thus there is no function $f(x)$, or part of a function, which cannot be expressed by a trigonometric series."[4]

It is difficult to exaggerate the mathematical richness of this idea. It has been convincingly argued by mathematical historians that the ensuing investigation into the validity of Fourier's conjecture was the fundamental catalyst for the pursuit of rigor that characterizes 19th century mathematics. Power series had been in wide use in the 150 years leading up to Fourier's work, largely because they behaved so well under the operations of calculus. A function expressed as a power series is continuous, differentiable an infinite number of times, and

---

[2]Exercise 8.4.1, as well as the insight of comparing the development of $x!$ to 2^x, are borrowed from this piece.

[3]Exercise 8.4.8 is borrowed from Fowler's treatment in [15].

[4]Quoted passages in this section are taken from [9].

can be integrated and differentiated as though it were a polynomial. In the presence of such agreeable behavior, there was no compelling reason for mathematicians to formulate a more precise understanding of "limit" or "convergence" because there were no arguments to resolve. Fourier's successful implementation of trigonometric series to the study of heat flow changed all of this. To understand what the fuss was really about, we need to look more closely at what Fourier was asserting, focusing individually on the terms "function," "express," and "trigonometric series."

Trigonometric Series

The basic principle behind any series representations is to express a given function $f(x)$ as a sum of simpler functions. For power series, the component functions are $\{1, x, x^2, x^3, \ldots\}$, so that the series takes the form

$$f(x) = \sum_{n=0}^{\infty} a_n x^n = a_0 + a_1 x + a_2 x^2 + a_3 x^3 + \cdots.$$

A *trigonometric series* is a very different type of infinite series where the functions

$$\{1, \cos(x), \sin(x), \cos(2x), \sin(2x), \cos(3x), \sin(3x), \ldots\}$$

serve as the components. Thus, a trigonometric series has the form

$$
\begin{aligned}
f(x) &= a_0 + a_1 \cos(x) + b_1 \sin(x) + a_2 \cos(2x) + b_2 \sin(2x) + a_3 \cos(3x) + \cdots \\
&= a_0 + \sum_{n=1}^{\infty} a_n \cos(nx) + b_n \sin(nx).
\end{aligned}
$$

The idea of representing a function in this way was not completely new when Fourier first publicly proposed it in 1807. About 50 years earlier, Jean Le Rond d'Alembert (1717–1783) published the partial differential equation

$$\text{(1)} \qquad\qquad \frac{\partial^2 u}{\partial x^2} = \frac{\partial^2 u}{\partial t^2}$$

as a means of describing the motion of a vibrating string. In this model, the function $u(x, t)$ represents the displacement of the string at time $t \geq 0$ and at some point x, which we will take to be in the interval $[0, \pi]$. Because the string is understood to be attached at each end of this interval, we have

$$\text{(2)} \qquad\qquad u(0, t) = 0 \quad \text{and} \quad u(\pi, t) = 0$$

for all values of $t \geq 0$. Now, at $t = 0$, the string is displaced some initial amount, and at the moment it is released we assume

$$\text{(3)} \qquad\qquad \frac{\partial u}{\partial t}(x, 0) = 0,$$

meaning that, although the string immediately starts to move, it is given no initial velocity at any point. Finding a function $u(x, t)$ that satisfies equations (1), (2), and (3) is not too difficult.

Exercise 8.5.1. (a) Verify that

$$u(x, t) = b_n \sin(nx) \cos(nt)$$

satisfies equations (1), (2), and (3) for any choice of $n \in \mathbf{N}$ and $b_n \in \mathbf{R}$. What goes wrong if $n \notin \mathbf{N}$?

(b) Explain why any finite sum of functions of the form given in part (a) would also satisfy (1), (2), and (3). (Incidentally, it is possible to hear the different solutions in (a) for values of n up to 4 or 5 by isolating the harmonics on a well-made stringed instrument.)

Now, we come to the truly interesting issue. We have just seen that any function of the form

$$(4) \qquad u(x, t) = \sum_{n=1}^{N} b_n \sin(nx) \cos(nt)$$

solves d'Alembert's *wave equation*, as it is called, but the particular solution we want depends on how the string is originally "plucked." At time $t = 0$, we will assume that the string is given some initial displacement $f(x) = u(x, 0)$. Setting $t = 0$ in our family of solutions in (4), the hope is that the initial displacement function $f(x)$ can be expressed as

$$(5) \qquad f(x) = \sum_{n=1}^{N} b_n \sin(nx).$$

What this means is that *if* there exist suitable coefficients $b_1, b_2, \ldots, b_N$ so that $f(x)$ can be written as a sum of sine functions as in (5), then the vibrating-string problem is completely solved by the function $u(x, t)$ given in (4). The obvious question to ask, then, is just what types of functions can be constructed as linear combinations of the functions $\{\sin(x), \sin(2x), \sin(3x), \ldots\}$. How general can $f(x)$ be? Daniel Bernoulli (1700–1782) is usually credited with proposing the idea that by taking an *infinite* sum in equation (5), it may be possible to represent *any* initial position $f(x)$ over the interval $[0, \pi]$.

Fourier was studying the propagation of heat when trigonometric series resurfaced in his work in a very similar way. For Fourier, $f(x)$ represented an initial temperature applied to the boundary of some heat-conducting material. The differential equations describing heat flow are slightly different from d'Alembert's wave equation, but they still involve the second derivatives that make expressing $f(x)$ as a sum of trigonometric functions the crucial step in finding a solution.

Periodic Functions

In the early stages of his work, Fourier focused his attention on even functions (i.e., functions satisfying $f(x) = f(-x)$) and sought out ways to represent them

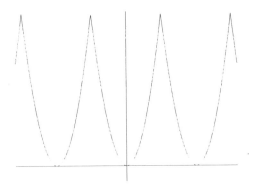

Figure 8.2: $f(x) = x^2$ **over** $(-\pi, \pi]$, **extended to be** 2π**-periodic.**

as series of the form $\sum a_n \cos(nx)$. Eventually, he arrived at the more general formulation of the problem, which is to find suitable coefficients (a_n) and (b_n) to express a function $f(x)$ as

$$(6) \qquad\qquad f(x) = a_0 + \sum_{n=1}^{\infty} a_n \cos(nx) + b_n \sin(nx).$$

As we begin to explore how arbitrary $f(x)$ can be, it is important to notice that every component of the series in equation (6) is periodic with period 2π. Turning our attention to the term "function," it now follows that any function we hope to represent by a trigonometric series will necessarily be periodic as well. We will give primary attention to the interval $(-\pi, \pi]$. What this means is that, given a function such as $f(x) = x^2$, we will restrict our attention to f over the domain $(-\pi, \pi]$ and then extend f periodically to all of **R** via the rule $f(x) = f(x + 2k\pi)$ for all $k \in \mathbf{Z}$ (Fig. 8.2).

   This convention of focusing on just the part of $f(x)$ over the interval $(-\pi, \pi]$ hardly seems controversial, but it did generate some confusion in Fourier's time. In Sections 1.2 and 4.1, we alluded to the fact that in the early 1800s the term "function" was used to mean something more like "formula." It was generally believed that a function's behavior over the interval $(-\pi, \pi]$ determined its behavior everywhere else, a point of view that follows naturally from an overly zealous faith in Taylor series. The modern definition of function given in Definition 1.2.3 is attributed to Dirichlet from the 1830s, although the idea had been suggested earlier by others. In *Théorie Analytique de la Chaleur*, Fourier clarifies his own use of the term by stating that a "function $f(x)$ represents a succession of values or ordinates, each of which is arbitrary... We do not suppose these ordinates to be subject to a common law; they succeed each other in any matter whatever, and each of them is given as if it were a single quantity."

   In the end, we will need to make a few assumptions about the nature of our functions, but the requirements we will need are quite mild, especially when compared with restrictions such as "infinitely differentiable," which are necessary—but not sufficient—for the existence of a Taylor series representation.

Types of Convergence

This brings us to a discussion of the word "expressed." The assumptions we must ultimately place on our function depend on the kind of convergence we aim to demonstrate. How are we to understand the equal sign in equation (6)? Our usual course of action with infinite series is first to define the partial sum

(7)
$$S_N(x) = a_0 + \sum_{n=1}^{N} a_n \cos(nx) + b_n \sin(nx).$$

To "express $f(x)$ as a trigonometric series" then means finding coefficients $(a_n)_{n=0}^{\infty}$ and $(b_n)_{n=1}^{\infty}$ so that

(8)
$$f(x) = \lim_{N \to \infty} S_N(x).$$

The question remains as to what kind of limit this is. Fourier probably imagined something akin to a *pointwise* limit because the concept of uniform convergence had not yet been formulated. In addition to pointwise convergence and uniform convergence, there are still other ways to interpret the limit in equation (8). Although it won't be discussed here, it turns out that proving

$$\int_{-\pi}^{\pi} |S_N(x) - f(x)|^2 \, dx \to 0$$

is a natural way to understand equation (8) for a particular class of functions. This is referred to as L^2 *convergence*. An alternate type of convergence that we will discuss, called *Cesaro mean convergence*, relies on demonstrating that the *averages* of the partial sums converge, in our case uniformly, to $f(x)$.

Fourier Coefficients

In the discussion that follows, we are going to need a few calculus facts.

Exercise 8.5.2. Using trigonometric identities when necessary, verify the following integrals.

(a) For all $n \in \mathbf{N}$,

$$\int_{-\pi}^{\pi} \cos(nx) \, dx = 0 \quad \text{and} \quad \int_{-\pi}^{\pi} \sin(nx) \, dx = 0.$$

(b) For all $n \in \mathbf{N}$,

$$\int_{-\pi}^{\pi} \cos^2(nx) \, dx = \pi \quad \text{and} \quad \int_{-\pi}^{\pi} \sin^2(nx) \, dx = \pi.$$

(c) For all $m, n \in \mathbf{N}$,

$$\int_{-\pi}^{\pi} \cos(mx) \sin(nx) \, dx = 0.$$

For $m \neq n$,

$$\int_{-\pi}^{\pi} \cos(mx) \cos(nx) \, dx = 0 \quad \text{and} \quad \int_{-\pi}^{\pi} \sin(mx) \sin(nx) \, dx = 0.$$

The consequences of these results are much more interesting than their proofs. The intuition from inner-product spaces is useful. Interpreting the integral as a kind of dot product, this exercise can be summarized by saying that the functions

$$\{1, \cos(x), \sin(x), \cos(2x), \sin(2x), \cos(3x), \dots\}$$

are all *orthogonal* to each other. The content of what follows is that they in fact form a *basis* for a large class of functions.

The first order of business is to deduce some reasonable candidates for the coefficients (a_n) and (b_n) in equation (6). Given a function $f(x)$, the trick is to *assume* we are in possession of a representation described in (6) and then manipulate this equation in a way that leads to formulas for (a_n) and (b_n). This is exactly how we proceeded with Taylor series expansions in Section 6.6. Taylor's formula for the coefficients was produced by repeatedly differentiating each side of the desired representation equation. Here, we integrate.

To compute a_0, integrate each side of equation (6) from $-\pi$ to π, brazenly take the integral inside the infinite sum, and use Exercise 8.5.2 to get

$$
\begin{aligned}
\int_{-\pi}^{\pi} f(x) \, dx &= \int_{-\pi}^{\pi} \left[a_0 + \sum_{n=1}^{\infty} a_n \cos(nx) + b_n \sin(nx) \right] dx \\
&= \int_{-\pi}^{\pi} a_0 \, dx + \sum_{n=1}^{\infty} \int_{-\pi}^{\pi} \left[a_n \cos(nx) + b_n \sin(nx) \right] dx \\
&= a_0(2\pi) + \sum_{n=1}^{\infty} a_n 0 + b_n 0 = a_0(2\pi).
\end{aligned}
$$

Thus,

(9)
$$a_0 = \frac{1}{2\pi} \int_{-\pi}^{\pi} f(x) \, dx.$$

The switching of the sum and the integral sign in the second step of the previous calculation should rightly raise some eyebrows, but keep in mind that we are really working backward from a hypothetical representation for $f(x)$ to get a *proposal* for what a_0 should be. The point is not to justify the derivation of the formula but rather to show that using this value for a_0 ultimately gives us the representation we want. That hard work lies ahead.

Now, consider a fixed $m \geq 1$. To compute a_m, we first multiply each side of equation (6) by $\cos(mx)$ and again integrate over the interval $[-\pi, \pi]$.

Exercise 8.5.3. Derive the formulas

$$(10) \qquad a_m = \frac{1}{\pi} \int_{-\pi}^{\pi} f(x) \cos(mx)\,dx \quad \text{and} \quad b_m = \frac{1}{\pi} \int_{-\pi}^{\pi} f(x) \sin(mx)\,dx$$

for all $m \geq 1$.

Let's take a short break and empirically test our recipes for (a_m) and (b_m) on a few simple functions.

Example 8.5.1. Let

$$f(x) = \begin{cases} 1 & \text{if } 0 < x < \pi \\ 0 & \text{if } x = 0 \text{ or } x = \pi \\ -1 & \text{if } -\pi < x < 0. \end{cases}$$

The fact that f is an odd function (i.e., $f(-x) = -f(x)$) means we can avoid doing any integrals for the moment and just appeal to a symmetry argument to conclude

$$a_0 = \frac{1}{2\pi} \int_{-\pi}^{\pi} f(x)\,dx = 0 \quad \text{and} \quad a_n = \frac{1}{\pi} \int_{-\pi}^{\pi} f(x) \cos(nx)\,dx = 0$$

for all $n \geq 1$. We can also simplify the integral for b_n by writing

$$\begin{aligned} b_n &= \frac{1}{\pi} \int_{-\pi}^{\pi} f(x) \sin(nx)\,dx &= \frac{2}{\pi} \int_0^{\pi} \sin(nx)\,dx \\ &&= \frac{2}{\pi} \left(\frac{-1}{n} \cos(nx) \Big|_0^{\pi} \right) \\ &&= \begin{cases} 4/n\pi & \text{if } n \text{ is odd} \\ 0 & \text{if } n \text{ is even.} \end{cases} \end{aligned}$$

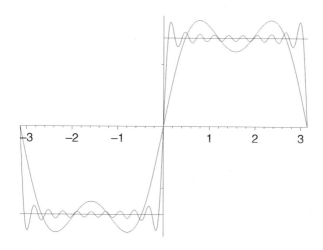

Figure 8.3: f, S_4, and S_{20} on $[-\pi, \pi]$.

Proceeding on blind faith, we plug these results into equation (6) to get the representation

$$f(x) = \frac{4}{\pi} \sum_{n=0}^{\infty} \frac{1}{2n+1} \sin((2n+1)x).$$

A graph of a few of the partial sums of this series (Fig. 8.3) should generate some optimism about the legitimacy of what is happening.

Exercise 8.5.4.   (a) Referring to the previous example, explain why we can be sure that the convergence of the partial sums to $f(x)$ is *not* uniform on any interval containing 0.

(b) Repeat the computations of Example 8.5.1 for the function $g(x) = |x|$ and examine graphs for some partial sums. This time, make use of the fact that g is even ($g(x) = g(-x)$) to simplify the calculations. By just looking at the coefficients, how do we know this series converges uniformly to something?

(c) Use graphs to collect some empirical evidence regarding the question of term-by-term differentiation in our two examples to this point. Is it possible to conclude convergence or divergence of either differentiated series by looking at the resulting coefficients? Theorem 6.4.3 is about the legitimacy of term-by-term differentiation. Can it be applied to either of these examples?

The Riemann–Lebesgue Lemma

In the examples we have seen to this point, the sequences of Fourier coefficients (a_n) and (b_n) all tend to 0 as $n \to \infty$. This is always the case. Understanding why this happens is crucial to our upcoming convergence proof.

We start with a simple observation. The reason

$$\int_{-\pi}^{\pi} \sin(x)\, dx = 0$$

is that the positive and negative portions of the sine curve cancel each other out. The same is true of

$$\int_{-\pi}^{\pi} \sin(nx)\, dx = 0.$$

Now, when n is large, the period of the oscillations of $\sin(nx)$ becomes very short—$2\pi/n$ to be precise. If $h(x)$ is a continuous function, then the values of h do not vary too much as $\sin(nx)$ ranges over each short period. The result is that the successive positive and negative oscillations of the product $h(x)\sin(nx)$ (Fig. 8.4) are nearly the same size so that the cancellation leads to a small value for

$$\int_{-\pi}^{\pi} h(x)\sin(nx)\, dx.$$

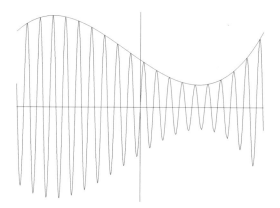

Figure 8.4: $h(x)$ **and** $h(x)\sin(nx)$ **for large** n.

Theorem 8.5.2 (Riemann–Lebesgue Lemma). *Assume $h(x)$ is continuous on $(-\pi, \pi]$. Then,*

$$\int_{-\pi}^{\pi} h(x)\sin(nx)\,dx \to 0 \quad \text{and} \quad \int_{-\pi}^{\pi} h(x)\cos(nx)\,dx \to 0$$

as $n \to \infty$.

Proof. Remember that, like all of our functions from here on, we are mentally extending h to be 2π-periodic. Thus, while our attention is generally focused on the interval $(-\pi, \pi]$, the assumption of continuity is intended to mean that the periodically extended h is continuous on all of **R**. Note that in addition to continuity on $(-\pi, \pi]$, this amounts to insisting that $\lim_{x \to -\pi^+} h(x) = h(\pi)$.

Exercise 8.5.5. Explain why h is uniformly continuous on **R**.

Given $\epsilon > 0$, choose $\delta > 0$ such that $|x - y| < \delta$ implies $|h(x) - h(y)| < \epsilon/2$. The period of $\sin(nx)$ is $2\pi/n$, so choose N large enough so that $\pi/n < \delta$ whenever $n \geq N$. Now, consider a particular interval $[a, b]$ of length $2\pi/n$ over which $\sin(nx)$ moves through one complete oscillation.

Exercise 8.5.6. Show that $\left| \int_a^b h(x)\sin(nx)\,dx \right| < \epsilon/n$, and use this fact to complete the proof. $\qquad \square$

Applications of Fourier series are not restricted to continuous functions (Example 8.5.1). Even though our particular proof makes use of continuity, the Riemann–Lebesgue lemma holds under much weaker hypotheses. It is true, however, that any proof of this fact ultimately takes advantage of the cancellation of positive and negative components. Recall from Chapter 2 that this type of cancellation is the mechanism that distinguishes conditional convergence from absolute convergence. In the end, what we discover is that, unlike power series,

Fourier series can converge conditionally. This makes them less robust, perhaps, but more versatile and capable of more interesting behavior.

A Pointwise Convergence Proof

Let's return once more to Fourier's claim that every "function" can be "expressed" as a trigonometric series. Our recipe for the Fourier coefficients in equations (9) and (10) implicitly requires that our function be integrable. This is the major motivation for Riemann's modification of Cauchy's definition of the integral. Because integrability is a prerequisite for producing a Fourier series, we would like the class of integrable functions to be as large as possible. The natural question to ask now is whether Riemann integrability is enough or whether we need to make some additional assumptions about f in order to guarantee that the Fourier series converges back to f. The answer depends on the type of convergence we hope to establish.

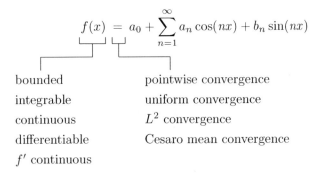

$$f(x) \;=\; a_0 + \sum_{n=1}^{\infty} a_n \cos(nx) + b_n \sin(nx)$$

bounded	pointwise convergence
integrable	uniform convergence
continuous	L^2 convergence
differentiable	Cesaro mean convergence
f' continuous	

There is no tidy way to summarize the situation. For pointwise convergence, integrability is not enough. At present, "integrable" for us means Riemann-integrable, which we have only rigorously defined for bounded functions. In 1966, Lennart Carleson proved (via an extremely complicated argument) that the Fourier series for such a function converges pointwise at every point in the domain excluding possibly a set of *measure zero*. This term surfaced in our discussion of the Cantor set (Section 3.1) and is defined rigorously in Section 7.6. Sets of measure zero are small in one sense, but they can be uncountable, and there are examples of continuous functions with Fourier series that diverge at uncountably many points. Lebesgue's modification of Riemann's integral in 1901 proved to be a much more natural setting for Fourier analysis. Carleson's proof is really about Lebesgue-integrable functions which are allowed to be unbounded but for which $\int_{-\pi}^{\pi} |f|^2$ is finite. One of the cleanest theorems in this area states that, for this class of square Lebesgue-integrable functions, the Fourier series always converges to the function from which it was derived if we interpret convergence in the L^2 sense described earlier. As a final warning about how fragile the situation is, there is an example due to A. Kolmogorov (1903–1987) of a Lebesgue-integrable function where the Fourier series fails to converge at any point.

Although all of these results require significantly more background to pursue in any rigorous way, we are in a position to prove some important theorems that require a few extra assumptions about the function in question. We will content ourselves with two interesting results in this area.

Theorem 8.5.3. *Let $f(x)$ be continuous on $(-\pi, \pi]$, and let $S_N(x)$ be the Nth partial sum of the Fourier series described in equation (7), where the coefficients (a_n) and (b_n) are given by equations (9) and (10). It follows that*

$$\lim_{N \to \infty} S_N(x) = f(x)$$

pointwise at any $x \in (-\pi, \pi]$ where $f'(x)$ exists.

Proof. Cataloging a few preliminary facts makes for a smoother argument.

Fact 1: (a) $\cos(\alpha - \theta) = \cos(\alpha)\cos(\theta) + \sin(\alpha)\sin(\theta)$.
(b) $\sin(\alpha + \theta) = \sin(\alpha)\cos(\theta) + \cos(\alpha)\sin(\theta)$.

Fact 2: $\frac{1}{2} + \cos(\theta) + \cos(2\theta) + \cos(3\theta) + \cdots + \cos(N\theta) = \dfrac{\sin((N + 1/2)\theta)}{2\sin(\theta/2)}$ for any $\theta \neq 2n\pi$.

Facts 1(a) and 1(b) are familiar trigonometric identities. Fact 2 is not as familiar. Its proof (which we omit) is most easily derived by taking the real part of a geometric sum of complex exponentials. The function in Fact 2 is called the *Dirichlet kernel* in honor of the mathematician responsible for the first rigorous convergence proof of this kind. Integrating both sides of this identity leads to our next important fact.

Fact 3: Setting

$$D_N(\theta) = \begin{cases} \frac{\sin((N+1/2)\theta)}{2\sin(\theta/2)}, & \text{if } \theta \neq 2n\pi \\ 1/2 + N, & \text{if } \theta = 2n\pi \end{cases}$$

from Fact 2, we see that

$$\int_{-\pi}^{\pi} D_N(\theta)d\theta = \pi.$$

Although we will not restate it, the last fact we will use is the Riemann–Lebesgue Lemma.

Fix a point $x \in (-\pi, \pi]$. The first step is to simplify the expression for $S_N(x)$. Now x is a fixed constant at the moment, so we will write the integrals in equations (9) and (10) using t as the variable of integration. Keeping an eye on Facts 1(a) and (2), we get that

$$S_N(x) = a_0 + \sum_{n=1}^{N} a_n \cos(nx) + b_n \sin(nx)$$

$$= \left[\frac{1}{2\pi} \int_{-\pi}^{\pi} f(t)\,dt \right] + \sum_{n=1}^{N} \left[\frac{1}{\pi} \int_{-\pi}^{\pi} f(t)\cos(nt)\,dt \right] \cos(nx)$$

$$+ \sum_{n=1}^{N} \left[\frac{1}{\pi} \int_{-\pi}^{\pi} f(t)\sin(nt)\,dt \right] \sin(nx)$$

$$= \frac{1}{\pi} \int_{-\pi}^{\pi} f(t) \left[\frac{1}{2} + \sum_{n=1}^{N} \cos(nt)\cos(nx) + \sin(nt)\sin(nx) \right] dt$$

$$= \frac{1}{\pi} \int_{-\pi}^{\pi} f(t) \left[\frac{1}{2} + \sum_{n=1}^{N} \cos(nt - nx) \right] dt$$

$$= \frac{1}{\pi} \int_{-\pi}^{\pi} f(t) D_N(t - x)\,dt.$$

As one final simplification, let $u = t - x$. Then,

$$S_N(x) = \frac{1}{\pi} \int_{-\pi-x}^{\pi-x} f(u+x) D_N(u)\,du = \frac{1}{\pi} \int_{-\pi}^{\pi} f(u+x) D_N(u)\,du.$$

The last equality is a result of our agreement to extend f to be 2π-periodic. Because D_N is also periodic (it is the sum of cosine functions), it does not matter over what interval we compute the integral as long as we cover exactly one full period.

To prove $S_N(x) \to f(x)$, we must show that $|S_N(x) - f(x)|$ gets arbitrarily small when N gets large. Having expressed $S_N(x)$ as an integral involving $D_N(u)$, we are motivated to do a similar thing for $f(x)$. By Fact 3,

$$f(x) = f(x) \frac{1}{\pi} \int_{-\pi}^{\pi} D_N(u)\,du = \frac{1}{\pi} \int_{-\pi}^{\pi} f(x) D_N(u)\,du,$$

and it follows that

$$(11) \qquad S_N(x) - f(x) = \frac{1}{\pi} \int_{-\pi}^{\pi} (f(u+x) - f(x)) D_N(u)\,du.$$

Our goal is to show this quantity tends to zero as $N \to \infty$. A sketch of $D_N(u)$ (Fig. 8.5) for a few values of N reveals why this might happen. For large N, the Dirichlet kernel $D_N(u)$ has a tall, thin spike around $u = 0$, but this is precisely where $f(u+x) - f(x)$ is small (because f is continuous). Away from zero, $D_N(u)$ exhibits the fast oscillations that hearken back to the Riemann–Lebesgue Lemma (Theorem 8.5.2). Let's see how to use this theorem to finish the argument.

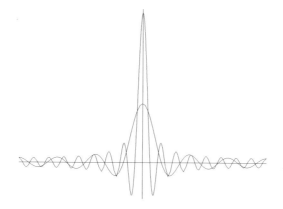

Figure 8.5: $D_6(u)$ and $D_{16}(u)$.

Using Fact 1(b), we can rewrite the Dirichlet kernel as

$$D_N(u) = \frac{\sin((N+1/2)u)}{2\sin(u/2)} = \frac{1}{2}\left[\frac{\sin(Nu)\cos(u/2)}{\sin(u/2)} + \cos(Nu)\right].$$

Then, equation (11) becomes

$$
\begin{aligned}
S_N(x) - f(x) &= \frac{1}{2\pi}\int_{-\pi}^{\pi}(f(u+x)-f(x))\left[\frac{\sin(Nu)\cos(u/2)}{\sin(u/2)} + \cos(Nu)\right]du \\
&= \frac{1}{2\pi}\int_{-\pi}^{\pi}(f(u+x)-f(x))\left(\frac{\sin(Nu)\cos(u/2)}{\sin(u/2)}\right) \\
&\qquad\qquad + (f(u+x)-f(x))\cos(Nu)\,du \\
&= \frac{1}{2\pi}\int_{-\pi}^{\pi}p_x(u)\sin(Nu)\,du + \frac{1}{2\pi}\int_{-\pi}^{\pi}q_x(u)\cos(Nu)\,du,
\end{aligned}
$$

where in the last step we have set

$$p_x(u) = \frac{(f(u+x)-f(x))\cos(u/2)}{\sin(u/2)} \quad \text{and} \quad q_x(u) = f(u+x)-f(x).$$

Exercise 8.5.7. (a) First, argue why the integral involving $q_x(u)$ tends to zero as $N \to \infty$.

(b) The first integral is a little more subtle because the function $p_x(u)$ has the $\sin(u/2)$ term in the denominator. Use the fact that f is differentiable at x (and a familiar limit from calculus) to prove that the first integral goes to zero as well. □

This completes the argument that $S_N(x) \to f(x)$ at any point x where f is differentiable. If the derivative exists everywhere, then we get $S_N \to f$

pointwise. If we add the assumption that f' is continuous, then it is not too difficult to show that the convergence is uniform. In fact, there is a very strong relationship between the speed of convergence of the Fourier series and the smoothness of f. The more derivatives f possesses, the faster the partial sums S_N converge to f.

Cesaro Mean Convergence

Rather than pursue the proofs in this interesting direction, we will finish this very brief introduction to Fourier series with a look at a different type of convergence called Cesaro mean convergence.

Exercise 8.5.8. Prove that if a sequence of real numbers (x_n) converges, then the arithmetic means

$$y_n = \frac{x_1 + x_2 + x_3 + \cdots + x_n}{n}$$

also converge to the same limit. Give an example to show that it is possible for the sequence of means (y_n) to converge even if the original sequence (x_n) does not.

The discussion preceding Theorem 8.5.3 is intended to create a kind of reverence for the difficulties inherent in deciphering the behavior of Fourier series, especially in the case where the function in question is not differentiable. It is from this humble frame of mind that the following elegant result due to L. Fejér in 1904 can best be appreciated.

Theorem 8.5.4 (Fejér's Theorem). *Let $S_n(x)$ be the nth partial sum of the Fourier series for a function f on $(-\pi, \pi]$. Define*

$$\sigma_N(x) = \frac{1}{N+1} \sum_{n=0}^{N} S_n(x).$$

If f is continuous on $(-\pi, \pi]$, then $\sigma_N(x) \to f(x)$ uniformly.

Proof. This argument is patterned after the proof of Theorem 8.5.3 but is actually much simpler. In addition to the trigonometric formulas listed in Facts 1 and 2, we are going to need a version of Fact 2 for the sine function, which looks like

$$\sin(\theta) + \sin(2\theta) + \sin(3\theta) + \cdots + \sin(N\theta) = \frac{\sin\left(\frac{N\theta}{2}\right) \sin\left((N+1)\frac{\theta}{2}\right)}{\sin\left(\frac{\theta}{2}\right)}.$$

Exercise 8.5.9. Use the previous identity to show that

$$\frac{1/2 + D_1(\theta) + D_2(\theta) + \cdots + D_N(\theta)}{N+1} = \frac{1}{2(N+1)} \left[\frac{\sin\left((N+1)\frac{\theta}{2}\right)}{\sin\left(\frac{\theta}{2}\right)} \right]^2.$$

The expression in Exercise 8.5.9 is called the *Fejér kernel* and will be denoted by $F_N(\theta)$. Analogous to the Dirichlet kernel $D_N(\theta)$ from the proof of Theorem 8.5.3, F_N is used to greatly simplify the formula for $\sigma_N(x)$.

Exercise 8.5.10. (a) Show that

$$\sigma_N(x) = \frac{1}{\pi} \int_{-\pi}^{\pi} f(u+x) F_N(u)\, du.$$

(b) Graph the function $F_N(u)$ for several values of N. Where is F_N large, and where is it close to zero? Compare this function to the Dirichlet kernel $D_N(u)$. Now, prove that $F_N \to 0$ uniformly on any set of the form $\{u : |u| \geq \delta\}$, where $\delta > 0$ is fixed (and u is restricted to the interval $(-\pi, \pi]$).

(c) Prove that $\int_{-\pi}^{\pi} F_N(u)\, du = \pi$.

(d) To finish the proof of Fejér's Theorem, first choose a $\delta > 0$ so that

$$|u| < \delta \quad \text{implies} \quad |f(x+u) - f(x)| < \epsilon.$$

Set up a single integral that represents the difference $\sigma_N(x) - f(x)$ and divide this integral into sets where $|u| \leq \delta$ and $|u| \geq \delta$. Explain why it is possible to make each of these integrals sufficiently small, independently of the choice of x. $\qquad\square$

Weierstrass Approximation Theorem

The hard work of proving Fejér's Theorem has many rewards, one of which is access to a relatively short argument for a profoundly important theorem discovered by Weierstrass in 1885. The Weierstrass Approximation Theorem (WAT) is studied in depth in Section 6.7 and is restated here for ease of reference.

Theorem 6.7.1 (Weierstrass Approximation Theorem). *Let $f : [a, b] \to \mathbf{R}$ be continuous. Given $\epsilon > 0$, there exists a polynomial $p(x)$ satisfying*

$$|f(x) - p(x)| < \epsilon$$

for all $x \in [a, b]$.

Proof. We have actually seen a few special cases of this result before in Section 6.6 on Taylor series. For instance, we showed that

$$\sin(x) = x - \frac{x^3}{3!} + \frac{x^5}{5!} - \frac{x^7}{7!} + \frac{x^9}{9!} - \cdots,$$

where the series converges uniformly on any bounded subset of $\mathbf{R}$. Uniform convergence of a series means the partial sums converge uniformly, and the partial sums in this case are polynomials. Notice that this is precisely what WAT asks us to prove, only we must do it for an arbitrary, continuous function in place of $\sin(x)$.

Using Taylor series does not work in general. To construct a Taylor series we need the function to be infinitely differentiable—not just continuous—and even in this case we might get a series that either does not converge or converges to the wrong thing. Taylor series are a valuable tool, however. In Section 6.7 we used the Taylor series for $\sqrt{1-x}$ as the starting point for a proper proof of WAT. Fejér's Theorem, in conjunction with the Taylor series for $\sin(x)$ and $\cos(x)$, provides a significant shortcut to the same result.

Exercise 8.5.11.   (a) Use the fact that the Taylor series for $\sin(x)$ and $\cos(x)$ converge uniformly on any compact set to prove WAT under the added assumption that $[a, b]$ is $[0, \pi]$.

  (b) Show how the case for an arbitrary interval $[a, b]$ follows from this one.

□

A comment from Section 6.7 that bears repeating relates to the striking contrast between this result and Weierstrass's demonstration of a continuous nowhere-differentiable function. Although there exist continuous functions that oscillate so wildly that they fail to have a derivative at any point, these unruly functions are always uniformly within ϵ of an infinitely differentiable polynomial.

Approximation as a Unifying Theme

Viewing the last section of this chapter as a kind of appendix (included to clear up some loose ends from Chapter 1 regarding the definition of the real numbers), the Weierstrass' Approximation Theorem makes for a fitting close to our introductory survey of some of the gems of analysis.

The idea of approximation permeates the entire subject. Every real number can be approximated with rational ones. The value of an infinite sum is approximated with partial sums, and the value of a continuous function can be approximated with its values nearby. A function is differentiable when a straight line is a good approximation to the curve, and it is integrable when finite sums of rectangles are a good approximation to the area under the curve. Now, we learn that every continuous function can be approximated arbitrarily well with a polynomial. In every case, the approximating objects are tangible and well-understood, and the issue is how well these properties survive the limiting process. By viewing the different infinities of mathematics through pathways crafted out of finite objects, Weierstrass and the other founders of analysis created a paradigm for how to extend the scope of mathematical exploration deep into territory previously unattainable. Although our journey ends here, the road is long and continues to be written.

8.6 A Construction of R From Q

This entire section is devoted to constructing a proof for the following theorem:

Theorem 8.6.1 (Existence of the Real Numbers). *There exists an ordered field in which every nonempty set that is bounded above has a least upper bound. In addition, this field contains* **Q** *as a subfield.*

There are a few terms to define before this statement can be properly understood and proved, but it can essentially be paraphrased as "the real numbers exist." In Section 1.1, we encountered a major failing of the rational number system as a place to do analysis. Without the square root of 2 (and uncountably many other irrational numbers) we cannot confidently move from a Cauchy sequence to its limit because in **Q** there is no guarantee that such a number exists. (A review of Sections 1.1 and 1.3 is highly recommended at this point.) The resolution we proposed in Chapter 1 came in the form of the Axiom of Completeness, which we restate.

Axiom of Completeness. *Every nonempty set of real numbers that is bounded above has a least upper bound.*

Now let's be clear about how we actually proceeded in Chapter 1. This is the property that distinguishes **Q** from **R**, but by referring to this property as an *axiom* we were making the point that it was not something to be proved. The real numbers were defined simply as an extension of the rational numbers in which bounded sets have least upper bounds, but no attempt was made to demonstrate that such an extension is actually possible. Now, the time has finally come. By explicitly building the real numbers from the rational ones, we will be able to demonstrate that the Axiom of Completeness does not need to be an axiom at all; it is a theorem!

There is something ironic about having the final section of this book be a construction of the number system that has been the underlying subject of every preceding page, but there is something perfectly apt about it as well. Through eight chapters stretching from Cantor's Theorem to the Baire Category Theorem, we have come to see how profoundly the addition of completeness changes the landscape. We all grow up believing in the existence of real numbers, but it is only through a study of classical analysis that we become aware of their elusive and enigmatic nature. It is because completeness matters so much, and because it is responsible for such perplexing phenomena, that we should now feel obliged—compelled really—to go back to the beginning and verify that such a thing really exists.

As we mentioned in Chapter 1, proceeding in this order puts us in good historical company. The pioneering work of Cauchy, Bolzano, Abel, Dirichlet, Weiestrass, and Riemann preceded—and in a very real sense led to—the host of rigorous definitions for **R** that were proposed in the last half of the 19th century. Georg Cantor is a familiar name responsible for one of these definitions, but alternate constructions of the real number system also came from Charles

Meray (1835–1911), Eduard Heine (1821–1881), and Richard Dedekind (1831–1916). The formulation that follows is the one due to Dedekind. In a sense it is the most abstract of the approaches, but it is the most appropriate for us because the verification of completeness is done in terms of least upper bounds.

Dedekind Cuts

We begin this discussion by assuming that the rational numbers and all of the familiar properties of addition, multiplication, and order are available to us. At the moment, there is no such thing as a real number.

Definition 8.6.2. A subset A of the rational numbers is called a *cut* if it possesses the following three properties:

(c1) $A \neq \emptyset$ and $A \neq \mathbf{Q}$.

(c2) If $r \in A$, then A also contains every rational $q < r$.

(c3) A does not have a maximum; that is, if $r \in A$, then there exists $s \in A$ with $r < s$.

Exercise 8.6.1.   (a) Fix $r \in \mathbf{Q}$. Show that the set $C_r = \{t \in \mathbf{Q} : t < r\}$ is a cut.

The temptation to think of all cuts as being of this form should be avoided. Which of the following subsets of $\mathbf{Q}$ are cuts?

(b) $S = \{t \in \mathbf{Q} : t \leq 2\}$

(c) $T = \{t \in \mathbf{Q} : t^2 < 2 \text{ or } t < 0\}$

(d) $U = \{t \in \mathbf{Q} : t^2 \leq 2 \text{ or } t < 0\}$

Exercise 8.6.2. Let A be a cut. Show that if $r \in A$ and $s \notin A$, then $r < s$.

To dispel any suspense, let's get right to the point.

Definition 8.6.3. Define the *real numbers* $\mathbf{R}$ to be the set of all cuts in $\mathbf{Q}$.

This may feel awkward at first—real numbers should be numbers, not sets of rational numbers. The counterargument here is that when working on the foundations of mathematics, sets are about the most basic building blocks we have. We have defined a set $\mathbf{R}$ whose elements are subsets of $\mathbf{Q}$. We now must set about the task of imposing some algebraic structure on $\mathbf{R}$ that behaves in a way familiar to us. What exactly does this entail? If we are serious about constructing a proof for Theorem 8.6.1, we need to be more specific about what we mean by an "ordered field."

Field and Order Properties

Given a set F and two elements $x, y \in F$, an *operation* on F is a function that takes the ordered pair (x, y) to a third element $z \in F$. Writing $x + y$ or xy to represent different operations reminds us of the two operations that we are trying to emulate.

Definition 8.6.4. A set F is a *field* if there exist two operations—addition $(x + y)$ and multiplication (xy)—that satisfy the following list of conditions:

(f1) (commutativity) $x + y = y + x$ and $xy = yx$ for all $x, y \in F$.

(f2) (associativity) $(x+y)+z = x+(y+z)$ and $(xy)z = x(yz)$ for all $x, y, z \in F$.

(f3) (identities exist) There exist two special elements 0 and 1 with $0 \neq 1$ such that $x + 0 = x$ and $x1 = x$ for all $x \in F$.

(f4) (inverses exist) Given $x \in F$, there exists an element $-x \in F$ such that $x + (-x) = 0$. If $x \neq 0$, there exists an element x^{-1} such that $xx^{-1} = 1$.

(f5) (distributive property) $x(y + z) = xy + xz$ for all $x, y, z \in F$.

Exercise 8.6.3. Using the usual definitions of addition and multiplication, determine which of these properties are possessed by **N**, **Z**, and **Q**, respectively.

Although we will not pursue this here in any depth, all of the familiar algebraic manipulations in **Q** (e.g., $x + y = x + z$ implies $y = z$) can be derived from this short list of properties.

Definition 8.6.5. An *ordering* on a set F is a relation, represented by $\leq$, with the following three properties:

(o1) For arbitrary $x, y \in F$, at least one of the statements $x \leq y$ or $y \leq x$ is true.

(o2) If $x \leq y$ and $y \leq x$, then $x = y$.

(o3) If $x \leq y$ and $y \leq z$, then $x \leq z$.

We will sometimes write $y \geq x$ in place of $x \leq y$. The strict inequality $x < y$ is used to mean $x \leq y$ but $x \neq y$.

A field F is called an *ordered field* if F is endowed with an ordering $\leq$ that satisfies

(o4) If $y \leq z$, then $x + y \leq x + z$.

(o5) If $x \geq 0$ and $y \geq 0$, then $xy \geq 0$.

Let's take stock of where we are. To prove Theorem 8.6.1, we are accepting as given that the rational numbers are an ordered field. We have defined the real numbers **R** to be the collection of cuts in **Q**, and the challenge now is to invent addition, multiplication, and an ordering so that each possesses the properties outlined in the preceding two definitions. The easiest of these is the ordering. Let A and B be two arbitrary elements of $\mathbf{R}$.

$$\text{Define} \quad A \leq B \quad \text{to mean} \quad A \subseteq B.$$

Exercise 8.6.4. Show that this defines an ordering on $\mathbf{R}$ by verifying properties (o1), (o2), and (o3) from Definition 8.6.5.

Algebra in R

Given A and B in $\mathbf{R}$, define

$$A + B = \{a + b : a \in A \text{ and } b \in B\}.$$

Before checking properties (f1)–(f4) for addition, we must first verify that our definition really defines an operation. Is $A + B$ actually a cut? To get the flavor of how these arguments look, let's verify property (c2) of Definition 8.6.2 for the set $A + B$.

Let $a + b \in A + B$ be arbitrary and let $s \in \mathbf{Q}$ satisfy $s < a + b$. Then, $s - b < a$, which implies that $s - b \in A$ because A is a cut. But then

$$s = (s - b) + b \in A + B,$$

and (c2) is proved.

Exercise 8.6.5. (a) Show that (c1) and (c3) also hold for $A + B$. Conclude that $A + B$ is a cut.

(b) Check that addition in $\mathbf{R}$ is commutative (f1) and associative (f2).

(c) Show that property (o4) holds.

(d) Show that the cut

$$O = \{p \in \mathbf{Q} : p < 0\}$$

successfully plays the role of the additive identity (f3). (Showing $A + O = A$ amounts to proving that these two sets are the same. The standard way to prove such a thing is to show two inclusions: $A + O \subseteq A$ and $A \subseteq A + O$.)

What about additive inverses? Given $A \in \mathbf{R}$, we must produce a cut $-A$ with the property that $A + (-A) = O$. This is a bit more difficult than it sounds. Conceptually, the cut $-A$ consists of all rational numbers less than $-\sup A$. The problem is how to define this set without using suprema, which are strictly off limits at the moment. (We are building the field in which they exist!)

Given $A \in \mathbf{R}$, define

$$-A = \{r \in \mathbf{Q} : \text{ there exists } t \notin A \text{ with } t < -r\}.$$

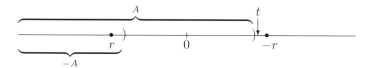

Exercise 8.6.6. (a) Prove that $-A$ defines a cut.

(b) What goes wrong if we set $-A = \{r \in \mathbf{Q} : -r \notin A\}$?

(c) If $a \in A$ and $r \in -A$, show $a + r \in O$. This shows $A + (-A) \subseteq O$. Now, finish the proof of property (f4) for addition in Definition 8.6.4.

Although the ideas are similar, the technical difficulties increase when we try to create a definition for multiplication in **R**. This is largely due to the fact that the product of two negative numbers is positive. The standard method of attack is first to define multiplication on the non-negative cuts.

Given $A \geq O$ and $B \geq O$ in **R**, define the product

$$AB = \{ab : a \in A, b \in B \text{ with } a, \ b \geq 0\} \cup \{q \in \mathbf{Q} : q < 0\}.$$

Exercise 8.6.7. (a) Show that AB is a cut and that property (o5) holds.

(b) Propose a good candidate for the multiplicative identity (1) on **R** and show that this works for all cuts $A \geq O$.

(c) Show the distributive property (f5) holds for non-negative cuts.

Products involving at least one negative factor can be defined in terms of the product of two positive cuts by observing that $-A \geq 0$ whenever $A \leq O$. (Given $A \leq O$, property (o4) implies $A + (-A) \leq O + (-A)$, which yields $O \leq -A$.) For any A and B in **R**, define

$$AB = \begin{cases} \text{as given} & \text{if } A \geq O \text{ and } B \geq O \\ -[A(-B)] & \text{if } A \geq O \text{ and } B < O \\ -[(-A)B] & \text{if } A < O \text{ and } B \geq O \\ (-A)(-B) & \text{if } A < O \text{ and } B < O. \end{cases}$$

Verifying that multiplication defined in this way satisfies all the required field properties is important but uneventful. The proofs generally fall into cases for when terms are positive or negative and follow a pattern similar to those for addition. We will leave them as an unofficial exercise and move on to the punch line.

Least Upper Bounds

Having proved that **R** is an ordered field, we now set our sights on showing that this field is complete. We defined completeness in Chapter 1 in terms of least upper bounds. Here is a summary of the relevant definitions from that discussion.

Definition 8.6.6. A set $\mathcal{A} \subseteq \mathbf{R}$ is *bounded above* if there exists a $B \in \mathbf{R}$ such that $A \leq B$ for all $A \in \mathcal{A}$. The number B is called an *upper bound* for $\mathcal{A}$.

A real number $S \in \mathbf{R}$ is the *least upper bound* for a set $\mathcal{A} \subseteq \mathbf{R}$ if it meets the following two criteria:

  (i) S is an upper bound for $\mathcal{A}$ and

  (ii) if B is any upper bound for $\mathcal{A}$, then $S \leq B$.

Exercise 8.6.8. Let $\mathcal{A} \subseteq \mathbf{R}$ be nonempty and bounded above, and let S be the *union* of all $A \in \mathcal{A}$.

  (a) First, prove that $S \in \mathbf{R}$ by showing that it is a cut.

  (b) Now, show that S is the least upper bound for $\mathcal{A}$.

   This finishes the proof that $\mathbf{R}$ is complete. Notice that we could have proved that least upper bounds exist immediately after defining the ordering on $\mathbf{R}$, but saving it for last gives it the privileged place in the argument it deserves. There is, however, still one loose end to sew up. The statement of Theorem 8.6.1 mentions that our complete ordered field contains $\mathbf{Q}$ as a subfield. This is a slight abuse of language. What it should say is that $\mathbf{R}$ contains a subfield that looks and acts exactly like $\mathbf{Q}$.

Exercise 8.6.9. Consider the collection of so-called "rational" cuts of the form

$$C_r = \{t \in \mathbf{Q} : t < r\}$$

where $r \in \mathbf{Q}$. (See Exercise 8.6.1.)

  (a) Show that $C_r + C_s = C_{r+s}$ for all $r, s \in \mathbf{Q}$. Verify $C_r C_s = C_{rs}$ for the case when $r, s \geq 0$.

  (b) Show that $C_r \leq C_s$ if and only if $r \leq s$ in $\mathbf{Q}$.

Cantor's Approach

As a way of giving Georg Cantor the last word, let's briefly look at his very different approach to constructing $\mathbf{R}$ out of $\mathbf{Q}$. One of the many equivalent ways to characterize completeness is with the assertion that "Cauchy sequences converge." Given a Cauchy sequence of rational numbers, we are now well aware that this sequence may converge to a value not in $\mathbf{Q}$. Just as before, the goal is to create something, which we will call a *real number*, that can serve as the limit of this sequence. Cantor's idea was essentially to define a real number to be the entire Cauchy sequence. The first problem one encounters with this approach is the realization that two different Cauchy sequences can converge to the same real number. For this reason, the elements in $\mathbf{R}$ are more appropriately defined as *equivalence classes* of Cauchy sequences where two sequences (x_n) and (y_n) are in the same equivalence class if and only if $(x_n - y_n) \to 0$.

As with Dedekind's approach, it can be momentarily disorienting to supplant our relatively simple notion of a real number as a decimal expansion with something as unruly as an equivalence class of Cauchy sequences. But what exactly do we mean by a decimal expansion? And how are we to understand the number $1/2$ as both $.5000\ldots$ and $.4999\ldots$? We leave it as an exercise.

Bibliography

[1] Robert G. Bartle, *The Elements of Real Analysis*. Second Edition. John Wiley and Sons, New York, 1964.

[2] Robert G. Bartle, "Return to the Riemann Integral." American Mathematical Monthly, October, 1996.

[3] Robert G. Bartle, *A Modern Theory of Integration*. Graduate Studies in Mathematics, Vol. 2, American Mathematical Society, Providence, Rhode Island, 2001.

[4] R.P. Boas, "Counterexamples to L'Hôpital's Rule." American Mathematical Monthly, October, 1986.

[5] Carl B. Boyer, *A History of Mathematics*. Princeton University Press, Princeton, New Jersey, 1969.

[6] David Bressoud, *A Radical Approach to Lebesgue's Theory of Integration*. The Mathematical Association of America, Washington D.C., 2008.

[7] David Bressoud, *A Radical Approach to Real Analysis*. The Mathematical Association of America, Washington D.C., 1994.

[8] Soo Bong Cha, *Lebesgue Integration*. Monographs and Textbooks in Pure and Applied Mathematics, Marcel Dekker, New York, 1980.

[9] W.A. Coppel, "J.B. Fourier—On the Occasion of his Two Hundredth Birthday." American Mathematical Monthly, 76, 1969.

[10] Roger Cooke, "Uniqueness of Trigonometric Series and Descriptive Set Theory." Offprint from *Archive for History of Exact Sciences*, Volume 45, number 4, Springer–Verlag, New York, 1993.

[11] Philip Davis, "Leonard Euler's Integral: A Historical Profile of the Gamma Function." American Mathematical Monthly, December, 1959.

© Springer Science+Business Media New York 2015
S. Abbott, *Understanding Analysis*, Undergraduate Texts in Mathematics, DOI 10.1007/978-1-4939-2712-8

[12] William Dunham, *The Master of Us All*. Dociani Mathematical Expositions no. 22. Mathematical Association of America, Washington D.C., 1999.

[13] Peter Duren, *Invitation to Classical Analysis*. John Wiley and Sons, New York, 2012.

[14] H. Dym and H.P. McKean, *Fourier Series and Integrals*. Academic Press, Inc., New York, 1972.

[15] David Fowler, "A Simple Approach to the Factorial Function." The Mathematical Gazette, vol. 80, July 1996.

[16] David Fowler, "A Simple Approach to the Factorial Function: The Next Step." The Mathematical Gazette, vol. 83, March 1999.

[17] David Fowler, "The Factorial Function: Convex Functions, the Bohr–Mollerup–Artin Theorem, and Some Formulae." The Mathematical Gazette, vol. 84, November 2000.

[18] E. Hairer and G. Wanner, *Analysis by Its History*. Undergraduate Texts in Mathematics, Springer–Verlag, New York, 1996.

[19] Paul R. Halmos, *Naive Set Theory*. Undergraduate Texts in Mathematics, Springer–Verlag, New York, 1974.

[20] G.H. Hardy, *A Mathematician's Apology*. Cambridge University Press (Canto Edition), Cambridge, 1992.

[21] E.W. Hobson, *The Theory of Functions of a Real Variable and the Theory of Fourier's Series*. Volume 1, Third Edition. Harren Press, Washington D.C., 1950.

[22] T.W. Körner, *A Companion to Analysis. A Second First and First Second Course in Analysis*. Graduate Studies in Mathematics, Vol. 62. American Mathematical Society, Providence, Rhode Island, 2004.

[23] James Propp, "Real Analysis in Reverse," American Mathematical Monthly, Volume 120, May 2013, pp. 392–408.

[24] Walter Rudin, *Principles of Mathematical Analysis*. International Series in Pure and Applied Mathematics, McGraw–Hill, New York, 1964.

[25] George Simmons, *Calculus Gems: Brief Lives and Memorable Mathematics*. McGraw–Hill, New York, 1992.

Index

© Springer Science+Business Media New York 2015
S. Abbott, *Understanding Analysis*, Undergraduate Texts
in Mathematics, DOI 10.1007/978-1-4939-2712-8